Nearrings, Nearfields and Related Topics

Nearrings, Nearfields and Related Topics

Editors

Kuncham Syam Prasad
Manipal Institute of Technology, Manipal University, India

Kedukodi Babushri Srinivas
Manipal Institute of Technology, Manipal University, India

Harikrishnan Panackal
Manipal Institute of Technology, Manipal University, India

Bhavanari Satyanarayana
Acharya Nagarjuna University, India

Associate Editors

Kent Neuerburg
Southeastern Louisiana University, USA

G. L. Booth
Nelson Mandela Metropolitan University, South Africa

Bijan Davvaz
Yazd University, Iran

Mark Farag
Fairleigh Dickinson University, USA

S. Juglal
Nelson Mandela Metropolitan University, South Africa

Ayman Badawi
American University of Sharjah, UAE

NEW JERSEY • LONDON • SINGAPORE • BEIJING • SHANGHAI • HONG KONG • TAIPEI • CHENNAI • TOKYO

Published by

World Scientific Publishing Co. Pte. Ltd.
5 Toh Tuck Link, Singapore 596224
USA office: 27 Warren Street, Suite 401-402, Hackensack, NJ 07601
UK office: 57 Shelton Street, Covent Garden, London WC2H 9HE

Library of Congress Cataloging-in-Publication Data
Names: Prasad, Kuncham Syam, editor. | Srinivas, Kedukodi Babushri, 1977– editor. |
Harikrishnan, Panackal, editor. | Satyanarayana, Bhavanari, editor.
Title: Nearrings, nearfields, and related topics / edited by Kuncham Syam Prasad (Manipal University, India), Kedukodi Babushri Srinivas (Manipal University, India), Panackal Harikrishnan (Manipal University, India), Bhavanari Satyanarayana (Manipal University, India).
Other titles: Near rings, near fields, and related topics
Description: New Jersey : World Scientific, 2017.
Identifiers: LCCN 2016047615 | ISBN 9789813207356 (hardcover : alk. paper)
Subjects: LCSH: Near-rings. | Associative rings. | Near-fields. | Algebraic fields.
Classification: LCC QA251.5 .N435 2017 | DDC 512/.46--dc23
LC record available at https://lccn.loc.gov/2016047615

British Library Cataloguing-in-Publication Data
A catalogue record for this book is available from the British Library.

Printed in Singapore

Dedicated to

Founder of Manipal

Dr T. M. A. Pai

Padma Shri Awardee

(1898–1979)

"Education at all levels is a great factor to elevate an individual, a society or a nation"

Dr T. M. A. Pai

Manipal - Nearrings Conference 2015
WILEY
euclid

Note from the editors

DST-DRDO-ISRO-KSTA sponsored 24th International Conference on Nearrings, Nearfields and Related Topics held at MIT, Manipal University

Algebra is often described as the language of mathematics. Historically, the terms algebra and algorithm originate from the same source. Algebra is so beautiful that many people naturally possess the ability to understand and apply algebra. The theory of nearrings is now a sophisticated theory in algebra. Nearrings are generalized rings: commutativity of addition is not assumed, and — more important — only one distributive law is required. Compared with a standard class of rings, endomorphism rings of Abelian groups, one sees that rings describe "linear" maps on groups, while nearrings handle the general non-linear case. Nearrings in which the nonzero elements form a multiplicative group are called nearfields. The connections between nearrings (especially nearfields) and geometry are well-known. Nearring conferences are held approximately every two years to discuss new ideas, share results and identify new trends. The conferences were held previously in countries namely Germany, Italy, USA, UK, Canada, South Africa, Austria etc. starting from the year 1968. In Asia the conference took place earlier in India (in the year 1985) and in Taiwan (in the year 2005). The 23rd International conference on Nearrings was organized by Prof. J. H. Meyer, at University of the Free State (South Africa) during 07–12 July 2013. The business meeting held during the conference at South Africa offered consent for the next conference to be held at MIT, Manipal University, India. Prof. Pilz (Johannes Kepler University, Austria) gave total encouragement and conveyed full support on behalf of all nearring researchers.

The Department of Mathematics, Manipal Institute of Technology, Manipal organized the 24th International Conference on Nearrings, Nearfields and Related Topics during 05–12 July 2015. The Chancellor of Manipal University Dr Ramdas M. Pai (Padma Bhushan awardee, Govt. of India) was the chief patron of the conference. The patrons of the international conference were Dr H. S. Ballal (Pro-Chancellor), Dr H. Vinod Bhat (Vice Chancellor), Dr V. Surendra Shetty (Pro-Vice Chancellor), and Dr G. K. Prabhu (Former Registrar, present Director, MIT). The honorary committee members were Dr Vinod V. Thomas (Former Director MIT, present Registrar Evaluation), Dr Raghuvir Pai (Former Director Research-Tech.), and Dr D. Srikanth Rao (Former Joint Director MIT, present Director Admissions).

Dr H. Vinod Bhat, Vice Chancellor, Manipal University was the chief guest for the inaugural function of the conference. Dr I. B. S. Passi, IISER Mohali, former president of the Indian Mathematical Society was the guest of honor for the inaugural session. Dr Vinod V. Thomas, then Director of Manipal Institute of Technology presided over the function and spoke about the achievements of the institute. Dr Pradeep G. Bhat, Head of the Department of Mathematics, MIT welcomed the gathering. Dr H. Vinod Bhat, Vice Chancellor,

Manipal University released abstract booklet of Manipal Nearrings Conference 2015. The conference committee acknowledged Prof. Pilz G. (Johannes Kepler University, Austria), Prof. D. Ramakotaiah (Former Vice Chancellor, Acharya Nagarjuna University, India) and Dr Narasimha Bhat (Manipal Dot Net) for their messages and advises. The plenary / invited speakers who delivered talks at the conference are Prof. WenKe Fong (National Cheng Kung University, Taiwan), Prof. Johan Meyer (University of the Free State, South Africa), Prof. I. B. S. Passi (IISER Mohali, India), Prof. Sudhir R. Ghorpade (IIT Bombay, India), Prof. Bhavanari Satyanarayana, (Acharya Nagarjuna University, India), Prof. Booth G. L., Prof. Groenewald, Prof. Stefan Veldsman and Prof. Juglal — all from Nelson Mandela Metropolitan University, South Africa, Prof. Gary F. Birkenmeier, (University of Louisiana at Lafayette, USA), Prof. Ambedkar Dukkipati (IISc, India), Prof. Shankar B. R. (NITK, India), Prof. S. D. Scott (University of Auckland, New Zealand), Prof. Tim Boykett (Johannes Kepler University, Austria), Prof. Bijan Davvaz (Yazd University, Iran), Prof. Ayman Badawi (American University of Sharjah, UAE), Prof. Rajan A. R. (University of Kerala, India), Prof. Kent Neuerburg, (Southeastern Louisiana University, USA), Prof. Mark Farag (Fairleigh Dickinson University, USA), Prof. Y. V. Reddy (Acharya Nagarjuna Univesity, India), Prof. B. S. Kiranagi (University of Mysore, India), Prof. S. Parameshwara Bhatta (Mangalore University, India), Prof. T. Srinivas (Kakatiya University), Prof. Manjunatha Prasad (Manipal University, India), and Dr Srikanth Prabhu (Manipal University, India). Besides, there were 34 short communication talks which highlighted recent developments on nearrings and nearfields.

One of the highlight of the conference was that an entire solution to the famous and mammoth n-gen problem in nearrings was given in a series of three plenary talks by Prof. S. D. Scott (University of Auckland, New Zealand). High quality talks were a feature of this conference. The conference was well received by the participants and some of the participants from abroad mentioned that they have been overwhelmed, excited and intrigued. The conference also had evening programs — classical in nature — featuring mathemagi music by Raghavendra Bhat, Violin and Classical instruments program by Pandit Ravi Kumar, traditional dance by Shruti snd team, and Yoga demonstration by Prajwal.

The valedictory function was held with dignitaries on dais Prof. Raviraja A. (Associate Director Research and Consultancy, MIT), Prof. Veldsman (Nelson Mandela Metropolitan University, South Africa). The organizing committee acknowledged the scientific organizations namely Department of Science and Technology (DST, Govt. of India), Defence Research and Development Organization (DRDO, Govt. of India), Indian Space Research Organization (ISRO, Govt. of India), Karnataka Science and Technology Academy (KSTA, Govt. of Karnataka) and Manipal Institute of Technology, Manipal University for sponsoring the conference. It is worth mentioning that as a precursor to the conference a National Workshop entitled Computer Applications based on Modern Algebra was organized jointly by the Department of Mathematics and the Department of Computer Science and Engineering, MIT, Manipal during 03–05 January 2013. The National Workshop was sponsored by National Board for Higher Mathematics (NBHM, Govt. of India), DST, DRDO, Council of Scientific and Industrial Research (CSIR, Govt. of India) and Infosys.

This lecture notes consists of 25 chapters (a mix of research articles and exploratory articles) reviewed by experts and editorial team. The editorial team thanks chapter contributors and reviewers for their efforts. The editorial team thank Dr G. K. Prabhu (Director, MIT Manipal), Dr B. H. V. Pai (Joint Director, MIT Manipal), Dr P. G. Bhat (HOD Mathematics, MIT Manipal) and all the faculty and staff members of the Department of Mathematics, MIT Manipal for their support. The committee specifically thanks Dr Jagadeesha B. (St. Joseph Engineering College, Mangalore), Mohana K. S. (Department of Mathematics, Mangalore Univeristy), Kavitha K. (Department of Mathematics, MIT Manipal), Hamsa Nayak (Research Scholar, Department of Mathematics, MIT Manipal), for their support.

Editors
Kuncham Syam Prasad
Kedukodi Babushri Srinivas
Harikrishnan Panackal
Bhavanari Satyanarayana

Contents

The development of near-rings and near-fields with greetings to the 24th near-ring conference

Günter Pilz

Linz (Austria)
E-mail: Guenter.Pilz@jku.at

We are connected by our interest in the fascinating structures of near-rings (and related structures). Up to now, I attended all near-ring conferences (except the one in January 1985 at Nagarjuna University), so I am by far the most frequent participant at these conferences. But this time I regret that I cannot attend for health reasons. Let me wish all of you a most interesting conference; I am sure that the industrious and professional organizers, Professors Prasad, Srinivas, and Harikrishnan, will provide a fine environment for this conference.

The interest in near-rings and near-fields started at the beginning of the 20th century when Leonard E. Dickson wanted to know if the list of axioms for skew-fields is redundant. He found in Ref. 1 that there do exist "near-fields", which fulfill all axioms for skew fields except one distributive law. Dickson achieved this by starting with a field and changing the multiplication in a skillful way. To his honor, these types of near-fields are now called "Dickson near-fields". In a monumental paper, Hans Zassenhaus later showed[2] that all finite near-fields are Dickson near-fields, with the exception of 7 cases of order p^2 with $p = 5, 7, 11$ (two cases), $23, 29$, and 59 (which are well understood). In the infinite case, there do exist non-Dickson near-fields, even of every characteristic[4].

Very soon later, it became clear that near-fields are very useful in geometry. If one wants to represent points in a geometry as pairs of "numbers in a structure S" and the (non-vertical) lines by the usual equations $y = ax + b$, one clearly needs something like an addition and a multiplication in S. Soon later, Veblen and Wedderburn[3] found out that for this "coordinatization" of important geometric planes, S has to be exactly a near-field. In the sequel, many more connections to geometry were found — see, e.g., Ref. 5.

The next big success was the characterization of all finite sharply 2-transitive groups. These are permutation groups G, acting on a set X, such that for all $x_1, x_2, y_1, y_2 \in X$ with $x_1 \neq x_2$ and $y_1 \neq y_2$, there is precisely one $g \in G$ with $g(x_1) = y_1$ and $g(x_2) = y_2$. W. Kerby[6] could show that for all finite sharply 2-transitive permutation groups G there is a finite near-field F such that $G \cong \{x \mapsto ax + b \,|\, a, b \in F, a \neq 0\}$. Then all finite sharply 3-transitive permutation groups could be found in a similar way[6]. Since sharply 1- and sharply n-transitive permutation groups for $n \geq 4$ were "known" already[7], these results closed the "painful" gap, and all finite sharply transitive groups are known now.

General near-rings (which are not near-fields) had to wait until 1950, when Donald W. Blackett intensively investigated their structure in his dissertation[8] at Princeton University (before, Helmut Wielandt had done thorough studies, but published only the short note[9]). For near-rings, one also does not require the commutativity of addition (in near-fields, this is a consequence of the remaining axioms, see, e.g., Ref. 10). The main examples of near-rings are the ones of the type $M(G) = G^G$ of all functions from an additive group G into itself, with component-wise addition and with composition as "multiplication". These $M(G)$ are "catholic" examples in the sense that every near-ring can be embedded in one of the $M(G)$'s (see Ref. 11). In contrast, every ring can be embedded in the endomorphism ring of a suitable Abelian group, with the same operations as in $M(G)$. So rings can be viewed as systems of "linear" functions on groups, while near-rings describe the "non-linear" case.

After Blackett's dissertation[8], a large number of papers on near-rings came out. Gerhard Betsch started in the early 60's with a radical theory for near-rings[12] and reached a few years later the famous density theorem for 2-primitive near-rings[13]. Especially in the beginning, it was completely natural that near-ringers investigated which results from the "linear" ring theory could be transferred (by suitable modifications) to the non-linear world, so these people usually have a good knowledge of ring theory. Soon, however, more "genuine" research was done; in the late 60's, James R. Clay started a close investigation of "planar" near-rings in Ref. 14, which had a close connection to geometry from the beginning. But already in 1970, Celestina and Giovanni Ferrero[15] found that planar near-rings give rise to balanced incomplete block designs, which in turn can excellently be used in the design of statistical experiments. So, in the past years, I had a most interesting connection to the Department of Agriculture in Austria, where planar near-rings could be applied to improve the yield of farmers (cf Refs. 17, 16). This was one of the surprising moments when abstract algebra can be used for applications in daily life.

As mentioned above, near-rings "describe functions on groups", so a good knowledge of group theory is necessary in this respect. Especially John D. P. Meldrum showed us how much one can learn about near-rings with a superb mastery of group theory. Conversely, near-ring theory was soon strong enough to yield new insights into groups. The characterization of sharply transitive groups was mentioned already. It is well-known that the sum of two endomorphisms of a non-Abelian group $(G,+)$ is usually not an endomorphism any more, so the endomorphisms are not even closed w.r.t. $+$ in general. The additive closure $E(G)$ consists of all sums/differences of endomorphisms and is an important type of near-rings. Usually, the maps in $E(G)$ "behave better" than ordinary maps from G to G: they map 0 to 0, have "a little bit of distributivity", and so on. So they might be called "civilized". A consequence of the density theorem mentioned above is that every map from G to G is civilized iff G is a finite, simple, non-Abelian group (see Ref. 18, Theorem 7.46). From this, one can deduce that every subnear-ring N of $M(G)$, G finite, can be embedded in some $E(H)$, so that the maps in N can be civilized. It is a big open question if this also holds in the infinite case.

Near-rings of polynomials and polynomial functions over rings, groups, etc. have a lot to do with equations over rings, groups, ..., since these polynomials, as well as the polynomial functions form near-rings w.r.t. addition and composition $\circ$. The value $p(c)$ of a polynomial (function) p at c can be expressed within the near-ring as $p \circ \bar{c}$, where $\bar{c}$ is the constant polynomial (function) with value c. For more, see Ref. 31.

Many more connections of near-rings to the "outside world" were found: to automata, dynamical systems, graphs, homological algebra, universal algebra, category theory, and so on. At the same time, near-ring theory has become a deep and sophisticated discipline. After my book[18], several more books have appeared: by John Meldrum[19], Jim Clay[22], Hans Whling[20]), Stuart Scott[24], Giovanni Ferrero[40], Veljko Vukovic[21] and Bhavanari Satyanarayana / Kuncham Syam Prasad[23]. A large number of impressive results can be found in the numerous papers by Carl Maxson (e.g., Ref. 27 or 28), and the (long) papers of Stuart D. Scott (see, e.g., Refs. 24, 25, and 26).

But there are still huge areas which are almost untouched. I think the following are some examples of natural areas of further investigations:

- Non-linear functional analysis: when non-linear operators are studied, addition and composition are natural operations, and we are right in near-ring theory. See, e.g., Ref. 35.
- Topological near-rings (in which the operations are continuous) are a largely untouched area. Examples come from continuous mappings on topological groups, and the cooperation between topology and near-rings should shed some new light on this subject — see for instance Ref. 34 or 35.
- If one studies functions on a ring rather than on a group, one get the concept of a "composition ring" with 3 operations $+, \cdot, \circ$ and it makes sense to study differentiations as maps which fulfill the sum-, product-, and the chain rule. Its "inverse" is a formal integration. One then can treat "exponential elements" (= maps which coincide with their derivation), differential- and integral equations, and so on. This idea goes back to Karl Menger[36] in the early 40's, and was taken up much later by Winfried Mller[37,38], and others.
- For ideals in polynomial near-rings or composition rings, is there something like a "Groebner basis" to get something like a canonical generating set? Which consequences does this have for the theory of equations? See Refs. 39 and 31.
- If G is an ordered group, $M(G)$ is lattice-ordered by he component-wise order. What about lattice-ordered near-rings in general?

For computations in the area of near-rings, an excellent software "SONATA" was developed by E. Aichinger *et al.*[30], based on GAP.

As mentioned in the beginning, this is the 24th near-ring conference. Aside from them, there has been a number of "inofficial" conferences. I happily recall a "sub-conference" of the general conference of the American Mathematical Society in January 1970 in San Antonio (Texas), where I first met "near-ring giants" like Henry Heatherly and Jim Malone, and the lovely couple of Hanna and Bernhard Neumann. It was a BIG experience for me as a 24-year-old researcher to meet these people in person, and it confirmed my desire to make near-rings to the main object of my mathematical life. And I never felt any regret to have done this. So I wish all participants a great time in Manipal and in the future life as researchers!

Bibliography

1. DICKSON, Leonard E., *Definitions of a group and a field by independent postulates*, Trans. Amer. Math. Soc. **6** (1905), 198–204.
2. ZASSENHAUS, Hans, *ber endliche Fastkrper*, Abh. Math. Sem. Univ. Hamburg **11** (1935/36), 187–220.
3. VEBLEN, Oscar and WEDDERBURN, J. H. M, *Non-Desarguesian and Non-Pascalian Geometries*, Trans. Amer. Math. Soc. **8** (1907), 379–388.
4. GRUNDHFER, Theo and ZASSENHAUS, Hans, *A Criterion for Infinite Non-Dickson Near-Fields of Dimension One*, Res. Math. **15** (1989), 221–226.
5. KARZEL, Helmut, *Bericht berprojektive Inzidenzgruppen*, Jahresber. dt. Math., Vereinigung **67** (1965), 58–92.
6. KERBY, William E., *On infinite sharply multiply transitive groups*, Vandenhoeck & Ruprecht, Gttingen (1974).
7. HALL, Marshal Jr., *Combinatorial Theory*, Ginn/Blaisdell, Waltham (1967).
8. BLACKETT, Donald W., *Simple and Semi-Simple Near-Rings*, Doctoral Dissertation, Princeton Univ. (1950).
9. WIELANDT, Helmut, *ber Bereiche aus Gruppenabbildungen*, Dt. Mathematik **3** (1938), 9–10.
10. NEUMANN, Bernhard H., *On the Commutativity of Addition*, J. London Math. Soc. **15** (1940), 203–208.
11. HEATHERLY, Henry E. and MALONE, Joseph J., *Some Near-Ring Embeddings*, Quart. J. Math. Oxford Ser. **20** (1969), 81–85.
12. BETSCH, Gerhard, *Ein Radikal fr Fastringe*, Math. Z. **78** (1962), 86–90.
13. BETSCH, Gerhard, *Primitive near-Rings*, Math. Z. **130** (1973), 351–361.
14. CLAY, James R., *Some Algebraic Aspects of Planarity*, Atti del Convegno di Geometria Combinatoria e Sui Applicationi, Univ. degli Studi, Perugia (1971), 163–172.
15. FERRERO, Giovanni, *Stems planari e BIB-disegni*, Riv. Mat. Univ. Parma (2) **11** (1970), 79–96.
16. KE, Wenfong and PILZ, Günter, *Abstract Algebra in Statistics*, Journal of Algebraic Statistics **1** (2010), 6–12.
17. BCK, Martin, KPPL, Hubert, PILZ, Günter and WENDT, Gerhard, *Influence of various Parameters on Mycotoxins in Winter Wheat*, ALVA-Tagung (2008), 87–88.
18. PILZ, Günter, *Near-Rings*, North-Holland, Amsterdam, rev. ed. (1983).
19. MELDRUM, John D. P., *Near-Rings and their Links with Groups*, Pitman, London (1985).
20. WHLING, Heinz, *Theorie der Fastkrper*, Thales-Verlag, Essen (1987).
21. VUKOVIC, Veljko, *Nonassociative Near-Rings*, Univ. of Kragujevac - Studio Plus, Belgrade, 1996.
22. CLAY, James R., *Nearrings: Geneses and Applications*, Oxford Univ. Press, Oxford, 1992.
23. SATYANARAYANA, Bhavanari and PRASAD, Kuncham Syam, *Near-Rings, Fuzzy Ideals, and Graph Theory*, CRC Press, New York (2013).
24. SCOTT, Stuart, *Tame Theory*, Amo Publishing, Auckland (1983).
25. SCOTT, Stuart, *The Z-constraint conjecture*, in: Near-rings and Near-fields, Springer, Dordrecht (2005), 69–168
26. SCOTT, Stuart, *The Structure of Ω-groups*, in: "Nearrings, Nearfields, and K-Loops", Kluwer, Dordrecht (1997), 47–137.
27. MAXSON, Carlton J., *On Local Near-Rings*, Math. Z. **106** (1968), 197–205.

28. MAXSON, Carlton J., *Maximal near-rings of polynomial functions on groups*, Math. Pannon. **21** (2010), 1–6.
29. FERRERO, Giovanni and FERRERO-COTTI, Celestina, *Nearrings. Some Developments Linked to Semigroups and Groups*, Kluwer, Dordrecht (2002).
30. AICHINGER, Erhard, BINDER, Franz, ECKER, Jrgen, EGGETSBERGER, Roland, MAYR, Peter and NBAUER, Roland, *SONATA - A System of Near-rings and Their Applications* and *9 Easy Pieces for SONATA (Tutorial)*, Near-ring Newsletter # **17** (1998).
31. AICHINGER, Erhard and PILZ, Günter, *A survey on polynomials and polynomial functions*, Proceedings of The Third International Algebra Conference, Chang Jung University, Tainan, Taiwan, in: Y. Fong, L.-S. Shiao, E. Zelmanov, Kluwer (2003), 1–16.
32. SCOTT, Stuart, *The Z-constraint Conjecture*, in: Near-rings and Near-fields, Springer, Dordrecht (2005), 69–168.
33. SCOTT, Stuart, The Structure of Ω-groups, in: Nearrings, Nearfields, and K-Loops, Kluwer, Dordrecht (1997), 47–137.
34. THAMARATNAM, Vellupillai, Regular Topological Distributively Generated Near-Rings, Bull. Austral. Math. Soc. **35** (1987), 59–72.
35. MAGILL, Kenneth D., Nerrings of Continuous Functions from Topological Spaces into Topological Nearings, Canadian Math. Bull. **39** (1996), 316–329.
36. MENGER, Karl, Algebra of Analysis, Notre Dame Mathematical Lectures **3** (1944).
37. MLLER, Winfried, Differentiations-Kompositionsringe, Acta Sci. Math. Szeged **40** (1978), 157–161.
38. AICHINGER, Erhard, A Note on Simple Composition Rings, Near-Rings, Near-Fields, and K-Loops, Kluwer, Dordrecht (1997), 167–174.
39. HONG, Hoon, Groebner Bases under Composition I, J. of Symbolic Computation **25** (1998), 643–663.
40. FERRERO, Giovanni and FERRERO-COTTI, Celestina, Nearrings. Some Developments Linked to Semigroups and Groups, Kluwer, Dordrecht (2002).

Near-rings of mappings: Past, present and future

G. L. Booth

Nelson Mandela Metropolitan University,
Port Elizabeth,
South Africa
E-mail: geoff.booth@nmmu.ac.za

It has long been known that the set $M(G)$ of self-maps of an additive (but not necessarily Abelian) group G is a near-ring with respect to pointwise addition and composition of functions. Moreover, the set $M_0(G)$ of zero-preserving self-maps of G is a zero-symmetric subnear-ring of $M(G)$. Such near-rings and their subnear-rings have been extensively studied. Various authors have given this study a topological flavour by requiring that G be a topological group and considering the sets $N(G)$ and $N_0(G)$ of continuous and continuous, zero-preserving self maps of G, respectively. We will survey the main results of this study, and also some more recent ones concerning certain subnear-rings of $N(G)$ and $N_0(G)$.

1. Preliminaries

In this exposition, all near-rings will be right distributive, and will only be zero-symmetric when explicitly so stated. Implicitly, this means that all functions will act from the left. For the basics on near-rings, we refer to any of the standard texts, e.g. Refs. 23 and 25. If N is a near-ring, the notation "$I \lhd N$ "will mean "I is an ideal of N ". Let $(G,+)$ be a (not necessarily Abelian) group, and let $M(G)$ denote the set of all self-maps of G. As is well-known, $M(G)$ is a near-ring with respect to pointwise addition and composition of maps. The set $M_0(G) := \{a \in M(G) : a(0) = 0\}$ is a subnear-ring of $M(G)$ which is zero-symmetric, i.e. $a0 = 0$ for all $a \in M_0(G)$. $M(G)$ and $M_0(G)$ provide prototypes for all near-rings (resp. all zero-symmetric near-rings) in the sense that every near-ring (resp. zero-symmetric near-ring) is isomorphic to a subnear-ring of $M(G)$ (resp. $M_0(G)$) for a suitable choice of G (cf Ref. 25, Corollary 1.18). The following result is well-known.

Theorem 1.1. *(Reference 23, Theorems 1.42 and 1.43) Let $(G,+)$ be a group. Then:*

(1) $M_0(G)$ is a simple near-ring.
(2) If $|G| \neq 2$, then $M(G)$ is a simple near-ring.

A number of different kinds of subnear-rings of $M(G)$ and $M_0(G)$ have been studied, for example near-rings of homogeneous functions[22], centraliser near-rings (Ref. 23, p. 31) and Wielandt near-rings[21,29]. In Section 3, we will be concerned with Wielandt near-rings and near-rings of homogeneous maps.

From the late 1960's various authors began to add a topological flavour to the study of near-rings of self-maps by requiring that the group be topological and that the self-maps be continuous. In the sequel, all topological groups will be T_0, and hence completely regular.

Definition 1.1. Let $(G,+)$ be a topological group. Then

(1) $N(G) := \{a \in M(G) : a \text{ is continuous}\}$;
(2) $N_0(G) := \{a \in M_0(G) : a \text{ is continuous}\}$.

It is clear that $N(G)$ and $N_0(G)$ are subnear-rings of $M(G)$ and $M_0(G)$, respectively. Moreover, if the topology on G is discrete, then $N(G) = M(G)$ and $N_0(G) = M_0(G)$. For surveys of early work done on these near-rings, we refer to Refs. 19 and 20. For information on the theory of topological groups, any of the standard texts may be consulted, for example Ref. 12.

We conclude this introduction by listing some concepts which will be of importance in the sequel. Several notions of primeness for near-rings exist in the literature. We will consider three of these.

Definition 1.2. A near-ring N is

- 0-prime if $A, B \lhd N$, $AB = \{0\}$ implies $A = \{0\}$ or $B = \{0\}$;
- 3-prime if $a, b \in N$, $aNb = \{0\}$ implies $a = 0$ or $b = 0$[11];
- equiprime (e-prime) if $a, x, y \in N$, $anx = any$ for all $n \in N$ implies $a = 0$ or $x = y$[7].

We remark that an equiprime near-ring is necessarily zero-symmetric[7]. The radicals $\mathcal{P}_\upsilon$ ($\upsilon = 0, 3, e$) are defined by $\mathcal{P}_\upsilon(N) = \cap\{I \lhd N : N/I \text{ is a } \upsilon\text{-prime near-ring}\}$. The equiprime radical $\mathcal{P}_e$ is of special interest in that it is the only known Kurosh-Amitsur prime radical for both zero-symmetric and all near-rings[7].

Definition 1.3. Let N be a zero-symmetric near-ring and let G be a nonzero left N-module. Then

- G is of type 2 if $NG \neq \{0\}$ and G has no nontrivial N-subgroups;
- G is of type 3 if G is of type 2 and $g, g' \in G$, $ng = ng'$ for all $n \in N$ implies $g = g'$[17].

Definition 1.4. N is υ-primitive ($\upsilon = 2, 3$) if N has a faithful N-module G (i.e. the left annihilator of G in N is the zero ideal) of type υ.

These definitions enable us to define two Jacobson-type radicals for zero-symmetric near-rings.

Definition 1.5. Let N be a zero-symmetric near-ring. Then $\mathcal{J}_\upsilon(N) = \cap\{P \lhd N : N/P \text{ is } \upsilon\text{-primitive}\}$ for $\upsilon = 2, 3$.

In general, $\mathcal{J}_2(N) \subseteq \mathcal{J}_3(N)$, and this inclusion can be strict. However, if N has a unity, then $\mathcal{J}_2(N) = \mathcal{J}_3(N)$.

2. The Near-rings $N(G)$ and $N_0(G)$

In this section, we will consider properties of the near-rings $N(G)$ and $N_0(G)$ and explore how these are affected by the topology on G.

2.1. *Simplicity*

We know from Theorem 2.1 that for any group $(G,+)$, $M_0(G)$ is simple. The question arises: is this true in general for $N_0(G)$? The following example provides a negative answer to this question.

Example 2.1. Let $(G,+)$ be a topological group. Then define $P_G := \{a \in N_0(G) :$ there exists a neighbourhood U of 0 such that $a(U) = 0\}$. Then $P_G \lhd N_0(G)$. There are many examples where the ideal P_G is non-trivial, for example, $G = \mathbb{R}$ with the usual topology.

As we shall see, the instances where $N_0(G)$ is simple seem to be the exception rather than the rule.

Definition 2.1. Let $(G,+)$ be a topological group. Suppose that for every proper closed subset F of G, $x \in G \backslash F$ and $0 \neq y \in G$, there exists a continuous function $f : G \to G$ such that $f(F) = 0$ and $f(x) = y$. Then G is called an S^*-group.

We remark that the class of S^*-groups includes the arcwise connected groups, as well as the 0-dimensional ones. A topological space X is arcwise connected if for all $x, y \in X$, there exists a continuous mapping $f : [0,1] \to X$ such that $f(0) = x$ and $f(1) = y$. X is 0-dimensional if its topology has a basis consisting of clopen (i.e. both open and closed) sets.

Theorem 2.1.[15] *Let $(G,+)$ be an S^*-group or be disconnected. Then $N_0(G)$ is simple if and only if the topology on G is discrete.*

It seems easier to find cases where $N(G)$ is simple, as the next result shows.

Theorem 2.2.[15] *Let $(G,+)$ be the additive group of a topological division ring, such that $|G| > 2$. Then $N(G)$ is simple.*

2.2. *Isomorphisms*

Let G, H be topological groups and let $f : G \to H$ be a topological group isomorphism. It is easily seen that f induces a near-ring isomorphism $\varphi_f : N(G) \to N(H)$ whose restriction to $N_0(G)$ is an isomorphism of $N_0(G)$ onto $N_0(H)$, defined by $\varphi_f(a) := f \circ a \circ f^{-1}$ for all $a \in N(G)$. The question arises: Is the converse true, i.e. if there exists an isomorphism $\varphi : N(G) \to N(H)$ or $\varphi : N_0(G) \to N_0(H)$, does it induce an isomorphism $f : G \to H$? This does not appear to be true in general, but Magill and Hofer have provided partial answers.

Definition 2.2. Let X be a topological space. Then X is called an equiliser space if it is Hausdorff and the sets of the form $E(f,g) := \{x \in X : f(x) = g(x)\}$, where f and g are continuous self-maps of X, form a sub-basis for the closed sets of X.

The class of equiliser spaces is quite large, as it contains all completely regular Hausdorff spaces which contain an arc, as well as all 0-dimensional Hausdorff spaces.

Theorem 2.3.[18] *Let G and H be topological groups both with equiliser topologies, and let $\varphi : N(G) \to N(H)$ be a near-ring isomorphism. Then there exists a unique topological group isomorphism: $f : G \to H$ such that $\varphi(a) = f \circ a \circ f^{-1}$ for all $a \in N(G)$.*

Theorem 2.4.[13,14] *Let G and H be topological groups which are either both 0-dimensional or both arcwise connected, and let $\varphi : N_0(G) \to N_0(H)$ be a near-ring isomorphism. Then there exists a unique topological group isomorphism $f : G \to H$ such that $\varphi(a) = f \circ a \circ f^{-1}$ for all $a \in N_0(G)$.*

2.3. *Primeness*

As Veldsman has noted[28], it is easily seen that $M_0(G)$ is equiprime for any group G. In general, this is not the case for $N_0(G)$, as the following example shows.

Example 2.2. Let $G := \mathbb{R} \times \mathbb{Z}_2$, where G has the product topology with respect to the usual and discrete topologies on $\mathbb{R}$ and $\mathbb{Z}_2$, respectively. Let $I := \{a \in N_0(G) : a(\mathbb{R} \times \{0\}) = \{0_G\}\}$ and $J := \{a \in N_0(G) : a(G) \subseteq \mathbb{R} \times \{0\}\}$. Then I and J are ideals of $N_0(G)$ and $I \cap J \neq \{0\}$. However, $(I \cap J)^2 = \{0\}$, so $N_0(G)$ is not 0-prime and hence not equiprime.

Nevertheless, $N_0(G)$ is equiprime in many cases.

Theorem 2.5. *Let G be an S^*-group. Then $N_0(G)$ is equiprime.*

In particular this implies that $N_0(G)$ is equiprime if G is arcwise connected or 0-dimensional. The next result provides a generalisation of Example 2.2.

Theorem 2.6.[3,6] *Let G be a disconnected topological group, with open components, each of which contains more than one element. Let H be the component of G which contains 0. Let $I := \{a \in N_0(G) : a(H) = 0\}$ and $J := \{a \in N_0(G) : a(G) \subseteq H\}$. Then $\mathcal{P}_0(G) = \mathcal{P}_e(G) = \mathcal{J}_3(G) = I \cap J$.*

2.4. *Strongly Prime Near-rings*

Strongly prime near-rings were introduced by Groenewald[10].

Definition 2.3. A near-ring N is

- strongly prime if for all $0 \neq a \in N$, there exists a finite subset F of N such that $x \in N$, $aFx = 0$ implies $x = 0$[10];
- strongly equiprime if for all $0 \neq a \in N$, there exists a finite subset F of N such that $x, y \in N$, $afx = afy$ for all $f \in F$ implies $x = y$[8].

If the finite subset F is independent of the choice of a in 1 or 2 above, then N is said to be uniformly strongly prime (resp. uniformly strongly equiprime).

We remark that strongly equiprime $\Longrightarrow$ strongly prime $\Longrightarrow$ 3-prime, and strongly equiprime $\Longrightarrow$ equiprime. Moreover, if N is finite, then the concepts of strongly prime and 3-prime (resp. strongly equiprime and equiprime) coincide.

Theorem 2.7.[9] *Let G be an arcwise connected topological group with more than one element. Then $N_0(G)$ is not strongly prime.*

Theorem 2.8.[9] *Let G be a 0-dimensional topological group. Then:*

(1) *$N_0(G)$ is strongly prime if and only if the topology on G is discrete;*
(2) *$N_0(G)$ is strongly equiprime if and only if G is finite.*

Recall that an ideal I of a near-ring N is strongly prime (resp. uniformly strongly prime) if the factor near-ring N/I is strongly prime, (resp. uniformly strongly prime). The strongly prime radical $\mathcal{P}_s(N)$ (resp. uniformly strongly prime radical $\mathcal{P}_u(N)$) of N is the intersection of the strongly prime (resp. uniformly strongly prime) ideals of N.

In what follows, $\mathbb{R}^n$ ($n \in \mathbb{N}$) will be endowed with the usual (product) topology.

Theorem 2.9.[4] *$P_{\mathbb{R}^n} := \{a \in N_0(\mathbb{R}^n) :$ there exists a neighbourhood U of 0 such that $a(U) = 0\}$ is a uniformly strongly prime ideal of $N_0(\mathbb{R}^n)$ and is contained in every strongly prime ideal of $N_0(\mathbb{R}^n)$.*

We remark that, for $n \geq 2$, this result makes use of the Peano space-filling curves.

Corollary 2.1.[2,4] $\mathcal{P}_s(N_0(\mathbb{R}^n)) = \mathcal{P}_u(N_0(\mathbb{R}^n)) = P_{\mathbb{R}^n}$.

Now we investigate strongly prime ideals in $N_0(\mathbb{R}^\omega)$, where ω is the first transfinite cardinal, and $\mathbb{R}^\omega$ has the usual (Tychonoff) product topology.

Recall that $\mathbb{R}^\omega$ metrizable, with metric d defined by $d(x,y) := \sum_{i=1}^{\infty} \frac{|x_i - y_i|}{2^i(1+|x_i-y_i|)}$, where $x := (x_i)_{i\in\mathbb{N}}$ and $y := (y_i)_{i\in\mathbb{N}}$.

Theorem 2.10.[4] *$P_{\mathbb{R}^\omega}$ is a strongly prime ideal of $N_0(\mathbb{R}^\omega)$, and is contained in every strongly prime ideal of $N_0(\mathbb{R}^\omega)$.*

We remark that $P_{\mathbb{R}^\omega}$ is not uniformly strongly prime in this case.

Corollary 2.2. $\mathcal{P}_s(N_0(\mathbb{R}^\omega)) = P_{\mathbb{R}^\omega}$.

At this stage, we have not been able to characterise $\mathcal{P}_u(N_0(\mathbb{R}^\omega))$.

3. Important Subnear-rings of $N(G)$

In this section, we will consider two classes of subnear-rings of $N(G)$, namely Wielandt near-rings and near-rings of homogeneous self-maps. We will be particularly concerned with the effect of the topology on the structure of the near-rings. Except where otherwise stated, the results in this section come from the doctoral thesis of Mogae[24].

3.1. *Wielandt Near-rings*

This class of near-rings was defined by Wielandt[29], and was studied in Refs. 21 and 28, inter alia. We start by introducing a concept from group theory, which will be required in the sequel.

Definition 3.1. Let B_1, B_2 be subgroups of a group G, let $\overline{B}_1, \overline{B}_2$ be normal subgroups of B_1, B_2, respectively, and let $\kappa : B_1/\overline{B}_1 \to B_2/\overline{B}_2$ be a group isomorphism. Let α be an ordinal, and let $\{b_{1\eta} : \eta < \alpha\}$ be a class of coset representatives of $B_1/\overline{B}_1$. Then define $H = B_1/\overline{B}_1 \times_\kappa B_2/\overline{B}_2 := \bigcup_{\eta<\alpha} (b_{1\eta} + \overline{B}_1) \times \kappa(b_{1\eta} + \overline{B}_1)$. The product $B_1/\overline{B}_1 \times_\kappa B_2/\overline{B}_2$ is a subgroup of G^2 and is called a 2-fold meromorphic product.

Remak[26] showed that every subgroup of G^2 is expressible as a 2-fold meromorphic product. However, no analogy of this result for G^n exists if $n > 2$.

Definition 3.2. Let G be a group, let c be a cardinal and let H be a subgroup of G^c. If $a \in M(G)$, we define $\widehat{a} : G^c \to G^c$ as follows. Let I be an indexing set of cardinality c. If $x := (x_i)_{i\in I} \in G^c$, then let $\widehat{a}(x) := (a(x_i))_{i\in I}$. Let $M(G,c,H) := \{a \in M(G) : \widehat{a}(H) \subseteq H\}$ and $M_0(G,c,H) := \{a \in M_0(G) : \widehat{a}(H) \subseteq H\}$. Then $M(G,c,H)$ and $M_0(G,c,H)$ are subnear-rings of $M(G)$ and $M_0(G)$, respectively, and are known as Wielandt near-rings. If G is a topological group, let $N(G,c,H) := M(G,c,H) \cap N(G)$ and $N_0(G,c,H) := M_0(G,c,H) \cap N_0(G)$.

Theorem 3.1.[21] *Let G be a finite group and let $H := G/A \times_\kappa G/B$, where $A \cap B = 0$. Then $N_0(G,2,H)$ is simple if and only if it is $\mathcal{J}_2$-semisimple.*

This result cannot be extended to 0-dimensional topological groups in general.

Example 3.1. Let $\{0\} \neq G$ be any non-discrete 0-dimensional topological group, and let $H := (G/\{0\}) \times_\iota (\{G\}/\{0\})$, where ι is the identity. It is easily verified that $H = \{(g,g) : g \in G\}$. Then $\widehat{a}(H) \subseteq H$ for all $a \in N_0(G)$, so $N_0(G,2,H) = N_0(G)$. Since the topology on G is not discrete, $N_0(G)$ is not simple by Theorem 2.1. However, it can be shown to be 2-primitive (and hence $\mathcal{J}_2$-semisimple) on G.

We now consider primeness in Wielandt near-rings.

Theorem 3.2. *Let G be a 0-dimensional topological group, k a positive integer and let H be a proper open subgroup of G^k. If $H = \prod_{i=1}^{k} H_i$, then the following are equivalent:*

(1) $N(G,k,H)$ is 3-prime.
(2) For each j, either $H_j = G$ or $H_j = \{0\}$.
(3) $N(G,k,H)$ is equiprime.

We remark that this result was proved for the discrete case by Veldsman[28]. Moreover, if $N(G,k,H)$ satisfies the premises of this theorem it must be zero-symmetric and it holds that $N(G,k,H) = N_0(G,k,H) = N_0(G)$. Another similar result is the following.

Theorem 3.3. *Let G be a topological group, k a positive integer and let H be a proper closed, arcwise connected subgroup of G^k such that $H = \prod_{i=1}^{k} H_i$, and $\bigcap_{i=1}^{n} H_i$ is arcwise connected. Then the following are equivalent:*

(1) $N(G,k,H)$ is 3-prime.
(2) For each j, either $H_j = G$ or $H_j = \{0\}$.
(3) $N(G,k,H)$ is equiprime.

Once again, $N(G,k,H) = N_0(G,k,H) = N_0(G)$ in this case. The requirement that H be arcwise connected in the above result cannot be omitted.

Example 3.2. Consider $N(\mathbb{R},2,\mathbb{Z}\times\{0\})$, where $\mathbb{R}$ has the usual topology. Clearly $\mathbb{Z}\times\{0\}$ is a closed subgroup of $\mathbb{R}^2$. It may be shown that $N(\mathbb{R},2,\mathbb{Z}\times\{0\}) = N_0(\mathbb{R},1,\mathbb{Z})$ is equiprime, and hence that conditions 1 and 3 of the above result are satisfied, but clearly condition 2 is not.

3.2. *Near-rings of Homogeneous Functions*

In this sub-section, we will consider near-rings of homogeneous functions, and in particular the effects of topologising the theory.

Definition 3.3. Let R be a ring and let G be a right R-module. A function $f : G \to G$ is said to be homogeneous if $f(gr) = f(g)r$ for all $g \in G$ and $r \in R$.

The set of all homogeneous functions of G into itself will be denoted $M_R(G)$.

If G is a topological R-module, the set of all continuous homogeneous functions of G into itself will be denoted $N_R(G)$.

It is clear that $M_R(G)$ and $N_R(G)$ are zero-symmetric near-rings and that $N_R(G) \subseteq M_R(G)$. However, equality does not hold in general.

Example 3.3. Let $G := \mathbb{R}^2$ be considered as an $\mathbb{R}$-module with its usual topology. Define $f(a,b) := \begin{cases} (-a,0) & \text{if } b \neq 0 \\ (a,0) & \text{otherwise} \end{cases}$. Then $f \in M_{\mathbb{R}}(\mathbb{R}^2)$. Let $x_n := \frac{1}{n}$. Then $\lim_{n\to\infty} f(1,x_n) = (-1,0)$ and $f\left(\lim_{n\to\infty} f(1,x_n)\right) = (1,0)$, so f is discontinuous at $(1,0)$, and hence $f \notin N_{\mathbb{R}}(\mathbb{R}^2)$.

For the remainder of this section we will be concerned with $M_R(R^2)$ and $N_R(R^2)$, where R is a topological ring with unity.

Theorem 3.4. *The following are equivalent:*

(1) R is a prime ring.
(2) $N_R(R^2)$ is 3-prime.
(3) $N_R(R^2)$ is equiprime.

This result was proved for the discrete case by Veldsman[28].

Definition 3.4. A near-ring N is

- 3-semiprime if $x \in N$, $xNx = \{0\}$ implies $x = 0$;
- semi-equiprime if $x, y \in N$, $(x-y)nx = (x-y)ny$ for all $n \in N$ implies $x = y$.

Theorem 3.5. *The following are equivalent:*

(1) R is a semiprime (resp. strongly prime) ring.
(2) $N_R(R^2)$ is 3-semiprime (resp. strongly prime).
(3) $N_R(R^2)$ is semi-equiprime (resp. strongly equiprime).

Making use of the Wedderburn-Artin theorem, we can prove the following.

Corollary 3.1. *Let R be an Artinian ring. Then the following are equivalent:*

(1) R is a semiprime ring.
(2) $\mathcal{P}_e\left(N_R(R^2)\right) = 0$.
(3) $N_R(R^2)$ is semi-equiprime.
(4) $N_R(R^2)$ is 3-semiprime.

4. Conclusion

The theory of near-rings of self-maps has been extensively studied for almost as long as near-rings themselves. This exposition is not exhaustive, nor could it be. For further information, we refer the reader to the references and their references. In this final section we pose some questions for further study, and introduce some similar structures which lie outside the scope of the present exposition.

4.1. *Some Questions for Further Study*

(1) Determine the ideal structure of $N_0(G)$ for familiar topological groups G, e.g. $G = \mathbb{R}^n$, $\mathbb{T}^m$ (the m-dimensional torus), $\mathbb{R}^n \times \mathbb{T}^m$.[19]
(2) Find interesting results about the topological properties of G which result from algebraic properties of $N(G)$ and $N_0(G)$.
(3) Given that $N_0(G)$ and $N_0(H)$ are isomorphic, determine the isomorphisms between them in the case that G and H are not necessarily 0-dimensional or arcwise connected. Will G and H necessarily be isomorphic in such cases, and, if so, will the isomorphisms between $N(G)$ and $N(H)$ be induced by isomorphisms between G and H as per Section 2.2?[19]
(4) Does there exist a non-discrete topological group G such that $N_0(G)$ is strongly prime?
(5) Does there exist a disconnected topological group G which is not 0-dimensional such that $N_0(G)$ is equiprime?

4.2. *Similar Structures*

In this section, we discuss some structures related to near-rings of self-maps.

Definition 4.1. Let X and G be a nonempty set and a group, respectively, and let $\theta : G \to X$ be a mapping. Let $M_0(X,G,\theta) := \{a : X \to G : a\theta(0_G) = 0_G\}$. Then $M_0(X,G,\theta)$ is a zero-symmetric near-ring with respect to pointwise addition and multiplication defined by $a \cdot b := a\theta b$ for all $a,b \in M_0(X,G,\theta)$. Such near-rings are called sandwich near-rings.

The study of sandwich near-rings can be topologised by requiring X and G to be a topological space and a topological group, respectively, and all functions to be continuous. For a sample of the work done on sandwich near-rings in this setting, we refer to Refs. 2, 3, 5, 9, 16 and 24.

Definition 4.2. A Γ-near-ring is a triple $(M,+,\Gamma)$ where

(1) $(M,+)$ is a (not necessarily abelian) group;
(2) Γ is a nonempty set of binary operators on M such that $(M,+,\gamma)$ is a near-ring for each $\gamma \in \Gamma$;
(3) $a\gamma(b\mu c) = (a\gamma b)\mu c$ for all $a,b,c \in M$ and $\gamma,\mu \in \Gamma$.
If in addition $a\gamma 0 = 0$ for all $a \in M$ and $\gamma \in \Gamma$, then the Γ-near-ring $(M,+,\Gamma)$ is said to be zero-symmetric.

This definition is due to Satyanarayana[27]. For some of the basic properties of Γ-near-rings, the reader is referred to Ref. 1.

Now let X and G be a nonempty set and a group, respectively and let Γ be a nonempty set of mappings from G to X. Let $M = M_0(X,\Gamma,G) = \{a : X \to G : a\gamma(0_G) = 0_G$ for all $\gamma \in \Gamma$. Then M is clearly a zero-symmetric Γ-near-ring with respect to pointwise addition and composition of functions. As for sandwich near-rings, the study of Γ-near-rings of continuous functions can be topologised in the natural way. For some results in this setting, we refer to Ref. 24, Chapter 4.

Bibliography

1. G. L. Booth, A note on Γ-near-rings, *Studia Sci. Math. Hungar.* **23** (1988), 471–475.
2. G. L. Booth, Primeness and radicals in near-rings of continuous functions, *Nearrings and Nearfields, Proceedings on the Conference on Nearrings and Nearfields, Hamburg, Germany, July 27–August 2, 2003*, Springer, Berlin, 2005, pp. 171–176.
3. G. L. Booth, Primeness in near-rings of continuous functions 2, *Beiträge Alg. Geom.* **46** (2005), 207–214.
4. G. L. Booth, Strongly prime ideals of near-rings, *Advances in Ring Theory, Proceedings of the International Conference on Algebra and Applications, June 18–21, 2008, Athens, Ohio*, Birkhäuser/Springer, Basel, 2010, pp. 63–68.
5. G. L. Booth, Prime radicals in sandwich near-rings, *Acta Math. Hungar.* **131** (2011), 25–34.
6. G. L. Booth, Primitivity in near-rings of continuous functions, *Topology and its Applications* **159**(9) (2012), 2274–2279.
7. G. L. Booth, N. J. Groenewald and S. Veldsman, A Kurosh-Amitsur prime radical for near-rings, *Comm. in Algebra* **18** (1990), 3111–3122.

8. G. L. Booth, N. J. Groenewald and S. Veldsman, Strongly equiprime near-rings, *Quaestiones Math.* **14** (1991), 483–489.
9. G. L. Booth and P. R. Hall, Primeness in near-rings of continuous functions, *Beiträge Alg. Geom.* **45** (2004), No. 1, 21–27.
10. N. J. Groenewald, Strongly prime near-rings, *Proc. Edinburgh Math. Soc.* **31** (1988), 337–343.
11. N. J. Groenewald, Different prime ideals in near-rings, *Comm. in Algebra* **19** (1993), 2667–2675.
12. P. J. Higgins, *Introduction to topological groups*, London Math. Soc. Lect. notes, Cambridge University Press, London, 1974.
13. R. D. Hofer, Restrictive semigroups of continuous functions on 0-dimensional spaces, *Canad. J. Math.* **24** (1972), 598–611.
14. R. D. Hofer, Restrictive semigroups of continuous selfmaps on arcwise connected spaces, *Proc. London Math. Soc.* **25** (1972), 358–384.
15. R. D. Hofer, Near-rings of continuous functions on disconnected groups, *J. Austral. Math. Soc. (Ser. A)* **28** (1979), 433–451.
16. R. D. Hofer and K. D. Magill, On the simplicity of sandwich near-rings, *Acta Math. Hungar.* **60** (1992), 51–60.
17. W. M. Holcombe, A hereditary radical for near-rings, *Studia Sci. Math. Hungar.* **17** (1982), 453–456.
18. K. D. Magill, A survey of semigroups of continuous self-maps, *Semigroup Forum* **3** (1975/76) 189–282.
19. K. D. Magill, Near-rings of continuous self-maps: a brief survey and some open problems, *Proc. Conf. San Bernadetto del Tronto, 1981*, 25–47, 1982.
20. K. D. Magill, A survey of topological nearrings and nearrings of continuous functions, *Proc. Tenn. Top. Conf.*, World Scientific Pub. Co., Singapore, 1997, 121–140.
21. C. J. Maxson and K.C. Smith, Simple-near-rings associated with meromorphic products, *Proc. Amer. Math. Soc.* **105** (1989), No. 3, 564–574.
22. C. J. Maxson and L. van Wyk, The lattice of ideals of $M_R(R^2)$, R a commutative PIR, *J. Austral. Math. Soc. (Series A)* **52** (1992), 268–282.
23. J. D. P. Meldrum, *Near-rings and their links with groups*, Pitman, London, 1985.
24. K. Mogae, *Primeness in near-rings of continuous maps*, doctoral thesis, Nelson Mandela Metropolitan University, Port Elizabeth, South Africa, 2013.
25. G. Pilz, *Near-rings*, 2nd ed., North-Holland, Amsterdam, 1983.
26. R. Remak, Über die darstelling der eindlichen gruppen als untergruppen direkte produkte, *J. Reine Angew. Math.* **163** (1930), 1–44.
27. Bh. Satyanarayana, *Contributions to near-ring theory*, doctoral thesis, Nagarjuna University, India, 1984.
28. S. Veldsman, On equiprime near-rings, *Comm. in Algebra* **20** (1992), 2569–2587.
29. H. Wielandt, How to single out function near-rings, *Oberwolfach Abstracts*, 1972.

Planar nearrings: Ten years after

Wen-Fong Ke

Department of Mathematics and Research Center for Theoretical Sciences,
National Cheng Kung University, Tainan 701, Taiwan
E-mail: wfke@mail.ncku.edu.tw

In this article, we review the development of planar nearrings in the past ten years.

Keywords: planar nearring; fixed point free group; balanced incomplete block design; semi-homogeneous map.

1. Introduction

An algebraic structure $N = (N,+,\cdot)$ is called a (left) *nearring* if $(N,+)$ is a group, which may be nonabelian, and $(N,\cdot)$ a semigroup such that $a(b+c) = ab+ac$ for all $a,b,c \in N$. Further, if $(N \setminus \{0\},\cdot)$ is a group itself, N is called a *nearfield.* It is always true that $a \cdot 0 = 0$ for all $a \in N$. If $0 \cdot a = 0$ for all $a \in N$, we call N *zero-symmetric.* On N, an equivalence relation $\equiv_m$ on N can be defined:

$$a \equiv_m b \Leftrightarrow ax = bx \text{ for all } x \in N,$$

and we say that $(N,+,\cdot)$ is *planar* if (1) there are at least three equivalence classes in $N/{\equiv_m}$, and (2) for each triple $a,b,c \in N$ with $a \not\equiv_m b$, the equation $ax = bx + c$ has a unique solution for x in N. Geometrically, imagining that ax and $bx+c$, $x \in N$, are "lines" passing 0 and c, and of slops a and b, respectively, the given conditions say that (1) there are at least three lines with different slops, and (2) for any two lines of different slops, there is a unique intersection points.

One immediately realizes that all fields having more than 2 elements are planar nearrings. Also, all finite nearfields are planar nearrings. The first examples of planar nearrings which are not rings were given by Anshel and Clay[3] in 1968, which we record them here again.

For $a,b \in \mathbb{C}$, $a = (a_1,a_2) \in \mathbb{R}^2$, define $*_i$, $i = 1,2,3$, as following.

$$a *_1 b = \begin{cases} a_1 b, & \text{if } a_1 \neq 0, \\ a_2 b, & \text{if } a_1 = 0; \end{cases}$$

$$a *_2 b = |a| \cdot b;$$

$$a *_3 b = \begin{cases} \frac{ab}{|a|}, & \text{if } a \neq 0, \\ 0, & \text{if } a \neq 0. \end{cases}$$

Then $(\mathbb{C},+,*_1)$, $(\mathbb{C},+,*_2)$, and $(\mathbb{C},+,*_3)$ are planar nearrings which are not rings. The following observations were the source for a long and fruitful research line in planar nearrings.

Remark 1.1. Let $a, b \in \mathbb{C}$, $a \neq 0$. Then

(1) $\mathbb{C} *_1 a + b$ is the straight line through b along the direction of a;
(2) $\mathbb{C} *_2 a + b$ is the ray starting from b along the direction of a;
(3) $\mathbb{C} *_3 a + b$ is the circle centered at b with radius $|a|$, plus the center b.

The characterization for integral nearrings given by Ferrero[10] turns out to be one for planar nearrings since all finite integral nearrings are planar.

Let N be a nearring. For $a \in N$, define

$$\varphi_a : N \to N; x \mapsto ax \text{ for all } x \in N.$$

Then φ_a is an endomorphism of the additive group $(N, +)$. When N is planar, $a \in N$ not a left identity and $\not\equiv_m 0$, φ_a turns out to be an automorphism with the property that the map $(-1 + \varphi_a) : N \to N; x \mapsto -x + ax$ is bijective, and is refer to as a regular automorphism. The collection $\{\varphi_a \mid a \in N,\ a \not\equiv_m 0\}$ is itself of a regular group of automorphisms of $(N, +)$. Conversely, if $(N, +)$ is a group with a given regular group of automorphisms Φ, then (N, Φ) is called a *Ferrero pair*, and planar nearrings can be constructed from it easily. First, take any complete set of orbit representatives of Φ in N. Denote it by C. Next, choose a subset E of $C \setminus \{0\}$ with $|E| \geq 2$. Now, for each $x \in N$, if $x = \varphi(e)$ where $e \in E$ and $\varphi \in \Phi$, define $x * y = \varphi(y)$ for all $y \in N$; otherwise, set $x * y = 0$ for all $y \in N$. Then $N = (N, +, *)$ is a planar nearring. In this case, E is the set of the left identities of N, and the resulting nearring N is integral if and only if $E = C \setminus \{0\}$.

Example 1.1. We give the Ferrero pair correspondences of the three examples of Anshel and Clay here.

(1) For $(\mathbb{C}, +, *_1)$, the corresponding Ferrero pair is $(\mathbb{C}, \widehat{\mathbb{R}}^*)$,
where $\widehat{\mathbb{R}}^* = \{\varphi_r \mid r \in \mathbb{R} \setminus \{0\}\}$.
(2) For $(\mathbb{C}, +, *_2)$, the corresponding Ferrero pair is $(\mathbb{C}, \widehat{\mathbb{R}}^+)$,
where $\widehat{\mathbb{R}}^+ = \{\varphi_r \mid r > 0\}$.
(3) For $(\mathbb{C}, +, *_3)$, the corresponding Ferrero pair is $(\mathbb{C}, \widehat{C})$,
where $\widehat{C} = \{\varphi_c \mid |c| = 1\}$.

In general, let F be a field with $|F| > 2$. Then F is itself a planar nearring. We can get more planar nearrings out of it: Take $G \leq F^* = F \setminus \{0\}$ with $|G| \geq 2$, and put $\widehat{G} = \{\varphi_a \mid a \in G\}$; then $(F, \widehat{G})$ is a Ferrero pair. We call any nearring constructed from $(F, \widehat{G})$ a *field generated* planar nearring.

Remark 1.2. One can make a slight generation of the Ferrero construction. Let us consider a pair (N, Γ) where N is a group written additively and Γ is a group of automorphisms of N. Then N is a disjoint union of Γ-orbits. Let E be a subset of N satisfying the following two conditions:

(1) if $a \in G$, $\gamma \in \Gamma \setminus \{1\}$ and $\gamma a = a$ then $a \notin E$;
(2) if $a \in G$ then $|E \cap \Gamma a| \leq 1$.

In other words, E does not contain fixed points of any $\gamma \neq 1$ from Γ (in particular, $0 \notin E$) and there is at most one element from each Γ-orbit of G. It is easy to understand that the elements of the orbits $\Gamma e, e \in E$, cannot be fixed by some $\gamma \in \Gamma \setminus \{1\}$. Then the multiplication on G is defined as follows: $(\gamma e)a = \gamma a$ for all $\gamma \in \Gamma, e \in E$ and $a \in G$; all other products are zero. What we get in this way is a 0-symmetric near-ring called a *Ferrero nearring*[11].

It was shown that a sub-nearring of a Ferrero nearring is also a Ferrero nearring. This is not true for planar nearrings.

2. Balanced Incomplete Block Designs

As mentioned before in Remark 1.1, the three examples of Anshel and Clay were the source for much research on planar nearrings. The main object is the combinatorial structure called balanced incomplete block designs, BIBDs or 2-designs in short.

Definition 2.1. A set X with v elements together with a family $\mathcal{S}$ of k-subsets of X is called a *balanced incomplete block design* if

(1) each element belongs to exactly r subsets, and
(2) each pair of distinct elements belongs to exactly λ subsets.

The k-subsets in $\mathcal{S}$ are called *blocks*, and the integers $v, b = |\mathcal{S}|, r, k, \lambda$ are referred to as the *parameters* of the BIBD, and $(X, \mathcal{S})$ is also called a 2-(v,k,λ) design.

Example 2.1. Let N be a finite planar nearring. Then following collections of subsets of N are 2-designs.

(1) Collection $\mathcal{B}$: all $Na + c$ with $a, c \in N$ and $a \neq 0$, provided that either all or none of the Na are additive subgroups.
(2) Collection $\mathcal{B}^-$: all $(Na \cup N(-a)) + c$ with $a, c \in N$ and $a \neq 0$, provided that for all nonzero a, $Na \cap N(-a) = \{0\}$, $Na \cup N(-a)$ is not an additive subgroup of N, and the map $t_a : N \to N; t_a(x) = x + ax$ is bijective.
(3) Collection $\mathcal{B}^*$: all $N^*a + c$ with $a, c \in N$ and $a \neq 0$. Here $N^* = N \setminus \{z \in N \mid z \equiv_m 0\}$.
(4) Collection $\mathcal{S}$: all intersections $N(c-a)+a$ and $N(a-c)+a$ with $a \neq c$.

We can refer to the above examples as classical ones. Sun in 2010 developed the following construction of BIBDs from finite nearfields. Let $(F, +, \cdot)$ be a finite nearfield (or just a finite field) with $|F| = q$ and $\mathrm{char} F = p$. Take any proper subset S of F with $|S| = k \geq 2$. Define on F the following equivalence relations $b_1 \sim_c b_2$ if $Sb_1 = Sb_2 + a$ for some $a \in F$, and $a_1 \sim_r a_2$ if $S + a_1 = S + a_2$. Set $n = |F^* / \sim_c|$ and $\mu = |F / \sim_r|$.

Theorem 2.1 (Sun[22]). *With $\mathcal{B} = \{Sb + a \mid a, b \in F, b \neq 0\}$, $(F, \mathcal{B})$ is a BIBD with $b = n\mu$ blocks, and $\lambda = \frac{bk(k-1}{q(q-1)}$.*

When $\sum_{x \in S} x = 0$, more detailed information can be obtained.

A very important class of planar nearrings were motivated by the third Anshel-Clay example $(\mathbb{C}, +, *_3)$. Recall that for $a, c \in \mathbb{C}$ with $a \neq 0$, $\mathbb{C}^* a + c$ is the circle centered at c of radius $|a|$. This motivated the following definition.

Definition 2.2. Let N be a planar nearring. Then N is said to be circular if $(N^* a) \cap (N^* b + c)$ contains at most 2 elements for all $a, b, c \in N \setminus \{0\}$.

Thus, $(\mathbb{C}, +, *_3)$ is a circular planar nearring. Also, as $(N, \mathcal{B}^*)$ is a BIBD when N is finite, we can extend the above definition to single out a class of BIBDs.

Definition 2.3. Let $(X, \mathcal{B})$ be a BIBD. Then it is said to be circular if $A \cap B$ contains at most two elements whenever A and B are distinct.

Using the geometrical idea of circles, Benini, Frigeri and Morini[6] define the disk with center b and radius a in a circular planar nearring N as

$$D(a;b) = \cup\{N^* r + c \mid r \neq 0, b \in N^* r + c, |(N^* r + c) \cap (N^* a + b)| = 1\}.$$

Theorem 2.2. *If N is field generated with $N = \mathbb{Z}_p$, p a prime, and Φ is of order $2n$, then the collection $D(a;b)$, $a, b \in F$, $a \neq 0$, form a BIBD. Further, if $D(1;0) \setminus \{0\}$ is not a multiplicative group, then $k = p(p-1)/2n$ and $\lambda = n(2n^2+1)$.*

Another geometric idea to use circularity is the Modisett's lines[20]. Let N be a circular planar nearring. For $x \neq y$, define the set $A_{x,y} := \{n \in N \setminus \{x, y\} \mid x, y, n \in N^* a + c$ for some $a, c \in N$, $a \neq 0\}$. The complement of $A_{x,y}$ is the conceptual line in N containing x, y. Let $\mathcal{A} = \{A_{x,y} \mid x \neq y\}$ and $\mathcal{L} = \{A^c_{x,y} = N \setminus A_{x,y} \mid x \neq y\}$.

When $N = F$ is a finite field and Φ a subgroup of F^* with $k = |\Phi|$, $v = |F|$, then $b = v(v-1)/k$ and $\lambda = k - 1$. It is known that (N, Φ) is circular if and only if $|A_{0,1}| = (k-1)(k-2)$. Let G be the affine group over the field N and put $\mathcal{A} := \{A_{x,y}; x, y \in N, x \neq y\}$. Then $\mathcal{A}$ is the collection of orbits of $A_{0,1}$ under G and thus $(N, \mathcal{A})$ is a 2-design and $(N, \mathcal{L})$ is the complemented BIBD of $(N, \mathcal{A})$. Let $G_{A_{0,1}}$ be the setwise stabilizer of $A_{0,1}$ inside G, and denote $s = |G_{A_{0,1}}|$. Then we can show that

Lemma 2.1. *The parameters of $(N, \mathcal{A})$ are $v = |N|$,*

$$\begin{aligned}
k_{\mathcal{A}} &= (k-1)(k-2), \\
b_{\mathcal{A}} &= \frac{v(v-1)}{s}, \\
r_{\mathcal{A}} &= \frac{(v-1)(k-1)(k-2)}{s} = \frac{(v-1)}{s} \cdot k_{\mathcal{A}}, \\
\lambda_{\mathcal{A}} &= \frac{(k-1)(k-2)}{s} \cdot ((k-1)(k-2)-1) \\
&= \frac{k_{\mathcal{A}}(k_{\mathcal{A}}-1)}{s}.
\end{aligned}$$

Lemma 2.2. *The parameters of* $(N, \mathcal{L})$ *are* $v = |N|$,

$$\begin{aligned}
k_{\mathcal{L}} &= v - (k-1)(k-2), \\
b_{\mathcal{L}} &= b_{\mathcal{A}} = \frac{v(v-1)}{s}, \\
r_{\mathcal{L}} &= \frac{v-1}{s} \cdot (v - (k-1)(k-2)) = \frac{v-1}{s} \cdot k_{\mathcal{L}}, \\
\lambda_{\mathcal{L}} &= \frac{v-1}{s} \cdot (v - 2k_{\mathcal{A}}) + \frac{k_{\mathcal{A}}(k_{\mathcal{A}}-1)}{s} \\
&= \frac{(v-1)k_{\mathcal{A}}}{s} \cdot \left(\frac{v}{k_{\mathcal{A}}} + \frac{k_{\mathcal{A}}-1}{v-1} - 2 \right) \\
&= \frac{(v-1)(k-1)(k-2)}{s} \cdot \left(\frac{v}{(k-1)(k-2)} + \frac{k^2-3k+1}{v-1} - 2 \right).
\end{aligned}$$

The problem here is to find the number s. It seems to be not obvious what s is, and so one may try to restrict his/her search to a specific class of Galois fields in hope of getting some results.

Finally, related to this topic, Benini-Morini[1] and Benini-Pellegrini[7] considered partial balanced incomplete block designs (PBIBDs) and weakly divisible nearrings. Also, there were applications of circular planar nearrings to coding theory investigated in Benini-Frigeri-Morini[5]. Note that PBIBDs from planar nearrings had been considered by Clay (see Clay[9] (7.110) and Sun[21]).

3. Nearring of matrix maps

Given a ring R, we are so familiar with the ring of square matrices having entries taken from R, where the usual operations of matrix addition and multiplication are used. With square matrices having entries taken from a nearring, however, under the same operations, one obtains a nearring of matrices only when the given nearring is distributive, i.e. the nearring satisfies both distributive laws. Moreover, the resulting nearring of matrices is also distributive (see Heatherly[12]).

Meldrum and van der Walt[18] took an alternative view of matrices, namely, as mappings rather than square arrays of elements from some algebraic structure, and successfully arrived at the notion of a matrix nearring. They used certain elementary maps to generate these matrix nearrings. These elementary maps imitate the well-known elementary matrices

$$rE_{ij} = \begin{pmatrix} 0 \cdots 0\,0\,0 \cdots 0 \\ \vdots \quad\; \vdots\,\vdots\,\vdots \quad\; \vdots \\ 0 \cdots 0\,0\,0 \cdots 0 \\ 0 \cdots 0\,r\,0 \cdots 0 \\ 0 \cdots 0\,0\,0 \cdots 0 \\ \vdots \quad\; \vdots\,\vdots\,\vdots \quad\; \vdots \\ 0 \cdots 0\,0\,0 \cdots 0 \end{pmatrix},$$

where r (from a ring R) occupies the (i,j)-th entry of a square $n \times n$ array, and the other entries are zero. The idea of Meldum and van der Walt was to consider the elementary matrices rE_{ij} as maps $f_{ij}^r : N^n \to N^n$; $f_{ij}^r v = \iota_i(r\pi_j v)$, where, in this case, N is a nearring with identity, N^n denotes the direct sum of n copies of the additive group of N, and ι_i and π_j denote the usual i-th co-ordinate injection function and the j-th co-ordinate projection function, respectively. The $n \times n$ matrix nearring over N, denoted $\mathcal{M}_n(N)$, is then defined to be the subnearring of the nearring $M(N^n)$, generated by all the f_{ij}^r. A substantial amount of research has been done on the structure $\mathcal{M}_n(N)$ since its origin in 1986. See Meyer[19] for a general account on the development of matrix nearrings and related nearrings.

This idea can be carried over to the nearrings with no identity with a bit care.

Definition 3.1. Let N be a nearring, not necessarily with identity. For a positive integer n, the *nearring of $n \times n$ matrix maps over N*, denoted $\mathrm{Mat}_n(N)$, is defined to be the subnearring of $M(R^n)$ generated by the mappings $f_{ij}^r : R^n \to R^n$, $1 \le i \le n$, $1 \le j \le n$, and $r \in R$, where each f_{ij}^r is defined as in our discussion above.

Remark 3.1. (1) If R happens to possess an identity element, then $\mathrm{Mat}_n(R) = \mathcal{M}_n(R)$, the $n \times n$ matrix nearring over R, as defined by Meldum and van der Walt.

(2) In $\mathrm{Mat}_n(N)$, it may happen that for two different elements $r, s \in R$, the elementary matrix maps f_{ij}^r and f_{ij}^s are the same mapping on R^n while the $n \times n$ elementary matrices having r and s, respectively, as the (i,j)-entries and 0 elsewhere are different "matrices." See Meldum-van der Walt[18] for more detail on this issue.

Theorem 3.1 (Ke-Meyer-Wendt[14]). *Let N be a nearring, and $\mathrm{Mat}_n(N)$ the nearring of matrix maps over N.*

(1) If N is integral planar, then $\mathrm{Mat}_n(N)$ is simple.
(2) If N is integral planar, then $\mathrm{Mat}_n(N)$ is 2-primitive.
(3) If N is 1-primitive and planar, then $\mathrm{Mat}_n(N)$ is 1-primitive.
(4) If N is 2-primitive and planar, then $\mathrm{Mat}_n(N)$ is 2-primitive.
(5) If N is 1-primitive, finite, planar, and not 2-primitive, then $\mathrm{Mat}_n(N)$ is 2-radical.
(6) If N is finite, planar, then $J_1(\mathrm{Mat}_n(N))$ is the greatest proper ideal in $\mathrm{Mat}_n(N)$.

The matrix maps in $\mathrm{Mat}_n(N)$ may be difficult to recognize. In particular, one wonders where U^{-1} is also a matrix map if $U \in \mathrm{Mat}_n(N)$ is bijective. This is true when N is finite[14] Corollary 4.2, but is not true when N if infinite[14] Example 4.4. The given example is in $\mathrm{Mat}_n(N)$ with N a non-planar nearfield. Thus, it is interesting to see what would be the situation when N is an infinite planar nearfield.

4. Ideals and characterization

Ideal structure of planar nearrings are relatively simple. G. Wendt has been studying the structure of nearrings since 2004 (cf. Section 6 of Ke[13]). He first studied the nearrings solving the identity $ax = c$, a class of nearrings containing planar nearrings. He also defined in Wendt[23] sandwich centralizer nearrings, which are centralizer nearrings where the usual function composition is replaced by a composition with a sandwich function in between. In particular, he applied his works to the structure of planar nearrings. In 2004, he showed that

Theorem 4.1 (Wendt[25]). *Every planar ring is isomorphic to some minimal left ideal of a ring of linear transformations of a vector space over a skew field and we see that the finite planar rings are just the minimal left ideals of matrix rings over fields.*

Then in 2010, he continued and showed that

Theorem 4.2 (Wendt[25]). *A finite minimal left ideal L of a zero symmetric nearring N is a planar nearring if L is not contained in the radical $J_2(N)$.*

Theorem 4.3 (Wendt[25]). *Let N be a zero symmetric nearring and M a finite minimal N-subgroup of N. If $|M/\equiv_m| \geq 3$, M is a planar nearring.*

Let N be a finite planar nearring with fixed point free action group Φ. It was shown in Beidar-Fong-Ke[2] that if N is circular, then all Sylow subgroups of Φ are cyclic, i.e., Φ is a metacyclic group. The converse is not true. That is, given a finite metacyclic group Φ, there may not exist any planar nearring N whose fixed point free action group is isomorphic to Φ. With this in mind, it is said that a given group Φ is a *group without fixed points* if there exists a group N such that (N,Φ) is a Ferrero pair, and that Φ is a *finite group without fixed points* if the group N can be chosen to be finite (of cause, Φ is then finite as well). This is equivalent to the existence of a fixed point free representation of Φ on a finite vector space, as was proved in Ref. 2 (2.7).

Now, it is natural to ask *when is Φ circular?*

Question 4.1. Let Φ be finite metacyclic group which is also a group without fixed points. Under what conditions is Φ circular?

Recall that metacyclic groups can be characterized by the following presentation:

$$\langle A,B \mid A^{\mu}=1,\, B^{\nu}=A^{t},\, BAB^{-1}=A^{\rho}\rangle,$$

where μ and ρ are two relatively prime numbers, $s=\gcd(\rho-1,\mu)$, $t=\mu/s$, and ν is the order of ρ modulo μ. Partial answer to the above question is answered in the following statement.

Theorem 4.4 (Beidar-Ke-Kiechle[4]). *Let*

$$\Phi=\langle A,B \mid A^{\mu}=1,\, B^{\nu}=A^{t},\, BAB^{-1}=A^{\rho}\rangle$$

be a finite metacyclic group and let ν be the order of ρ modulo μ. Suppose $\nu=2$ and that Φ is embeddable into the multiplicative group of some skew field. Then Φ is circular.

The general case is still open. In particular, the following conjecture still wait for a validation.

Conjecture 4.1. *Let Φ be a finite metacyclic group. If Φ is embeddable into the multiplicative group of some skew field, then Φ is circular.*

5. Nearrings on groups and Semi-homogeneous maps

A natural way of getting nearrings with certain properties is to define a multiplication on a given additive group. For small groups, one can exhaust all possible multiplications and obtain all possible nearrings with the small group as the additive group. For larger groups, this task is much more difficult, or impossible. In 1968, Clay[8] demonstrated how to define possible multiplications $*$ on a given additive group $(N,+)$ in order for the triple $(N,+,*)$ to be a nearring.

Theorem 5.1 (Clay[8]). *For a group $(N,+)$, let $\phi : N \to \mathrm{End}(N)$; $n \mapsto \phi_n$ be a map that satisfies $\phi_n \circ \phi_{n'} = \phi_{\phi_{n'}(n)}$ for all $n,n' \in N$. Then $(N,+,\cdot_\phi)$ is a nearring, where $n \cdot_\phi n' := \phi_n(n')$. Moreover, all possible multiplications to turn $(N,+)$ into a nearring can be obtained in this way. Further, in such a nearring N, an element e is an identity element for the nearring $(N,+,\cdot_\phi)$ if and only if ϕ_e is the identity map in $\mathrm{End}(N)$ and, for all $n \in N$, $\phi_n(e) = n$.*

In an attempt to find nearrings with Euclidean spaces as the additive groups, Magill have a matrix version of Clay's theorem.

Theorem 5.2 (Magill[17]). *Let $n \in \mathbb{N}$, $f_{st} : \mathbb{R}^n \to \mathbb{R}$, $1 \leq s \leq n$, $1 \leq t \leq n$. For $v \in \mathbb{R}^n$, let $A_v := (f_{st}(v)) \in M_n(\mathbb{R})$. Define $v * w = A_v w$ for all $v,w \in \mathbb{R}^n$. Then $(\mathbb{R}^n,+,*)$ is a left distributive system. Every left distributive system on $(\mathbb{R}^n,+)$ is obtained in exactly this manner. Furthermore, $(\mathbb{R}^n,+,*)$ is a nearring if and only if $A_v A_w = A_{A_v(w)}$ for all $v,w \in \mathbb{R}^n$.*

The final condition in the last theorem is difficult to achieve. A special situation is the following.

Theorem 5.3 (Magill[17]). *In the previous theorem, take $f_{ij} = 0$ for $i \neq j$ and $f_{ii} = f$ fixed. Then $(\mathbb{R}^n,+,*)$ is a nearring if and only if $f(av) = af(v)$ for all $v \in \mathbb{R}^n$ and all $a \in \mathrm{Im}(f)$. The obtained nearring is planar when $\mathrm{Im}(f)$ has at least three elements.*

The property $f(av) = af(v)$ for all $v \in \mathbb{R}^n$ and all $a \in \mathrm{Im}(f)$ makes f *semi-homogeneous.*

In Ke-Kiechle-Pilz-Wendt[15], certain topological planar nearrings on $(\mathbb{R}^2,+)$ were determined completely by determining the corresponding semi-homogeneous maps on $\mathbb{R}^2$. It turns out that semi-homogeneous maps themselves deserve some investigations.

Consider a monoid G acting on a nonempty set S, where we assume that the unit element 1 fulfills $1 \cdot s = s$ for all $s \in S$.

Definition 5.1 (Ke-Kiechle-Pilz-Wendt[16]). *A map $f : S \to G$ is called* semi-homogeneous *if*

$$f(g \cdot s) = gf(s) \quad \textit{for all } s \in S \textit{ and } g \in \mathrm{Im}(f).$$

Written as

$$f(f(t) \cdot s) = f(t)f(s) \quad \text{for all } s,t \in S$$

we see that $\mathrm{Im}(f)$ is a sub-semigroup of G.

Take any nonempty set S and put $G = (S^S, \circ)$, the monoid of all maps from S to itself. Then G acts naturally on S. For $f : S \to G$; $v \mapsto f_v$, define $s * t = f_s(t)$ for $s, t \in S$. Then$(S, *)$ is a semigroup if and only $f_{f_s(t)} = f_s \circ f_t$ for all $s, t \in S$, which is equivalent to saying that f is semi-homogeneous. Conversely, let $(S, *)$ be a monoid. Select some element $s_0 \in S$ and define a map $f : S \to S$ by $f(s) := s * s_0$. Then f is (semi-)homogeneous.

Theorem 5.4 (Ke-Kiechle-Pilz-Wendt[16]). *Let the group G act on the set S. Suppose that $f : S \to G$ is semi-homogeneous and put*

$$H = \{h \in G \mid f(h \cdot s) = hf(s) \textit{ for all } s \in S\}.$$

Then $\mathrm{Im}(f) = H$, *and this is a subgroup of G such that if $h \neq 1$, then $h \cdot s \neq s$ for all $s \in S$.*

In the above statement, the property that $h \in H$, $h \neq 1$, then $h \cdot s \neq s$ for all $s \in S$ makes H *fixed point free.* Let G be a group acting on a set S, and let H a fixed point free subgroup of G. Let R be a set of representatives of the orbits of H inside S. Then for every $s \in S$ there exist some $g_s \in H$ and $r_s \in R$ with $s = g_s \cdot r_s$.

Theorem 5.5 (Ke-Kiechle-Pilz-Wendt[16]). *Every map $\hat{f} : R \to H$ can be uniquely extended to a semi-homogeneous map $f : S \to G$ via $f(s) := h_s\hat{f}(r_s)$ if $s = h_s \cdot r_s$, and every semi-homogeneous map $f : S \to G$ arises in this way. We also have $H = \mathrm{Im}(f)$.*

In many cases, we encounter groups with zeros. Let G be a group with zero acting on a set S. Put $Z = 0 \cdot S$ and $S^\times = S \setminus Z$. Then the group $G^* = G \setminus \{0\}$ acts on $S^\times$.

Theorem 5.6 (Ke-Kiechle-Pilz-Wendt[16]). *Every semi-homogeneous map $f : S \to G$ can be constructed in one of the following ways:*

(1) f is a constant map with $\mathrm{Im}(f) = \{0\}$ or $\mathrm{Im}(f) = \{1\}$.

(2) Take a fixed point free subgroup H of G^*. Let R be a complete set of representatives of the orbits of H inside $S^\times$. The map f is uniquely determined by $\hat{f} = f|_R$ as by $f(s) = h_s\hat{f}(r_s)$ if $s = h_s \cdot r_s$ and $f(z) = 0$ for all $z \in Z$. Furthermore, $H \cup \{0\} = \mathrm{Im}(f)$.

Bibliography

1. A. Benini and F. Morini: *Partially balanced incomplete block designs from weakly divisible nearrings.* Discrete Math. **301** (2005), no. 1, 34–45.
2. K. I. Beidar, Y. Fong and W.-F. Ke: *On finite circular planar nearrings.* J. Algebra **185** (1996), 688–709.
3. M. Anshel and J. R. Clay: *Planar algebraic systems: some geometric interpretations.* J. Algebra **10** (1968), 166–173.
4. K. I. Beidar, W.-F. Ke and H. Kiechle: *Circularity of finite groups without fixed points.* Monatshefte fr Mathematik **144** (2005), 265–273.
5. A. Benini, A. Frigeri and F. Morini: *Codes and combinatorial structures from circular planar nearrings.* Algebraic informatics, 115–126, Lecture Notes in Comput. Sci., 6742, Springer, Heidelberg, 2011.
6. A. Benini, A. Frigeri and F. Morini: *BIB-designs from circular nearrings.* Results Math. **64** (2013), no. 1–2, 121–133.

7. A. Benini and S. Pellegrini: *Finite weakly divisible nearrings.* Riv. Mat. Univ. Parma (8) **2** (2009), 101–116.
8. J. R. Clay: *The near-rings on groups of low order.* Math. Z. **104** (1968), 364–371.
9. J. R. Clay: *Nearrings: Geneses and Applications*, Oxford University Press, 1992.
10. G. Ferrero: *Due generalizzazioni del concetto di anello e loro equivalenza nell'ambito degli "stems" finiti.* Riv. Mat. Univ. Parma **7** (1966), 145–150.
11. Y. Fong, K. Kaarli and W.-F. Ke: *On minimal varieties of near-rings.* in "Near-rings and Near-fields" (Fredericton, NB, 1993), pp. 123–131, Math. Appl. 336, Kluwer Acad. Publ., Dordrecht, 1995.
12. H. Heatherly: *Matrix near-rings.* J. London Math. Soc. (2) **7** (1973), 355–356.
13. W.-F. Ke: *On recent developments of planar nearrings.* In "Nearrings and nearfields" (Hamburg 2003), pp. 3–23. Springer 2005.
14. W.-F. Ke, J. H. Meyer and G. Wendt: *Matrix maps over planar near-rings.* Proc. Roy. Soc. Edinburgh Sect. A **140** (2010), no. 1, 83–99.
15. W.-F. Ke, H. Kiechle, G. Pilz and G. Wendt: *Planar nearrings on the Euclidean plane.* J. Geom. **105** (2014), no. 3, 577–599.
16. W.-F. Ke, H. Kiechle, G. Pilz and G. Wendt: *Semi-homogeneous maps,* Contemporary Mathematics **658**, Amer. Math. Soc., Providence, RI, 2016, to appear.
17. K. D. Magill, Jr.: *Topological nearrings on the Euclidean plane.* Papers on general topology and applications (Slippery Rock, PA, 1993), pp. 140–152 (Ann. New York Acad. Sci. **767**, New York Acad. Sci., New York (1995)).
18. J. D. P. Meldum and A. P. J. van der Walt: *Matrix near-rings.* Arch. Math. (Basel) **47** (1986), 312–319.
19. J. H. Meyer: *On the development of matrix nearrings and related nearrings over the past decade.* Near-rings and near-fields (Stellenbosch, 1997), 23–34, Kluwer Acad. Publ., Dordrecht, 2001.
20. M. C. Modisett: *A characterization of the circularity of balanced incomplete block designs.* Utilitas Math. **35** (1989), 83–94.
21. H.-M. Sun: *PBIB designs and association schemes obtained from finite rings.* Discrete Math. **252** (2002), no. 1–3, 267–277.
22. H.-M. Sun: *From planar nearrings to generating blocks.* Taiwanese J. Math. **14** (2010), no. 5, 1713–1739.
23. G. Wendt: *Characterisation results for planar near-rings.* Contributions to general algebra. **15**, 187–197, Heyn, Klagenfurt, 2004.
24. G. Wendt: *Planar near-rings, sandwich near-rings and near-rings with right identity.* in "Near-rings and nearfields," (Hamburg 2003), pp. 277–291, Springer, Dordrecht, 2005.
25. G. Wendt: *Minimal left ideals of near-rings.* Acta Math. Hungar. **127** (2010), no. 1–2, 52–63.

Goldie dimension and spanning dimension in modules and N-groups

Bhavanari Satyanarayana

AP Scientist Awardee,
Fellow, AP Akademy of Sciences,
Acharya Nagarjuna University, A. P., India
E-mail: davvaz@yazd.ac.ir

Dedication: In memory of Prof. Dr A. W. Goldie, University of Leeds

1. Introduction

It is well-known that the dimension of a vector space is defined as the number of elements in its basis. One can define a basis of a vector space as a maximal set of linearly independent vectors or a minimal set of vectors, which span the space. The former case when generalized to modules over rings, becomes the concept of Goldie dimension. A dualization of the concept 'finite Goldie dimension' is 'finite spanning dimension'. In this Chapter, we discuss some results and examples related to finite Goldie dimension (in short, FGD), fuzzy dimension, and finite spanning dimension in the algebraic systems: modules over rings and N-groups (that is, modules over near-rings).

2. Elementary concepts in vector spaces and modules over rings

Definition 2.1. (i) A subset S of a vector space V is called a basis of V if the elements of S are linearly independent, and $V = L(S)$; and
(ii) If V is a finite dimensional vector space, and S is a basis for V, $n = |S|$, then the integer n is called the dimension of V over F, and we write $n = dim\ V$.

Lemma 2.1. *If V is finite dimensional and if W is a sub space of V, then*
(i) W is finite dimensional,
(ii) $dim\ W \leq dim\ V$, and
(iii) $dim\ (V/W) = dim\ V - dim\ W$.

Corollary 2.1. *If A and B are finite dimensional subspaces of a vector space V. Then*
(i) $A+B$ is finite dimensional; and
(ii) $dim\ (A+B) = dim\ A + dim\ B - dim\ (A \cap B)$.

Definition 2.2. Let R be an associative ring. An Abelian group $(M,+)$ is said to be a module over R if there exists a mapping $f : R \times M \to M$ (the image of (r,m) is denoted by rm) satisfying the following three conditions:
(i) $r(a+b) = ra+rb$;
(ii) $(r+s)a = ra+sa$; and

(iii) $r(sa) = (rs)a$ for all $a, b \in M$ and $r, s \in R$.
Moreover if R is ring a with identity 1, and if $1m = m$ for all $m \in M$, then M is called a unital R-Module.

Example 2.1. (i) Every ring R is a module over itself;
(ii) Every group is a module over Z;
(iii) Every vector space over a field F, is a module over the ring F;
(iv) Let $(G,+)$ be an Abelian group. Write $R = \{f : G \to G \mid f \text{ is a group homomorphism}\}$. Define $(f+g)(x) = f(x) + g(x)$ for all $x \in G$ and $f, g \in R$. Define $(f.g)(x) = f(g(x))$ for all $f, g \in R$ and $x \in G$. Then $(R,+,.)$ becomes a ring with identity. For any $f \in R$ and $a \in G$, the element f(a) (the image of a under f) is in G. Now G becomes a module over R.
(v) Let R be a ring and L a left ideal of R. Define $a \sim b \Leftrightarrow a - b \in L$ for any $a, b \in R$. Then $\sim$ is an equivalence relation and the equivalence class containing a is $[a] = a + L$. Write $M = \{a + L \mid a \in R\}$. If we define $(a+L)+(b+L) = (a+b)+L$ on M, then $(M,+)$ is an Abelian group.
For any $r \in R$, $a + L \in M$, if we define $r(a+L) = ra + L$, then M is an R-module. It is called quotient module of R by L.

Definition 2.3. (i) Let M be an R-Module. A subgroup $(A,+)$ of $(M,+)$ is said to be a submodule of M if $r \in R$, $a \in A$ then $ra \in A$.
(ii) An R-module is said to be finitely generated if there exist elements $a_j \in M$, $1 \leq j \leq n$ such that $M = \{r_1 a_1 + ... + r_n a_n \mid r_j \in R, \text{ for } 1 \leq j \leq n\}$.

Definition 2.4. (i) If K, A are submodules of M, and K is a maximal submodule of M such that $K \cap A = (0)$, then K is said to be a complement of A (or a complement submodule in M).
(ii) A non-zero submodule K of M is called essential (or large) in M (or M is an essential extension of K) if A is a submodule of M and $K \cap A = (0)$, imply $A = (0)$. We denote as, $K \leq_e M$.

Remark 2.1. (i) If V is a vector space and W is a subspace of V, then W has no proper essential extensions.
(ii) If W, W^1 are two subspaces of V such that W is essential in W^1 , then $W = W^1$.
(iii) Every subspace W is a complement.

3. Finite Goldie dimension in modules

Henceforth, R denotes a fixed (not necessarily commutative) ring with 1.

Definition 3.1. (i) M has finite Goldie dimension (abbr. FGD) if M does not contain a direct sum of infinite number of non-zero submodules.
[Equivalently, M has FGD if for any strictly increasing sequence $H_0 \subseteq H_1 \subseteq ...$ of submodules of M, there exists an integer i such that H_k is an essential submodule in H_{k+1} for every $k \geq I$].

(ii) A non-zero submodule K of M is said to be an uniform submodule if every non-zero submodule of K is essential in K.

With the concepts defined above, Goldie proved the following theorem.

Theorem 3.1. *(Goldie[8]): If M is a module with finite Goldie dimension, then there exist uniform submodules $U_1, U_2, ..., U_n$ whose sum is direct and essential in M. The number 'n' is independent of the uniform submodules.*
The number 'n' of the above theorem is called the Goldie dimension of M, and is denoted by dim M.

Remark 3.1. (i) Let W be a subspace of V. Then W is uniform $\Leftrightarrow dim\ W = 1$. (ii) For any subspace W, we have that $dim\ W = 1 \Leftrightarrow W$ is indecomposable.

Note. (i) As in vector space theory, for any submodules K, H of M such that $K \cap M = (0)$, the condition $dim\ (K+H) = dim\ K + dim\ H$ holds.
(ii) If K and H are isomorphic, then $dim\ K = dim\ H$.
(iii) When we observe the following example, we will learn that the condition $dim\ (M/K) = dim\ M - dim\ K$ does not hold for a general submodule K of M.

Example 3.1. Consider $\mathbb{Z}$, the ring of integers. Since $\mathbb{Z}$ is uniform $\mathbb{Z}$-module, we have that $dim\ \mathbb{Z} = 1$. Suppose $p_1, p_2, ..., p_k$ are distinct primes and consider K, the submodule generated by the product of these primes. Now $\mathbb{Z}/K$ is isomorphic to the external direct sum of the modules $\mathbb{Z}/(p_i)$ where (p_i) denotes the submodule of $\mathbb{Z}$ generated by p_i (for $1 \leq i \leq k$) and so $dim\ \mathbb{Z}/K = k$. For $k \geq 2$, $dim\ \mathbb{Z} - dim\ K = 1 - 1 = 0 \neq k = dim(\mathbb{Z}/K)$.

Hence, there arise a type of submodules K which satisfy the condition $dim\ (M/K) = dim\ M - dim\ K$.

In this connection, Goldie obtained the following theorem.

Theorem 3.2. *(Goldie[8]): If M has finite Goldie dimension and K is a complement submodule, then $dim\ (M/K) = dim\ M - dim\ K$.*

On the way of getting the converse for Theorem 3.2., the concept 'E-irreducible submodule of M' was introduced in Reddy and Satyanarayana[15].

Definition 3.2. A submodule H of M is said to be E-irreducible if $H = K \cap J$ where K and J are submodules of M, and H is essential in K, imply $H = K$ or $H = J$.

Notation 3.1. Every complement submodule is an E-irreducible submodule, but the converse is not true.

Example 3.2. Consider $\mathbb{Z}$, the ring of integers and $\mathbb{Z}_{12}$ the ring of integers module 12. The principle submodule K of the $\mathbb{Z}$-module $\mathbb{Z}_{12}$ generated by 2, is E-irreducible submodule, but it is not a complement submodule.

It is proved in Reddy and Satyanarayana[15] that:

Theorem 3.3. *If K is a submodule of an R-module M and $f: M \rightarrow M/K$is the canonical epimorphism, then the conditions given below are equivalent: (i) $K = M$ or K is not essential, but E-irreducible; (ii) K has no proper essential extensions; (iii) K is a complement; (iv) For any submodule K^1 of M containing K, we have that K^1 is a complement in $M \Leftrightarrow f(K^1)$ is complement in M/K; and $(v) f(S)$ is essential in M/K for any essential submodule S of M.*
Moreover, if M has FGD, then each of the above conditions (i) to (v) are equivalent to (vi) M/K has FGD and dim $(M/K) = dim\ M - dim\ K$.

Note. The converse of the Theorem 3.2., is a part of the Theorem 3.3.

As consequence of Theorem 3.3., we have the following Theorem 3.4.

Theorem 3.4. *(Reddy and Satyanarayana[15]): If M is an R-module, then the following conditions are equivalent: (i) M is a completely reducible module; (ii) Every submodule of M is a complement submodule; (iii) Every proper submodule of M is not an essential submodule, but it is an E-irreducible submodule; (iv) Every proper submodule of M has no proper essential extensions; (v) For any submodule K of M with the canonical epimorphism $f: M \rightarrow M/K$, we have that: K^1 is a complement submodule in $M \Leftrightarrow f(K^1)$ is a complement submodule in M/K; and (vi) For any submodule K of M with the canonical epimorphism $f: M \rightarrow M/K$, we have that: S is an essential submodule in M imply $f(S)$ is an essential submodule in M/K.*

Moreover, if M has finite Goldie dimension, then the above conditions are equivalent to each of the following:
(vii) M has the descending chain condition on its submodules and M is completely reducible; and (viii) For any submodule K of M, we have that M/K has finite Goldie dimension and dim $(M/K) = dim\ M - dim\ K$.

Definition 3.3. A family $\{M_i\}_{i \in I}$ of submodules of M is said to be an E-direct system if, for any finite number of elements $i_1, i_2, ..., i_k$ of I there is an element $i_0 \in I$ such that $M_{i_0} \supseteq M_{i_1} + ... + M_{i_k}$ and M_{i_0} is non-essential submodule of M.

Theorem 3.5. *(Satyanarayana[20]): For an R-module M the following two conditions are equivalent:*
(i) M has FGD; and (ii) Every E-direct system of non-zero submodules of M is bounded above by a non-essential submodule of M.

Now we review the concepts of fuzzy demension in modules. We generalize the definition of fuzzy dimension given in Fu-Zheng PAN[5,6].

Definition 3.4. Let M be a unitary R-module and $\mu : M \to [0,1]$ is a mapping. μ is said to be a *fuzzy submodule* if the following conditions hold: (i) $\mu(m+m^1) \geq \min\{\mu(m), \mu(m^1)\}$ for all m and $m^1 \in M$; and (ii) $\mu(am) \geq \mu(m)$ for all $m \in M,\ a \in R$.

Proposition 3.1. *If M is a unitary R-module, $\mu : M \to [0,1]$ is a fuzzy set with $\mu(am) \geq \mu(m)$ for all $m \in M,\ a \in R$ then the following two conditions are true.*
(i) for all $0 \neq a \in R, \mu(am) = \mu(m)$ if a is left invertible; and (ii) $\mu(-m) = \mu(m)$.

Corollary 3.1. *If $\mu : M \to [0,1]$ is a fuzzy submodule and $m, m^1 \in M$, then $\mu(m - m^1) \geq$ min $\{\mu(m), \mu(m^1)\}$.*

Proposition 3.2. *If $\mu : M \to [0,1]$ is a fuzzy submodule, $m, m^1 \in M$ and $\mu(m) > \mu(m^1)$, then $\mu(m+m^1) = \mu(m^1)$.*

Corollary 3.2. *If $\mu : M \to [0,1]$ is a mapping satisfies the condition $\mu(am) \geq \mu(m)$ for all $m \in M$ and $a \in R$, then the following conditions are equivalent: (i) $\mu(m - m^1) \geq$ min $\{\mu(m), \mu(m^1)\}$; and (ii) $\mu(m+m^1) \geq$ min $\{\mu(m), \mu(m^1)\}$.*

Corollary 3.3. *If $\mu : M \to [0,1]$ is a fuzzy submodule and $m, m^1 \in M$ with $\mu(m) \neq \mu(m^1)$, then $\mu(m+m^1)$ = min $\{\mu(m), \mu(m^1)\}$.*

Proposition 3.3. *If $\mu : M \to [0,1]$ is a fuzzy submodule, then (i) $\mu(0) \geq \mu(m)$ for all $m \in M$; and (ii) $\mu(0) = sup_{m \in M} \mu(m)$.*

Theorem 3.6. *A fuzzy subset μ of a module M is a fuzzy submodule $\Leftrightarrow \mu_t = \{x \in M \mid \mu(x) \geq t\}$ is a submodule of M for all $t \in [0, \mu(0)]$.*

Definition 3.5. Let μ be any fuzzy submodule. The submodules $\mu_t,\ t \in [0,1]$ where $\mu_t = \{x \in M \mid \mu(x) \geq t\}$ are called *level submodules* of μ.

Note. Let $M_1 \subseteq M$. Define $\mu(x) = 1$ if $x \in M_1$ and $\mu(x) = 0$ otherwise.
Then the following conditions are equivalent:
(i) μ is a fuzzy submodule; and M_1 is a submodule of M.

Proposition 3.4. *Let μ be a fuzzy submodule of M and $\mu_t,\ \mu_s$ (with $t < s$) be two level submodules of μ. Then the following two conditions are equivalent:*
(i) $\mu_t = \mu_s$; and there is no $x \in M$ such that $t \leq \mu(x) < s$.

In the remaining part of this section, we discuss some results from Satyanarayana, Godloza and Shaw[29].

Definition 3.6. An element $x \in M$ is said to be a *minimal element* if the submodule generated by x is minimal in the set of all non-zero submodules of M.

Theorem 3.7. *If M has DCC on its submodules, then every nonzero submodule of M contains a minimal element.*

There are modules which do not satisfy DCC on its submodules, but contains a minimal element. For this we observe the following example.

Example 3.3. Write $M = \mathbb{Z} \oplus \mathbb{Z}_6$. Now M is a module over the ring $R = \mathbb{Z}$. Clearly M have no DCC on its submodules. Consider $a = (0,2) \in M$. Now the submodule generated by a, that is, $\mathbb{Z}a = \{(0,0),(0,2),(0,4)\}$ is a minimal element in the set of all non-zero submodules of M. Hence a is a minimal element.

Every minimal element is an u-element. The converse is not true, observe the Example 3.4.

When we consider M as a vector space over the field of real numbers $\mathbb{R}$, then every non-zero element is a minimal element as well as an u-element.

Example 3.4. Write $M = \mathbb{Z}$ as a module over the ring $R = \mathbb{Z}$. Since $\mathbb{Z}$ is a uniform module, and 1 is a generator, we have that 1 is an u-element. But $2\mathbb{Z}$ is a proper submodule of $1.\mathbb{Z} = \mathbb{Z} = M$. Hence 1 cannot be a minimal element. Thus 1 is an u-element but not a minimal element.

Theorem 3.8. *Suppose μ is a fuzzy submodule of M. (i) If $a \in M$, then for any $x \in Ra$ we have $\mu(x) \geq \mu(a)$; and (ii) If a is a minimal element, then for any $0 \neq x \in Ra$ we have $\mu(x) = \mu(a)$.*

Lemma 3.1. *If x is an u-element of a module M with DCC on submodules, then there exist minimal element $y \in Rx$ such that $Ry \leq_e Rx$.*

Theorem 3.9. *If M has DCC on its submodules, then there exist linearly independent minimal elements $x_1, x_2, ..., x_n$ in M where $n = \dim M$, and the sum $< x_1 > + ... + < x_n >$ is direct and essential in M. Also $B = \{x_1, x_2, ..., x_n\}$ forms a basis for M.*

Definition 3.7. Let M be a module and μ a fuzzy submodule of M. The elements $x_1, x_2, \ldots, x_n \in M$ are said to be *fuzzy $\mu-$linearly independent* (or *fuzzy linearly independent with respect to μ*) if (i) $x_1, x_2, ..., x_n$ are linearly independent; and (ii) $\mu(y_1 + ... + y_n)$ = min $\{\mu(y_1), ..., \mu(y_n)\}$ for any $y_i \in Rx_i, 1 \leq i \leq n$.

Theorem 3.10. *(Satyanarayana, Godloza and Shaw[29]): Let μ be a fuzzy submodule on M. If $x_1, x_2, ..., x_n$ are minimal elements in M with distinct μ-values, then $x_1, x_2, ..., x_n$ are (i) linearly independent, and (ii) fuzzy μ-linearly independent.*

Definition 3.8. (i) Let μ be a fuzzy submodule on M. A subset B of M is said to be a *fuzzy pseudo basis* for μ if B is a maximal subset of M such that $x_1, x_2, ..., x_k$ are fuzzy linearly independent for any finite subset $\{x_1, x_2, ..., x_k\}$ of B.
(ii) Consider the set $\mathscr{B} = \{k$: there exist a fuzzy pseudo basis B for μ with $|B| = k\}$. If $\mathscr{B}$ has no upper bound then we say that the *fuzzy dimension of* μ is infinite.

We denote this fact by $S - dim\ (\mu) = \infty$. If $\mathscr{B}$ has an upper bound, then the *fuzzy dimension* of μ is sup $\mathscr{B}$.
We denote this fact by $S - dim(\mu) = \sup \mathscr{B}$. If $m = S - dim(\mu) = \sup \mathscr{B}$, then a fuzzy pseudo basis B for μ with $|B| = m$, is called as *fuzzy basis* for the fuzzy submodule μ.

Note. Suppose M has FGD and μ is a fuzzy submodule on M. Then
(i) $|B| \leq dim\ M$ for any fuzzy pseudo basis B for μ; and (ii) $S - dim\ (\mu) \leq\ dim\ M$.

Definition 3.9. A module M is said to have a *fuzzy basis* if there exists an essential submodule A of M and a fuzzy submodule μ on A such that $S - dim\ (\mu) = dim\ M$.

The fuzzy pseudo basis of μ is called as fuzzy basis for M.

Remark 3.2. If M has FGD, then every fuzzy basis for M is a basis for M.

Theorem 3.11. *Let M be a module with DCC on submodules. Then M has a fuzzy basis (In other words, there exists an essential submodule A of M and a fuzzy submodule μ of A such that S-dim (μ) = dim M).*

4. Finite spanning dimension in modules

All the rings considered here are associative rings with unity, and all modules are unital.

Definition 4.1. A submodule A of M is said to be a small submodule if it satisfies the following condition:
$A + H = M$ and H is a submodule of $M \Rightarrow H = M$.

Example 4.1. (i) M is not a small submodule of M. (ii) (0) is a small submodule of M. (iii) Let M be a local module with unique maximal submodule H. Let X be a submodule of M such that $X \neq M$. Now we show that X is a small submodule of M. For this, take a submodule Y in M such that $X + Y = M$. If $Y \neq M$, then $Y \subseteq H$. Since X is also proper, $X \subseteq H$. So $M = X + Y \subseteq H + H = H$. $\Rightarrow H$ is not a proper submodule, a contradiction to the fact that H is unique maximal submodule. So we conclude that every proper submodule of a local module is a small submodule of M.
(iv) Let M be a completely reducible module. Then we can write $M = \Sigma_{i \in I} S_i$ (direct sum) of simple submodules $S_i, i \in I$ of M.
Then for any $k \in I$, we have that $M = S_k \oplus \Sigma\{S_j \mid j \in I, j \neq k\}$ and $\Sigma\{S_j \mid j \in I, j \neq k\} \neq M$. So each simple submodule S_i (for $i \in I$) is not a small submodule of M.

Note. Let X, Y be submodules of M such that $Y \subseteq X$ and Y is small in X. Then Y is also small in M.

Verification: Let $Y + H = M$ and H is a submodule of M. Then $X = M \cap X = (Y + H) \cap X = Y + (H \cap X)$. Since Y is small in X, we have that $X = H \cap X \Rightarrow X \subseteq H \Rightarrow Y \subseteq X \subseteq H \Rightarrow M = Y + H \subseteq H + H = H$.

Definition 4.2. We say that M is said to have finite spanning dimension (FSD, in short) if for every strictly decreasing sequence $U_0 \supseteq U_1 \supseteq \ldots$ of submodules, there is a k such that U_n is small in M for every $n \geq k$.

Example 4.2. (i) Any artinian module is a module with FSD. (ii) Consider the module $M = \mathbb{Z}$ over the ring $R = \mathbb{Z}$ of integers, and the submodule $2^n\mathbb{Z}$ generated by 2^n. Now $2^n\mathbb{Z} + 3\mathbb{Z} = M$ but $3\mathbb{Z} \neq \mathbb{Z}$. So $2^n\mathbb{Z}$ is not a small submodule for all $n \geq 1$.
So $2\mathbb{Z} \supseteq 2^2\mathbb{Z} \supseteq 2^3\mathbb{Z} \supseteq \ldots$ is an infinite strictly decreasing sequence of non-small submodules. Hence the module $M = \mathbb{Z}$ over the ring $R = \mathbb{Z}$ is not having FSD.

In the study of finite Goldie dimension, the concepts uniform submodule, and complement submodule play a vital role. The dual notions of uniform submodule and complement submodule are hollow submodule and supplement submodule respectively.

Definition 4.3. (i) M is said to be hollow if every proper submodule of M is small in M. (ii) Let U be a submodule of M. A submodule X of M is said to be a supplement of U in M if $U + X = M$ and $U + Y \neq M$ for any proper submodule Y of X. (In other words, X is a minimal submodule of M satisfying the property $U + X = M$).

Example 4.3. Let M be a finitely generated and completely reducible module. Then $M = S_1 \oplus \ldots \oplus S_n$ for some simple submodules S_i, $1 \leq i \leq n$ of M.
(i) Since S_i contains only the submodules (0) and S_i, we have that (0) is a small submodule in S_i. Hence S_i is a hollow submodule.
(ii) For any $1 \leq k \leq (n-1)$, we have that $S_1 \oplus \ldots \oplus S_k$ is a supplement of $S_{k+1} \oplus \ldots \oplus S_n$.

Theorem 4.1. *Let M be a module with FSD, and H be a non-small submodule of M. If every proper submodule of H is small in M, then H is hollow.*

Theorem 4.2. *Let M be a module with FSD and X is a non-small submodule of M. Then X contains a hollow submodule which is not small in M.*

Lemma 4.1. *Suppose M has FSD. (i) If Y and X are two submodules of M such that $Y + X = M$, then there exists a submodule Z of M such that $Z \subseteq X$, and Z is a supplement of Y in M. ii) If M has FSD, then every submodule of M has a supplement.*

Theorem 4.3. *(Fleury[4]) Let M have FSD. Then there is an integer p and hollow submodules N_i(for $1 \leq i \leq p$) such that $M = N_1 + \ldots + N_p$ and no N_i can be omitted (that is, the sum is irredundant). If there exists a summation $M = N_1^1 + \ldots + N_q^1$ with the same property, then $p = q$.*

Note. Let M be a module having FSD. The integer p determined in the Theorem 4.3 is called the spanning dimension of M, and it is denoted by $Sd(M)$.

Theorem 4.4. *(Fleury[4]) Let M be a module with FSD and $K \subseteq M$ be a supplement submodule. Then (i) K has FSD. (ii) If $Sd(K) = Sd(M)$, then $K = M$.*

Remark 4.1. For a general submodule, the condition (ii) of the Theorem 4.4 need not be true. That is, there exist a module M and a submodule K in M such that $Sd(K) = Sd(M)$ and $K \neq M$.

For example, consider the module $M = \mathbb{Z}_{24}$ over the ring $\mathbb{Z}$ of integers. Since $M = (8) + (3)$, the sum of two hollow submodules, we have that $Sd(M) = 2$. If $H = (2)$, then $H = (8) + (6)$ and so $Sd(H) = 2$. Thus we have $Sd(M) = Sd(H)$ but $M \neq H$.

Fluery[4] used the supplement submodules to obtain the condition $Sd(M/K) = Sd(M) - Sd(K)$, whereas Goldie used the complement submodules.

Theorem 4.5. *(Fleury[4]) Let M be a module with FSD and $K \subseteq M$ be a supplement. Then M/K has FSD and $Sd(M/K) = Sd(M) - Sd(K)$.*

Definition 4.4. Let K be a supplement of a submodule H in M. Then there exists a supplement H^1 of K with $H^1 \subseteq H$. Now H^1 and K are called mutual supplements.

Theorem 4.6. *(Satyanarayana[18]) If M is a module with FSD, H and K are mutual supplements, then $Sd(M) = Sd(H) + Sd(K)$.*

Theorem 4.7. *(Satyanarayana[18]) (The converse of Theorem 4.5) Let M be a module with FSD and H is a submodule of M also with FSD. If $Sd(M) = Sd(H) + Sd(M/H)$, then H is a supplement of some submodule in M.*

Definition 4.5. A family$\{M_i\}_{i \in I}$ of submodules of M is said to be an S-inverse system if for any finite number of elements $i_1, i_2, ..., i_k$ of I, there is an element $i_0 \in I$ such that $M_{i_0} \subseteq M_{i_1} + M_{i_2} + ... + M_{i_k}$ and M_{i_0} is non-small.

Theorem 4.8. *(Satyanarayana[20]) (i) If M has FSD, then every S-inverse system of submodules of M is bounded below by a non-small submodule of M. (ii) If every S-inverse system of submodules of M is bounded below by a non-small and non-hollow submodule, then M has FSD.*

5. Goldie dimension in N-Groups (where N is a near-ring)

Definition 5.1. An algebraic system $(N, +, .)$ is called a near-ring (or a right near-ring) if it satisfy: (i) $(N, +)$ is a group (not necessarily Abelian); (ii) $(N, .)$ is a semigroup; and (iii) $(n_1 + n_2)n_3 = n_1n_3 + n_2n_3$ (right distributive law) for all $n_1, n_2, n_3 \in N$.
$N_c = \{n \in N \mid nn^1 = n \text{ for all } n^1 \in N\} = \{n \in N \mid n0 = n\}$. N_c is called the constant part of N. If $N = N_c$ then we say that N is a constant near-ring.
$N_0 = \{n \mid n0 = 0\}$ is called the zero-symmetric part of N. If $N = N_0$ then we say that N is a zero-symmetric near-ring.
Throughout we consider N for a zero-symmetric right near-ring.

Definition 5.2. Let N be a near-ring. By an $N - group_N G$, we mean an additively written group G (but not necessarily Abelian), together with a mapping $N \times G \to G$ (denote the

image of (n,g) by $n.g$) satisfying the following conditions: (i) $(n_1+n_2)g = n_1g+n_2g$; and (ii) $n_1(n_2g) = (n_1n_2)g$ for all $g \in G$ and $n_1, n_2 \in N$.
It is clear that ${}_NN$ is an N-group.

Definition 5.3. (i) A subgroup H of an N-group G is said to be a $N-$*subgroup* of G if $NH \subseteq H$. It is denoted by $H \leq G$.
(ii) A normal subgroup K of G is said to be an *ideal* of G if $n(g+k)-ng \in K$ for all $n \in N, g \in G$ and $k \in K$. (equivalently, for all $n \in N, g \in G$ and $k \in K$, we have $n(k+g)-ng \in K$). Since N is zero symmetric, it is clear that every left ideal is an N-subgroup.

Note. (i) If H is an ideal of an N-group G, then this fact will be denoted by $H \trianglelefteq_N G$.
(ii) If H is an N- Subgroup of G then this fact will be denoted by $H \leq_N G$.
(iii) If $H \subseteq G$, then the intersection of all ideals of G containing H is called as the ideal generated by H (and it is denoted by $< H >$). If $H = \{a\}$, then we write $< a >$ in instead of $< \{a\} >$.

Definition 5.4. Let G and G^1 be two N-groups. A mapping $f : G \to G^1$ is called a $N-$ *homomorphism* if (i) $f(g_1+g_2) = f(g_1)+f(g_2)$; and (ii) $f(ng) = n(f(g))$ for all $g_1, g_2 \in G$ and $n \in N$.

Theorem 5.1. *(Pilz[13]): L is a left ideal of* $N \Rightarrow N_0L \subseteq L$.

Notation 5.1. For any two subsets H_1 and H_2 of G, we write $(H_1 : H_2) = \{n \in N \mid na \in H_1$ for all $a \in H_2\}$. For any element $x \in G, (0:x)$ is called the *annihilator* of x and for any subset H of $G, (0:H) = \{n \in N \mid nh = 0$ for all $h \in H\}$ is called the *annihilator* of H.

(Pilz[13]): Let G be an N-group. Then (i) for all $a \in G, (0:a)$ is left-ideal of N; (ii) for all N-subgroups H of G, $(0:H)$ is an ideal of N.

Definition 5.5. (i) An ideal I of G is said to be *finitely generated* if there exist $a_i \in G, 1 \leq i \leq k$ such that $I =< a_1 > + < a_2 > + ... + < a_k > =< \{a_1, a_2, ..., a_k\} >$. Let us recall that $< a_i >$ denotes the ideal of G generated by a_i, for each i.

(Pilz[13]): (i) Let I be an ideal of N and G be an N-group with $I \subseteq (0:G)$. Then $(n+I)g = ng$ makes G into an N/I- group G.
(ii) If G is an N/I-group then $ng = (n+I)g$ makes G into an N-group with $I \subseteq (0:G)_N$.
(iii) If $G = Na$ for some $a \in G$, then G is N-isomorphic to $N/(0:a)$.

The concept of Goldie dimension was generalized to N-groups by Reddy and Satyanarayana[16].

Now we list some definitions and results from Reddy and Satyanarayana[16].

Definition 5.6. Let H and K be ideals of G. H is said to be *essential* in K (written as, $H \leq_e K$) if it satisfies the following two conditions: (i) $H \subseteq K$; and (ii) $H \cap L = (0)$, L is an ideal of $G, L \subseteq K \Rightarrow L = (0)$. Moreover, an ideal H of G is said to be *essential* if it is essential in G.

Note. (i) From the above definition, it is clear that if I is an essential ideal of G and K is an ideal of G such that $I \subseteq K$, then K is essential.

Remark 5.1. (i) Intersection of a finite number of essential ideals is essential; (ii) If I,J,K are ideals of G such that $I \leq_e J$ and $J \leq_e K$ then $I \leq_e K$; (iii) $I \leq_e J \Leftrightarrow I \cap K \leq_e J \cap K$; and (iv) If $I \subseteq J \subseteq K$ then $I \leq_e K \Leftrightarrow I \leq_e J$ and $J \leq_e K$.

Definition 5.7. A non-zero ideal H of G is said to be *uniform* if for each pair of ideals K_1 and K_2 of G such that $K_1 \cap K_2 = (0), K_1 \subseteq H, K_2 \subseteq H$ implies $K_1 = (0)$ or $K_2 = (0)$.

Definition 5.8. An ideal H of G is said to have *finite Goldie dimension* (written as, FGD) if H does not contain an infinite number of non-zero ideals of G whose sum is direct.

Note. Let G be an N-group.Then every ideal of G contained in a uniform ideal of G is uniform.

Result 5.1. (Reddy and Satyanarayana[16]): Let G be an N-group with FGD then every non-zero ideal of G contains a uniform ideal.

Result 5.2. (Reddy & Satyanarayana[16]): Let H be an ideal of G.
(i) H has finite Goldie dimension if and only if for any sequence $H_0 \subseteq H_1 \subseteq H_2 \subseteq ...$ of ideals of G with $H_i \subseteq H$ for each i, there exists an integer k such that H_t is essential in H_{t+1} for all $t \geq k$;
(ii) If G has FGD then every ideal of G has FGD.

Notation 5.2. For any non-empty subset A of G, we write
$A_i^* = \{g+x-g \mid m \in N, x \in A_i\}$; $A_i^0 = \{x-y \mid x,y \in A_i\}$; and
$A_i^+ = \{n(g+x)-ng \mid n \in N, g \in G$ and $x \in A_i\}$.

Let X be a non-empty subset of G and write $X_0 = X$, and $X_{i+1} = X_i^* \cup X_i^+ \cup X_i^0$ for all integers $i \geq 0$. Then $X_0 \subseteq X_1 \subseteq X_2 \subseteq ...$ and clearly $\bigcup_{i=0}^{\infty} X_i$ is the ideal generated by X.

Result 5.3. Suppose $H_1, H_2, ..., H_t, K_1, K_2, ..., K_t$ are ideals of G such that the sum $K_1 + K_2 + ... + K_t$ is direct and $H_i \subseteq K_i$ for $1 \leq i \leq t$. Then the following two conditions are equivalent: (i) H_i is essential in K_i for $i = 1,2,...,t$; and (ii) $H_1 + H_2 + ... + H_t$ is essential in $K_1 \oplus K_2 \oplus ... \oplus K_t$.

Theorem 5.2. *(Reddy and Satyanarayana[16]) The following two conditions are equivalent: (i) G has finite Goldie dimension; and (ii) there exist uniform ideals $U_1, U_2, ..., U_n$ in G such that the sum $U_1 + U_2 + ... + U_n$ is direct and essential in G. This number n is independent of uniform ideals.*

Definition 5.9. Suppose G has finite Goldie dimension. Then the integer determined in Theorem 5.7. is called the *dimension* of G (or Goldie dimension of G) and it is denoted by $dim\ G$. We write, $dim\ G = n$.

Result 5.4. Suppose $dim\ G = n$. Then (i) The number of summands in any decomposition of H as the direct sum of non-zero ideals of G is at most n. (ii) An ideal H of G is essential if and only if H contains a direct sum of n uniform ideals.

Note. (i) Suppose G has FGD and H is an ideal of G. Then by Result 5.8, H has FGD. So there exist k uniform ideals in G whose sum is direct and essential in H. This number k is called the *dimension* of H and it is denoted by $dim\ H$. We write, $dim\ H = k$.
(ii) From (i), it is clear that H is uniform if and only if $dim\ H = 1$.

Result 5.5. (Reddy and Satyanarayana [16]): Suppose G has FGD and H, K are ideal of G. Then the following conditions holds: (i) H is essential if and only if $dim\ H = dim\ G$; (ii) If $H \cap K = (0)$ then $dim\ (H + K) = dim\ H + dim\ K$; and (iii) If $dim\ H < dim\ G$ then there exist uniform ideals $U_1, U_2, ..., U_k$ of G such that the sum $H + U_1 + ... + U_k$ is direct and essential in G. Moreover, $k = dim\ G - dim\ H$.

Theorem 5.3. *(Satyanarayana [25]) If G has finite Goldie dimension and K is a complement ideal of G, then* $dim\ (M/K) = dim\ M - dim\ K$.

Definition 5.10. A family $\{H_i\}_{i \in I}$ of ideals of G is said to be an E-direct system if, for any finite number of elements $i_1, i_2, ..., i_k$ of I there is an element $i_0 \in I$ such that $H_i \supseteq H_{i_1} + ... + H_{i_k}$ and H_{i_0} is non-essential submodule of M.

Theorem 5.4. *(Satyanarayana and Syam Prasad [33]): The following two conditions are equivalent: (i) G has FGD; and (ii) Every E-direct system of non-zero ideal of G is bounded above by a non-essential ideal of G.*

The concepts: "Linearly independent elements" and "uniform element (or u-element)" in the theory of N-groups were introduced and studied by Satyanarayana and Syam Prasad [35].
(i) Let X be a subset of G. X is said to be a *linearly independent* (l.i., in short) set if the sum $\sum_{a \in X} < a >$ is direct. If $\{a_i \mid 1 \le i \le n\}$ is a l.i. set, then we say that the elements $a_i, 1 \le i \le n$ are linearly independent. If X is not a l.i. set then we say that X is a linearly dependent (l.d., in short) set.
(ii) An element $0 \ne u \in G$ is said to be *uniform element* (u-element, in short) if $< u >$ is an uniform ideal of G.

Result 5.6. (a) Suppose G has FGD. If H is a non-zero ideal of G then H contains an u-element.
(b) (i) If $a_i, 1 \le i \le m$ are l.i. elements in G then $m \le n$ where $n = dim\ G$.
(ii) $dim\ G$ is equals to the least upper bound of the set A where $A = \{m \mid m$ is a natural number and there exist $a_i \in G, 1 \le i \le m$ such that $a_i, 1 \le i \le m\}$ are l.i..
(iii) If $n = dim\ G$ and $a_i, 1 \le i \le n$ are l.i., then each (a_i) is an uniform ideal (in other words, each a_i is an u-element).

Definition 5.11. If $n = dim\ G$ and $a_i, 1 \le i \le n$ are l.i., then $\{a_i : 1 \le i \le n\}$ is called an *essential basis* for G.

Lemma 5.1. *Let $f : G \to G^1$ be an isomorphism and $x_i \in G, 1 \leq i \leq k$. Then $x_1, x_2, ..., x_k$ are l.i. elements in G if and only if $f_{(x_1)}, f_{(x_2)}, ..., f_{(x_k)}$ are l.i. elements in G^1.*

Definition 5.12. A subset X of G is said to be $u-$ *linearly independent* (u.l.i., in short) set if every element of X is an u-element and X is a l.i. set. Let $a_i \in G$ for $1 \leq i \leq n$. $a_i, 1 \leq i \leq n$ are said to be u.l.i. if $\{a_i \mid 1 \leq i \leq n\}$ is an u.l.i. set.

Result 5.7. (a) Suppose $n = dim\ G$ and $a_i, 1 \leq i \leq n$ are l.i. elements. Then (i) $a_i, 1 \leq i \leq n$ are u.l.i. elements; (ii) $\{a_i \mid 1 \leq i \leq n\}$ forms an essential basis for G; and (iii) the conditions (i) and (ii) are equivalent.
(b) Suppose G has FGD. (i) If $b_i, 1 \leq i \leq k$ are l.i. elements, then there exists u-elements $a_i \in < b_i >, 1 \leq i \leq k$ such that $a_i, 1 \leq i \leq k$ are u.l.i. elements; (ii) If H is a non-zero ideal of G, then there exists an u.l.i set $X = \{a_i \mid 1 \leq i \leq k\}$ such that $< X >= \oplus_{i=1}^{k} < a_i > \leq_e H$. Moreover $dimH = k$.

Theorem 5.5. *(a) If G has FGD, then K is a complement ideal of G if and only if there exist u.l.i. elements $u_1 + K, u_2 + K, ..., u_m + K$ in G/K which spans G/K essentially with $m = dimG - dimK$.*
(b) Suppose G has FGD, $dimG = n, k < n$. If $u_1, u_2, ..., u_k$ are u.l.i. elements of G then there exists $u_{k+1}, ..., u_n$ in G such that $u_1, u_2, ..., u_k, u_{k+1}, ..., u_n$ are elements of G which spans G essentially.

Theorem 5.6. *If G has FGD then the following are equivalent: (i) dim $G = n$; (ii) there exist n uniform ideals $U_i, 1 \leq i \leq n$ whose sum direct and essential in G; (iii) the maximum number of u.l.i. elements in G is n; (iv) n is maximum with respect to the property that for any given $x_1, x_2, ..., x_k$ of u.l.i. elements with $k < n$, there exist $x_{k+1}, ..., x_n$ such that $x_1, x_2, ..., x_n$ are u.l.i. elements; (v) the maximum number of l.i. elements that can spans G essentially is n; (vi) the minimum number of u.l.i. elements that can span G essentially is n.*

The dimension concept in modules over matrix nearrings were introduced in Satyanarayana and Syam Prasad[36].

Notation 5.3. Consider a near-ring N with multiplicative identity 1. N^n denotes the direct sum of n-copies of $(N, +)$. For any $r \in N$, $1 \leq i \leq n$ and $1 \leq j \leq n$, define $f_{ij}^r : N^n \to N^n$ as $f_{ij}^r(a_1, a_2, ..., a_n) = (0, ..., ra_j, ..., 0)$ (here ra_j is in i^{th} place). If $f^r : N \to N$ defined by $f^r(x) = rx$ for all $x \in N, I_i : N \to N^n$ is the canonical monomorphism; and $\pi_j : N^n \to N$ is the jth projection map, then it is clear that $f_{ij}^r = I_i f^r \pi_j$ and $f_{ij}^r \in M(N^n)$ where $M(N^n)$ is the near ring of all mappings from $N^n \to N^n$. The subnear-ring $M_n(N)$ of $M(N^n)$ generated by $\{f_{ij}^r \mid r \in N, 1 \leq i \leq n, 1 \leq j \leq n\}$ is called the matrix near-ring over N. Now N^n becomes an $M_n(N)$-group and eventually obtained the following.

Theorem 5.7. *(Satyanarayana and Syam Prasad[36]) The Goldie dimension of the N-group N is equal to that of the $Mn(N)$-group N^n.*

6. Finite spanning dimension in N-groups

The concept of finite spanning dimension in N-groups was introduced by Reddy-Satyanarayana[17].

Definition 6.1. Let G be a N-group where N is a zero-symmetric right near-ring.
(i) A subset S of G is said to be small in G if $S+K=G$ and K is an ideal of G, imply $K=G$.
(ii) G is said to be hollow if every proper ideal of G is small in G.
(iii) G is said to have finite spanning dimension (FSD, in short) if for any decreasing sequence of N-subgroups $X_0 \supset X_1 \supset X_2 \supset X_3 \ldots$ of G such that X_i is an ideal of X_{i-1}, there exists an integer k such that X_j is small in G for all $j \geq k$.
Clearly every N-group with DCC on N-subgroups has FSD.
(iv) Suppose H and K are two N-groups of G. Then K is said to be a supplement for H if $H+K=G$ and $H+K^1 \neq G$ for any proper ideal K^1 of K.

Notation 6.1. A strictly decreasing sequence of N-subgroups $X_1 \supset X_2 \supset X_3 \ldots$ of G is said to be an ideal sequence if X_i is an ideal of X_{i-1} for $i \geq 2$. If there exists infinite number of N-subgroups such that $X_1 \supset X_2 \ldots$ is an ideal sequence, then we say the ideal sequence is an infinite ideal sequence.

Theorem 6.1. *(Reddy and Satyanarayana[17]) If G has FSD, then (i) there exists hollow N-subgroups $H_1, H_2, \ldots, H_i$ such that $G = H_1 + H_2 + \ldots + H_t$ and no H_i cannot be omitted (that is, $G \neq H_1 + \ldots + H_{j-1} + H_{j+1} + \ldots + H_t$ for every $1 \leq j \leq t$).*
(ii) If there exist hollow N-subgroups $H_1^1, H_2^1, \ldots., H_q^1$ such that $H_1^1 + H_2^1 + \ldots. + H_q^1 = G$ and none of the terms can be deleted from the summation, then $q = t$.

Theorem 6.2. *(Satyanarayana[25]) (i) If L is supplement of a non-small ideal K then $Sd(G/K) = Sd(L)$;*
(ii) If K is a supplement of an ideal H of G then K and G/K has FSD and $Sd(G) = Sd(K) + Sd(G/K)$;
(iii) If $(0) \neq K$ and $Sd(G) = Sd(G/K) + Sd(K)$ then K is non-small and supplement; and
(iv) K is a supplement $\Leftrightarrow$ K has FSD and $Sd(G) = Sd(G/K) + Sd(K)$.

Definition 6.2. (i) Let H, H^1 be an ideals of G such that $H^1 \subseteq H$. Then H^1 is said to be small-1 ideal of H if $H^1 + K = H$ and K is an ideal of G implies $K = H$.
(ii) An ideal H of G is said to be hollow-1 ideal if every ideal of G which is properly contained in H is small-1 in H.
(iii) G is said to have finite spanning dimension-1 (FSD-1, in short) if for any infinite strictly descending chain of ideals $H_1 \supset H_2 \supset \ldots$ of G there exists an integer k such that H_j is small in G for all $j \geq k$.

Theorem 6.3. *(Satyanarayana and Syam Prasad*[34]*) (i) If G has FSD-1 then every s-inverse system of ideals of G is bounded below by a non-small ideal of G. (ii) If every s-inverse system of ideals of G is bounded below by a non-small and non-hollow-1 ideal then G has FSD-1.*
Several related examples were presented in Satyanarayana-Syam Prasad[34].

Bibliography

1. CAMILLO V. and ZELMANOWITZ J., On the dimension of a sum of modules, *Communications in Algebra*, **6**(4) (1978), 345–352.
2. CAMILLO V. and ZELMANOWITZ J., Dimension modules, *Pacific Jour. Math.*, **91**(2) (1980), 249–261.
3. Chatters A. W. & Hajarnivas C. R., Rings with Chain Conditions, *Research Notes in Mathematics, Pitman Advanced publishing program*, Boston-London-Melbourne, 1980.
4. Fleury P., A Note on Dualizing Goldie Dimension, *Canad. Math. Bull.* **17**(4) 1974.
5. Fu-Zheng Pan, Fuzzy finitely generated Modules, *Fuzzy sets and Systems* **21** (1987), 105–113.
6. Fu-Zheng Pan, Fuzzy Quotient Modules, *Fuzzy sets and Systems* **28** (1988), 85–90.
7. Golan J. S., Making Modules Fuzzy, *Fuzzy sets and Systems* **32** (1989), 91–94.
8. Goldie A. W., The Structure of Noetherian Rings, *Lectures on Rings and Modules, Springer - Verlag*, New York, Lecture Notes, **246** (1974), 213–31.
9. JAIN S. K., LAM T. Y. and ANDRE LEROY, On uniform dimensions of ideals in right nonsingular rings, *Journal of Pure and Applied Algebra*, **133** (1998), 117–139.
10. Lambek J., Lectures on Rings and Modules, *Blaisdell Publishing Co.*, 1966.
11. Meldrum J. D. P. and Van der Walt A. P. J., Matrix Near-rings, *Arch. Math.* **47** (1986), 312–319.
12. Negotia C. A. and Ralescu D. A., Applications of Fuzzy sets to system Analysis, *Birkhauser, Basel*, 1975.
13. Pilz G., Near-rings, *North-Holland pub.*, 1983.
14. Ranga Swamy K. M., Modules with Finite Spanning dimension, *Candian Math. Bull.* **20** (1977), 255–262.
15. Reddy Y. V. and Satyanarayana Bh., A Note on Modules, *Proc. Japan Acad.*, **63-A** (1987), 208–211.
16. Reddy Y. V. and Satyanarayana Bh., A Note on N-groups, *Indian J. Pure & Appl. Math.* **19** (1988), 842–845.
17. Reddy Y. V. and Satyanarayana Bh., *Finite Spanning Dimension in N-groups*, The Mathematics Student, **56** (1988), 75–80.
18. Satyanarayana Bh., On Modules with Finite Spanning Dimension, *Proc. Japan Academy* **61-A** (1985), 23–25.
19. Satyanarayana Bh., On Modules with *FSD* and a property (P), *Proceedings of the international conf. on Number Theory (Ramanujan Centenniel International Conference)* Annamalai University, Dec. 15–18, 1987, pp. 137–140.
20. Satyanarayana Bh., A Note on E-direct and S-inverse Systems, *Proc. Japan Academy* **64A** (1988), 292–295.
21. Satyanarayana Bh., Lecture on Modules with Finite Goldie dimension and Finite Spanning dimension, *International Conference on General Algebra*, Krems, Vienna, Austria, August 21–27, 1988.
22. Satyanarayana Bh., The Injective Hull of a Module with *FGD*, *Indian J. Pure & Appl. Math.* **20** (1989), 874–883.
23. Satyanarayana Bh., On Modules with Finite Goldie Dimension *J.Ramanujan Math. Society.* **5** (1990), 61–75.

24. Satyanarayana Bh., Lecture on Modules with Finite Spanning Dimension, *Asian Mathematical Society Conference*, University of Hong Kong, Hong Kong, August 14–18, 1990.
25. Satyanarayana Bh., On Finite Spanning dimension in N-groups, *Indian J. Pure and Appl. Math.* **22** (1991), 633–636.
26. Satyanarayana Bh., Modules with Finite Spanning Dimension, *J. Austral, Math. Society.* (Series A) **57** (1994), 170–178.
27. Satyanarayana Bh., On Essential E-irreducible submodules, *Proc., 4th Ramanujan symposium on Algebra and its Applications*, University of Madras, Feb 1–3 (1995), pp. 127–129.
28. Satyanarayana Bh., Contributions to Near-ring Theory, *VDM Verlag Dr Muller*, Germany, 2010, (ISBN 978-3-639-22417-7).
29. Satyanarayana Bh., Godloza L. and Mohiddin Shaw Sk., On Fuzzy Dimension of a Module with DCC on Submodules, *Acharya Nagarjuna International Journal of Mathematics and Information Technology*, **01** (2004) 13–32.
30. Satyanarayan Bh. and Mohiddin Shaw Sk., Fuzzy Dimension of Modules over Rings, *VDM Verlag Dr Muller*, Germany, 2010, (ISBN 978-3-639-23197-7).
31. Satyanarayana Bh., Mohiddin Shah Sk., Eswaraiah Setty S. and Babu Prasad M., A generalization of Dimension of Vector Space to Modules over Associative Rings, *International Journal of Computational Mathematical Ideas*, Vol. 1., No. 2 (2009) 39–46 (India) (ISSN: 0974-8652).
32. Satyanarayana Bh. and Nagarju Dasari, Dimension and Graph Theoretic Aspects of Rings, *VDM Verlag Dr Muller*, Germany, 2011, (ISBN 978-3-639-30558-6).
33. Satyanarayana Bh. and Syam Prasad K., A Result on E-direct systems in N-groups, *Indian J. Pure and Appl. Math.* **29** (1998), 285–287.
34. Satyanarayana Bh. and Syam Prasad K., On Direct and Inverse Systems in N-groups, *Indian J. Math (BN Prasad Birth Commemoration Volume)* **42** (2000), 183–192.
35. Satyanarayana Bh. and Syam Prasad K., Linearly independent Elements in N-Groups with finite Goldie dimension, *Bull Korean Math Soc.*, **42** (2005), 433–441.
36. Satyanarayana Bh. and Syam Prasad K., On Finite Goldie Dimension of $M_n(N)$-group N^n, *Nearrings and Nearfields (Edited by H. Kiechel, A. Kreuzer & M.J. Thomsen) (Proc. 18th International Conference on Nearrings and Nearfields*, Universitat Bundeswar, Hamburg, Germany, July 27–Aug 03, 2003) Springer Verlag, Netherlands, 2005, pp. 301–310 (Zbl. 1078.16059).
37. Satyanarayana Bh. and Syam Prasad K., Discrete Mathematics and Graph Theory, *Printice Hall of India*, New Delhi, 2009 (ISBN: 978-81-203-3842-5).
38. Satyanarayana Bh. and Syam Prasad K., Near-rings, Fuzzy Ideals and Graph Theory, *CRC Press (Taylor and Francis Group), England/New York*, 2013 (ISBN: 978-1-4398-7310-6).
39. Satyanarayana Bh. and Syam Prasad K., Discrete Mathematics and Graph Theory, *Printice Hall of India*, New Delhi, 2014 (ISBN: 978-81-203-4948-3).
40. Satyanarayana Bh., Syam Prasad K. and Nagaraju D., A Theorem on Modules with Finite Goldie Dimension, *Soochow Journal of Mathematics*, **32**, No. 2 (2006), 311–315.
41. Satyanarayana Bh., Syam Prasad K. and Venkata Pradeep Kumar T., On Fuzzy Dimension of N-groups with DCC on Ideals, *East Asian Math. J.* **21** (2005), No. 2, pp. 205–216.
42. Sharpe D. W. and Vamaos P., Injective Modules, *Cambridge University Press*, 1972.
43. Syam Prasad K. and Satyanarayana Bh., Dimension of N-groups and Fuzzy ideals in Gamma Near-rings, *VDM Verlag Dr Muller, Germany*, 2011, (ISBN 978-3-639-30558-624851-7).
44. Varadarajan K., Dual Goldie Dimension, *Communications in Algebra*, **7** (1979), 565–610.
45. Zadeh L. A., Fuzzy Sets, *Information and Control*, **8** (1965), 338–353.

On the prime radicals of near-rings and near-ring modules

Nico Groenewald

Nelson Mandela Metropolitan University,
Port Elizabeth,
South Africa
E-mail: nico.groenewald@nmmu.ac.za

We introduce different prime radicals for near-rings and near-ring modules.

1. Near-rings

1.1. *Different prime ideals*

The study of prime ideals for rings or semi-groups is facilitated by the equivalence of the two conditions on an ideal I of a ring (semigroup) R:

(1) If A and B are ideals of R such that $AB \subseteq I$, then $A \subseteq I$ or $B \subseteq I$;
(2) If $x, y \in R$ are such that $xRy \subseteq I$, then $x \in I$ or $y \in I$.

These conditions are **not equivalent** in the class of near-rings. For near-rings there are many non equivalent definitions of prime near-rings. In this talk we discuss the impact on research in near-rings of these different prime near-rings.

All near-rings are **right near-rings**. We will use $\mathcal{R}$, $\mathcal{N}$ and $\mathcal{N}_0$ denote the variety of all **rings**, **near-rings** and **zero-symmetric near-rings** respectively.

From Birkenmeier *et al.*[3] we have the following:

Example 1.1. Let R be a zero-symmetric near-ring such that condition (1) above is satisfied for the ideal (0) i.e. if $AB = (0)$, then $A = (0)$ or $B = (0)$.

Suppose there is $0 \neq r \in R$ such that $Rr = (0)$.

In Ref. 3, many examples of near-rings satisfying these conditions are given.

Now, since R is a zero symmetric near-ring we have $rRr = (0)$ with $0 \neq r$ and R does not satisfy condition 2.

For $K \subseteq R$, $< K|_R$, $| K >_R$, $< K >_R$, $< K]_R$ and $[K >_R$ denote the **left ideal**, **right ideal**, **two-sided ideal**, **left** R**-subgroup** and **right** R**-subgroup** generated by K in R respectively. If it is clear in which near-ring we are working, the subscript R will be omitted.

Also $K \lhd_l R$, $K \lhd_r R$, $K \lhd R$ and $K \lhd_R R$ symbolize that K is a **left ideal, right ideal**, **two-sided ideal** or a **left** R**-subgroup** of R.

As we saw above in the study of near-rings one is quickly confronted by the fact that many conditions which are equivalent for rings are not necessarily equivalent for near-rings.

The following are possible candidates for a prime ideal in a near-ring.

Definition 1.1. Let R be a near-ring (not necessarily zero-symmetric) and P and ideal of R.

(1) P is a **0-prime** ideal if for every $A,B \lhd R$, $AB \subseteq P$ implies $A \subseteq P$ or $B \subseteq P$ (this is the same as the usual definition for a prime ideal in a ring).
(2) P is a **1-prime (r1-prime)** ideal if for every $A,B \lhd_l R$ $(A,B \lhd_r R)$, $AB \subseteq P$ implies $A \subseteq P$ or $B \subseteq P$
(3) P is a **2-prime (r2-prime)** ideal if for every A and B left R-subgroups (right $R-$subgroups) of R, $AB \subseteq P$ implies $A \subseteq P$ or $B \subseteq P$
(4) P is a **3-prime** ideal if for $a,b \in R$, $aRb \subseteq P$ implies $a \in P$ or $b \in P$.

R is called a **i-prime near ring** $(i = 0,1,r1,2,r2,3)$ if the **zero ideal** is a i-prime ideal. Denote the class of i-prime near-rings by $\mathcal{P}_i$.
Observe that the various prime conditions have the following relations:
3-prime $\Rightarrow$ 2-prime
2-prime$\Rightarrow$ 1-prime if R is zero-symmetric and
1-prime$\Rightarrow$ 0-prime.
We also have 3-prime $\Rightarrow$ $r2$-prime $\Rightarrow$ $r1$-prime $\Rightarrow$ 0-prime.

The notations for types 0 through 2 is due to Holcombe[16], while type 3 is due to the author[13], and types $r1$ and $r2$ are due to Birkenmeier[2].

Examples showing that types $0,1,2$ and 3 are distinct are provided in Fererro[11].

- Near-ring number 12 defined on $\mathbb{Z}_4$ Pilz[21] shows that $r1-$prime near-rings need not be $1-$prime and
- near-ring number 7 defined on S_3 Pilz[21] is an example of a $1-$prime near-ring which is not $r1-$prime.
- Near-ring number 10 defined on S_3 Pilz[21] is an example of a $2-$prime near-ring which is not $r2-$prime and
- near-ring number 6 defined on $\mathbb{Z}_6$ Pilz[21] is an example of an $r2-$prime near-ring which is not $2-$prime.
- Near-ring number 17 defined on S_3 Pilz[21] is an example of an $r1-$prime near-ring which is not $r2-$prime and
- near-ring number 20 on S_3 Pilz[21] is an example of an $r2-$prime near-ring which is not $3-$prime.

This is in sharp contrast to the ring case where $0-$prime and $3-$prime are equivalent.

1.2. *Different prime radicals*

The different definitions of prime ideals give rise to different prime radicals.

Definition 1.2. Let $R \in \mathcal{N}$. Then $\mathfrak{P}_\nu(R) = \cap\{P \lhd R : P \text{ is } \nu\text{-prime}\}$is the **$\nu$-prime radical** of R for $\nu \in \{0,1,r1,2,r2,3\}$.

Let ρ be a mapping which assigns to each near-ring R an ideal $\rho(R)$ of R. Such mappings will be called ideal -mappings

Definition 1.3. An ideal mapping ρ is a **Hoehnke radical** *(H-radical also called a radical map)* if it satisfies the following conditions:

(H1) $(\rho(R)+I)/I \subseteq \rho(R/I)$ for all $I \triangleleft R$;
(H2) $\rho(R/\rho(R)) = 0$ for all R.

The Hoehnke radicals are very general:

Let $\mathcal{M}$ be a class of near-rings and let ρ be the mapping which assigns to each near-ring R the ideal $\rho(R) = \cap\{I \triangleleft R : R/I \in \mathcal{M}\}$.The mapping ρ is an H-radical.

It is clear that such a radical only gives information on the relationships between the radical $\rho(R)$ of R and the radical of a homomorphic image of R.

Clearly all the ν-prime radical maps $\mathfrak{P}_\nu$ are Hoehnke radicals.

If $I \triangleleft R$ then a Hoehnke radical gives no information on the relationships between $\rho(R)$ and $\rho(I)$. But this is, amongst others, what the general theory of radicals is all about:

Given a near-ring R, then it should provide some information on the relationship between $\rho(R)$ and the radicals of near-rings related to R e.g., homomorphic images, ideals, extensions, etc. The following relationships between the radicals of a near-ring and its ideals play an important role in the general theory of radicals:

Definition 1.4. An H-radical ρ is:

(H3) **complete** if $\rho(I) = I \triangleleft R$ implies $I \subseteq \rho(R)$;
(H4) **idempotent** if $\rho(\rho(R)) = \rho(R)$;
(H5) **ideal-hereditary** if $\rho(I) = I \cap \rho(R)$.

If ρ is an H-radical which is **idempotent and complete**, then it is called a **Kurosh-Amitsur** *(KA) radical* map.

Since all the prime radicals are Hoehnke radicals, a natural question to ask is: **Which of the prime radicals are KA−radicals?**

In Ref. 4, Birkenmeier *et al.* proved that if S is a subnear-ring of R then $S \cap \mathfrak{P}_0(R) \subseteq \mathfrak{P}_0(S)$ and from Miltz and Veldsman[20] it now follows that $\mathfrak{P}_0$ is idempotent. In Ref. 19, an example was given by Kaarli to show that $\mathfrak{P}_0$ is not complete. Hence $\mathfrak{P}_0$ is not a KA-radical. There are examples to show that $\mathfrak{P}_1(\mathfrak{P}_1(R)) \neq \mathfrak{P}_1(R)$ and therefore $\mathfrak{P}_1$ is not idempotent. Thus $\mathfrak{P}_1$ is not a KA-radical. There are examples of finite near-rings in $\mathcal{N}$ for which $\mathfrak{P}_2$ is not complete and $\mathfrak{P}_3$ is not idempotent.What the situation is for $\mathfrak{P}_2$ and $\mathfrak{P}_3$ in $\mathcal{N}_0$ is not known For a long time it was believed that it is **not possible** to get a KA-**prime radical** for near-rings.

Initiated by Booth and then in conjunction with Veldsman and the present author, we could show that there is a KA-prime radical. In Ref. 9, we introduced a different generalization to near-rings of a prime ring. This generalization, it turns out, has some very satisfactory consequences and not only from a radical viewpoint.

Definition 1.5. A near-ring R is *equiprime* if for any $0 \neq a \in R$ and $x, y \in R$, $anx = any$ for all $n \in R$ implies $x = y$.

It is easy to check that an equiprime near-ring is zero-symmetric and 3-prime.

Let $\mathfrak{P}_e$ denote the **equiprime radical map**:

$$\mathfrak{P}_e(N) = \cap\{I \triangleleft N : N/I \text{ equiprime}\}$$

then $\mathfrak{P}_e$ is an **ideal-hereditary** ***KA*-radical** in the variety of all near-rings i.e., $\mathfrak{P}_e(N) \cap I = \mathfrak{P}_e(I)$ for every $I \triangleleft N \in \mathcal{N}$.

Equiprime near-rings are not too restrictive.

- For any group G, the simple near-ring with identity $\mathcal{M}_0(G)$ is equiprime.
- Any simple near-ring with identity which satisfies the descending chain condition on $R-$subgroups is equiprime.

In Ref. 25, van der Walt defined the notion of a s-prime near-ring (strong prime near-ring) and showed that the s-prime radical determined by the class of all s-prime near-rings is the same as the upper nil radical.

Hence if R is a near-ring then $\mathbb{N}(R)$ i.e., the sum of all nil ideals of R is equal to $s(R)$ the intersection of all the s-prime ideals of R (all ideals I such that R/I is an s-prime near-ring). In Ref. 18, Kaarli observed that the nil radical $\mathbb{N}(R)$ of the near-ring R is equal to the intersection of all the 0-prime ideals P of R such that R/P has no nonzero nil ideals. He mentioned that the proof of this result is essentially that given for rings by Divinsky, see Ref. 10 page 147. In Ref. 6, Birkenmeier *et al.* called an ideal I of the near-ring R **nilprime** if I is a 0-prime ideal and $\mathbb{N}(R/I) = 0$ i.e., R/I has no nonzero nil ideals. They then gave a self-contained proof within near-ring theory of the above result mentioned by Kaarli. In Ref. 6, it was proved that every $s-$prime near-ring is a nilprime near-ring and left it as an **open question** whether **every nilprime near-ring is an $s-$prime near-ring**.

We now introduce another notion of an s-prime near-ring which coincides with the notion of nilprime.

Definition 1.6.

(1) The subset M of the near-ring R is called an $m-$**system** if for every $a, b \in M$ there exists $c \in <a><b>$ such that $c \in M$.
(2) The subset N of the near-ring R is called an $sp-$**system** if for every $a \in N$ there exists $c \in <a><a>$ such that $c \in N$.
(3) The subset S of the near-ring R is called an $s-$**system** if for every $a, b \in S$ there exists $c \in <a><b>$ such that $c^n \in S$ for all $n \in \mathbb{N}$.
(4) The subset U of the near-ring R is called an $ss-$**system** if for every $a \in U$ there exists $c \in <a><a>$ such that $c^n \in U$ for all $n \in \mathbb{N}$.

Clearly an $s-$system is an $m-$system and also an $ss-$system. Furthermore an $ss-$system is an $sp-$system.

Let $C_R(Q)$ denote the complement of Q in R. An ideal Q of the near-ring R is an $s-$**prime**, **prime (0-prime)**, $s-$**semiprime** or **semiprime (0-semiprime)** ideal if $C_R(Q)$ is an $s-$system, $m-$system, $ss-$system or a $sp-$system respectively.

Definition 1.7. The $s-$radical (0-prime radical) of R, denoted by $s(R)$ ($\wp_0(R)$), consists of all those elements $r \in R$ such that every $s-$system ($m-$system) which contains r also contains 0.

From Ref. 25, it follows that $\wp_0(R)$ is equal to the intersection of all the 0-prime ideals of R. Hence $\wp_0(R) = \mathfrak{P}_0(R)$.

We now have:

Theorem 1.1. *The $s-$radical $s(R)$ of the near ring R is equal to the intersection of all the $s-$prime ideals of R and coincides with the upper nil radical $\mathbb{N}(R)$ of R*

Remark 1.1. We now have for our definition of an $s-$prime ideal that the notions of $s-$prime near-ring and nilprime near-ring coincide.

Theorem 1.2. *If $Q \lhd R$, then Q is an s- prime ideal if and only if Q is nilprime.*

Definition 1.8. An s-prime ideal P is a minimal $s-$prime ideal containing an ideal I if $I \subseteq P$ and there does not exist an $s-$prime ideal P' in R such that $I \subseteq P' \subsetneqq P$.

Theorem 1.3. *If $s'(R)$ is the intersection of all the minimal $s-$prime ideals of R then $\mathbb{N}(R) = s'(R) = s(R)$.*

We have that there are a number of non-equivalent notions of prime near-rings which coincide in the case of associative rings. We also have that the upper nil radical of the near-ring R is equal to the intersection of all the nil 0-prime ideals. Because of this we can now introduce the following:

Definition 1.9. A near-ring is **i-nilprime** if R is i-**prime** and R contains **no nonzero nilideals** for $i \in \{0, 1, r1, 2, r2, 3, equi\}$.

If R is an associative ring, this coincides with the notion prime nil-semisimple rings and the upper radical determined by this class of rings coincides with the nilradical $\mathbb{N}(R)$. We now show that in the case of near-rings this give rise to a number of nonequivalent nilradicals.

Examples:

0-nilprime but not 1-nilprime

Example 1.2. Let G be a finite group and let $0 \neq H$ be a proper subgroup of G.

Let $R = \{a \in M_0(G) : a(H) \subseteq H\}$.Then R is a zero-symmetric near-ring and its only ideals are R, $A = (0 : H) = \{a \in R : a(H) = 0\}$ and 0.

Let $a \in R$ be defined by:

$$a(x) = \begin{cases} g \text{ if } x = g \\ 0 \text{ if } x \neq g \end{cases} \quad \text{for } g \in G \backslash H.$$

Now $a \in A$ and $a^n(g) = a^{n-1}(g) = \cdots = a(g) = g$. Hence $a^n \neq 0$ for all $n \in \mathbb{N}$. Thus the only nil ideal of R is 0. R is $0-$prime since $A^2 \neq 0$ and it follows that R is $0-$nilprime. This near-ring is not $1-$nilprime since R is not $1-$prime because if $I = (0 : G \backslash H)$ then I is a left ideal of R and $AI = (0 : H)(0 : G/H) = 0$.

1-nilprime but not 2-nilprime

Example 1.3. Let G be a nonabelian simple group and let $0 \neq H$ be a proper subgroup of G. If $g \in G$, define multiplication by:

$$g \cdot x = \begin{cases} 0 \text{ if } & x \in H \\ g \text{ if } x \in G \backslash H \end{cases}.$$

$(G,+,\cdot)$ is a near-ring and 0 is a $1-$prime ideal i.e.$(G,+,\cdot)$ is a $1-$prime near-ring. Since H is a proper left $G-$subgroup and $H^2 = 0$, we have $(G,+,\cdot)$ is not a $2-$prime near-ring. Furthermore, for every $0 \neq x \in G \backslash H$ we have $x^n = x^{n-2} \cdot (x \cdot x) = x^{n-1} = \cdots = x \neq 0$. Hence $\mathbb{N}((G,+,\cdot)) = 0$. Thus $(G,+,\cdot)$ is a $1-$nilprime near-ring but not a $2-$nilprime near-ring.

2-nilprime but not 3-nilprime

Example 1.4. Let R be the near-ring on $\mathbb{Z}_3 = \{0,1,2\}$ multiplication defined by: $a \cdot b = \begin{cases} a \text{ if } b = 2 \\ 0 \text{ if } b \neq 2 \end{cases}$.

The only $R-$subgroups of R are 0 and R. We also have $R^2 \neq 0$. Hence R is $2-$prime. R is not $3-$prime since $1R1 = 0$. Furthermore we have $2^n = 2$ for every $n \in \mathbb{N}$. Thus R is a $2-$nilprime near-ring but not a $3-$nilprime near-ring.

3-nilprime but not equi-nilprime

Example 1.5. If $(R,+)$ is any cyclic group of prime order p $(p > 2)$, define a near-ring multiplication on R by:

$$ab = \begin{cases} a \text{ if } b \neq 0 \\ 0 \text{ if } b = 0 \end{cases}$$

Then R is a near-ring which is $3-$nilprime but not equi-nilprime.

Example 1.6. Near-ring number 17 defined on S_3[21] is an example of an $r1-$nilprime near-ring which is not $r2-$nilprime and near-ring number 20 on S_3[21] is an example of an $r2-$nilprime near-ring which is not $r3-$nilprime.

If R is any near-ring and $\rho_{n_i}(R)$ denotes the $H-$radical determined by the class of $i-$nilprime near-rings, then

- $\mathbb{N}(R) = \rho_{n_0}(R) \subsetneqq \rho_{n_1}(R) \subsetneqq \rho_{n_2}(R) \subsetneqq \rho_{n_3}(R) \subsetneqq \rho_{n_e}(R)$ and
- $\mathbb{N}(R) = \rho_{n_0}(R) \subsetneqq \rho_{n_{r1}}(R) \subsetneqq \rho_{n_{r2}}(R) \subsetneqq \rho_{n_3}(R) \subsetneqq \rho_{n_e}(R)$.

NOTATION:

If $a, b \in R$ we will use the following notation:

$$[a]^i[b]^i = \begin{cases} <a><b> & \text{for } i = 0 \\ <a|<b| & \text{for } i = 1 \\ |a>|b> & \text{for } i = r1 \\ [a>_R [b>_R & \text{for } i = r2 \\ <a]_R <b]_R & \text{for } i = 2 \\ aRb & \text{for } i = 3 \end{cases}$$

NOTE:

An ideal Q of R is i-prime, $i \in \{0,1,r1,2,r2,3\}$, if for $a,b \in R$, $[a]^i[b]^i \subseteq Q$ implies $a \in Q$ or $b \in Q$.

Definition 1.10. A subset T of the near-ring R is called a **complete system** if $a^n \in T$ for every $a \in T$ and every $n \in \mathbb{N}$.

In what follows let $i \in \{0,1,r1,2,r2,3\}$.

Definition 1.11. A subset $Z \subseteq R$ is called an n_i-**system** if Z contains a complete system U such that for every $t_1, t_2 \in Z$, it follows that $< [t_1]^i[t_2]^i > \cap U \neq \varnothing$.

Definition 1.12. An ideal Q is called $i-s-$**prime,** if for $a,b \in R$ and for all $x \in < [a]^i[b]^i >$, $x^m \in Q$ for some m implies $a \in Q$ or $b \in Q$.

Theorem 1.4. *$Q \lhd R$ is $i-s-$prime $\Leftrightarrow$ Q is i-nilprime $\Leftrightarrow$ $C_R(Q)$ is an n_i-system.*

In Ref. 24, Veldsman introduced the notion of $s-$equiprime near-rings and proved that in the variety of rings it coincides with the $s-$prime rings of Van der Walt[25] Veldsman proved that the class of $s-$equiprime near-rings determines an ideal-hereditary generalization of the nil radical.

Definition 1.13.[24] Veldsman A near-ring R is s-equiprime if it contains a nonempty multiplicative closed set S with $0 \notin S$ such that $0 \neq a \in R$ and $T_R(a,x,y) \cap S = \varnothing$ implies $x = y$ $(x,y \in R)$ where $T_R(a,x,y) = \{\text{all finite sums } \sum_i r_i(as_ix - as_iy)k_i \text{ with } r_i, s_i, k_i \in R\}$. In such a case S is called the kernel of R.

Theorem 1.5. *Every $s-$equiprime near-ring is equi-nilprime.*

We know that **equiprime radical map**: $\mathcal{P}_e(R) = \cap\{I \lhd R : R/I \text{ equiprime}\}$ is an **ideal-hereditary** ***KA*-radical map** in the variety of all near-rings i.e., $\mathcal{P}_e(N) \cap I = \mathcal{P}_e(I)$ for every $I \lhd N \in \mathcal{N}$.

We have the following:

QUESTION:

- **If $\mathcal{M}_{n_e}$ is the class of equi-nilprime near-rings, is the equi-nilprime radical map $\rho_{n_e}(R) = \cap\{I \lhd R : R/I \text{ equi-nilprime}\}$ a *KA*-radical map?**
- If R is a near-ring we know that $\rho_{n_e}(R) \subseteq \mathcal{P}_e(R)$. When will $\rho_{n_e}(R) = \mathcal{P}_e(R)$?

2. Near-ring modules

2.1. *Different prime submodules*

We attempt to generalise the various notions of primeness that were defined in R to the module M. We provide various characterizations of prime modules and show equivalences between these characterizations. Let R be a near-ring and let M, be any left R-module and P a subset of R. If P is an $R-$ideal ($R-$submodule) of M we denote it by $P \lhd_R M$ ($P \leq_R M$).

Definition 2.1. Let $P \lhd_R M$ such that $RM \nsubseteq P$. Then P is called:

- **0-prime** if $AB \subseteq P$ implies $AM \subseteq P$ or $B \subseteq P$ for all ideals, A of R, and all R-ideals, B of M.
- **1-prime** if $AB \subseteq P$ implies $AM \subseteq P$ or $B \subseteq P$ for all left ideals, A of R, and all R-ideals, B of M.
- **2-prime** if $AB \subseteq P$ implies $AM \subseteq P$ or $B \subseteq P$ for all R-subgroups, A of R, and all R-submodules, B of M.
- **3-prime** if $rRm \subseteq P$ implies that $rM \subseteq P$ or $m \in P$ for all $r \in R$ and $m \in M$.
- **completely prime** (c-prime) if $rm \in P$ implies that $rM \subseteq P$ or $m \in P$ for all $r \in R$ and $m \in M$.

Definition 2.2. M is said to be a ν-prime ($\nu = 0, 1, 2, 3, c$) R-module if $RM \neq 0$ and 0 is a ν-prime R-ideal of M.

In general, we cannot distinguish between 0-prime and 1-prime near-ring modules. Thus 1-prime modules were omitted from further investigations.

If $P \subseteq M$ then $(P : M)_R = \{r \in R : rM \subseteq P\}$.

Theorem 2.1. *Let $P \lhd_R M$. Then the following are equivalent:*

(1) *P is a 2-prime R-ideal.*
(2) *For all $a \in R$ and submodules B of M such that $aB \subseteq P$, it follows that $aM \subseteq P$ or $B \subseteq P$.*
(3) *For all $a \in R$ and $b \in M$ such that $a[b]_R \subseteq P$, it follows that $aM \subseteq P$ or $b \in P$. (Here $[b]_R$ is the submodule of M generated by b).*
(4) *For all R-submodules N of M such that $P \subset N$, we have that $(P : M) = (P : N)$.*

In a similar way we can construct and prove equivalent definitions for 0-prime and 1-prime R-ideals.

Theorem 2.2. *Let P be an R-ideal of M. Then the following are equivalent:*

(1) *P is a 0-prime (or 1-prime) R-ideal.*
(2) *For all $a \in R$ and for all R-ideals B of M such that $aB \subseteq P$, we have that $aM \subseteq P$ or $B \subseteq P$.*
(3) *For all $a \in R$ and $b \in M$ such that $a\langle b\rangle_R \subseteq P$, we have that $aM \subseteq P$ or $b \in P$. (Here $\langle b\rangle_R$ is the R-ideal of M generated by b).*
(4) *For all R-ideals N of M such that $P \subset N$, we have that $(P : M) = (P : N)$.*

Corollary 2.1. *An R-module M is:*

(1) 0-*prime if and only if for all non-zero R-ideals N of M, it follows that* $(0 : M) = (0 : N)$.
(2) 2-*prime if and only if for all non-zero submodules N of M, it follows that* $(0 : M) = (0 : N)$

Theorem 2.3. *Let M be an R- module and $P \lhd_R M$. Then the following are equivalent:*

(1) *P is 3- prime and $(P : m) \lhd R$ for every $m \in M \setminus P$.*
(2) *$RM \not\subseteq P$ and $(P : m) = (P : M)$ for every $m \in M \setminus P$.*

Theorem 2.4. *Let $P \lhd_R M$. Then the following are equivalent:*

(1) *P is a 3-prime R-ideal.*
(2) *$RM \not\subseteq P$ and $(P : Rm) = (P : M)$ for every $m \in M \setminus P$.*

Theorem 2.5. *Let $P \lhd_R M$. Then the following are equivalent:*

(1) *P is a completely prime R-ideal.*
(2) *$RM \not\subseteq P$ and $(P : m) = (P : M)$ for every $m \in M \setminus P$.*

Theorem 2.6. *Let $P \lhd_R M$. Then P is completely prime $\Rightarrow$ P is 3-prime $\Rightarrow$ P is 2-prime $\Rightarrow$ P is 0-prime.*

Remark 2.1. In general, a 0-prime R-ideal need not be 2-prime and a 2-prime R-ideal need not be 3-prime.

If $P \lhd_R M$, then we recall that $\tilde{P} = (P : M)$ is an **ideal** of R.
We have the following Question:
If P is a ν-prime ($\nu = 0, 1, 2, 3, c$) R-ideal does this imply that $\tilde{P}$ is a ν-prime ideal of R?

Theorem 2.7. *Let P be an R-ideal of M. Then:*

(1) *P is a 2-prime R-ideal of M implies that $\tilde{P}$ is a 2-prime ideal of R.*
(2) *P is a 3-prime R-ideal of M implies that $\tilde{P}$ is a 3-prime ideal of R.*
(3) *P is a completely prime R-ideal of M implies that $\tilde{P}$ is a completely prime ideal of R.*

That P is a 0-prime R-ideal implies that $\tilde{P}$ is a 0-prime ideal of R, unfortunately, does not follow as naturally as for the 2-prime, 3-prime and completely prime cases.

However, if we restrict M to a **tame** $R-$module or a **monogenic** R-module, we find that the relationship holds.

Theorem 2.8.

(1) *If M is a tame $R-$module and P be a 0-prime R-ideal of M,then $\tilde{P}$ is a 0-prime ideal of R.*
(2) *Let P be a 0-prime R-ideal of a monogenic R-module M. Then $\tilde{P}$ is a 0-prime ideal of R.*

(1) Now suppose that P is an R-ideal of M such that $(P : M)$ is a ν-prime ideal of R for $\nu = 0, 2, 3$ and c. Does this imply that P is a ν-prime R-ideal of M?

Fo the various types of prime R-ideals (modules) we were easily able to prove that if an R-ideal P of M satisfied a certain prime condition, then so did the corresponding ideal $\widetilde{P} = (P : M)$ of R.

However the converse relation turned out to be problematic in many situations, especially since it is difficult to construct an R-ideal of M by starting with an ideal of R.

To overcome this problem, we now introduce the notion of a multiplication near-ring module.

Definition 2.3. Let M be an $R-$module. Then:

(1) $C \subseteq M$ is called a **multiplication set** if $\widetilde{C}M = C$.

(2) $m \in M$ is called a multiplication element if the singleton set $\{m\}$ is a multiplication set.

Definition 2.4. Let M be an R-module. Then:

(1) *M is called a 0-multiplication module if every R-ideal is multiplication ideal.*

(2) *M is called a 2-multiplication module if every R-submodule is multiplication submodule.*

(3) *M is called a c-multiplication module if every $m \in M$ is a multiplication element.*

Theorem 2.9.

(1) Let P be an R-ideal of a 0-multiplication R-module M such that $\widetilde{P}$ is a 0-prime ideal of R. Then P is a 0-prime R-ideal of M.

(2) Let P be an R-ideal of a 2-multiplication R-module M such that $\widetilde{P}$ is a 2-prime ideal of R. Then P is a 2-prime R-ideal of M.

(3) Let P be an R-ideal of a c-multiplication R-module M such that $\widetilde{P}$ is a 3-prime (resp. c-prime) ideal of R. Then P is a 3-prime (resp. c-prime) R-ideal of M.

Corollary 2.2. *Suppose that M is a ν-multiplication R-module ($\nu = 0, 2, c$). Then M is ν-prime if and only if R is ν-prime. Furthermore, if M is a c-multiplication module, then M is 3-prime if and only if R is 3-prime.*

Definition 2.5. A submodule $P \leq M$ is called an s-prime submodule if for every $A \lhd R$ and for every $N \leq M$ if $x \in A$ and $x^n N \subseteq P$ for some $n \in N$, then $N \subseteq P$ or $AM \subseteq P$.

Example 2.1. Let R be the near-ring on $\mathbb{Z}_3 = \{0, 1, 2\}$ multiplication defined by: $a \cdot b = \begin{cases} a \text{ if } b = 2 \\ 0 \text{ if } b \neq 2 \end{cases}$. The only $R-$subgroups of R are 0 and R. We also have $R^2 \neq 0$. Hence R is 2$-$prime. Furthermore we have $2^n = 2$ for every $n \in \mathbb{N}$. Thus $M = {}_R R$ is an s-prime module.

Example 2.2. Every type 2-R-module M is an s-prime R-module: Suppose $N \neq 0$ and $AM \neq 0$ for some $A \lhd R$ and $N \leq {}_R M$. Since M is type 2, there exists $m \in M$ such that $M = Rm$. Since $AM \neq 0$, we have $0 \neq ARm \subseteq Am$. Since Am is a submodule of M and M is of type 2, $Am = M$ and there exists $a \in A$ such that $am = m$. Hence $a^k m = m$ for every $k \in \mathbb{N}$. Consequently $a^k N \neq 0$ for every $k \in \mathbb{N}$ and therefore M is $s-$prime.

Example 2.3. A prime module which is not an s-prime module. We use the construction and computation of S. U. Hwang, Y. C. Jeol and Y. Lee in Ref. 17, Example 1.2 and proposition 1.3. Let S be a domian, n be a positive integer and R_n be the 2^n by 2^n upper triangular matrix ring over S. Define a map $\delta : R_n \to R_{n+1}$ by $A \to \begin{bmatrix} A & 0 \\ 0 & A \end{bmatrix}$. Then R_n is considered as a subring of R_{n+1} via δ. $D_n = \{R_n, \delta_{nm}\}$, with $\delta_{nm} = \delta^{m-n}$ whenever $n \leq m$, is a direct system over $I = \{1,2,3,\cdots,\}$. Let $R = \varinjlim R_n$ be the direct limit of D. From Ref. 17, R is a prime near-ring with $\mathcal{N}(R) \neq 0$. Hence a prime near-ring which is not an s-prime near-ring. If we let $M = {}_R R$, then M is a prime R-module which is not an s-prime R-module.

Proposition 2.1. *A prime $R-$ideal $P \lhd_R M$ is s-prime if and only if:*

(1) P is a prime $R-$ideal

(2) for every $A \lhd R$ such that $A \not\subset (P:M)$ there exists $a \in A \setminus (P:M)$ such that $a^n M \nsubseteq P$ for all $n \in N$.

Corollary 2.3. *An $R-$ideal $P \lhd_R M$ is s-prime if and only if*

(a) P is a prime $R-$ideal and

(b) $R/(P:M)$ contains no nonzero nil ideals i.e. $\mathcal{N}(R/(P:M)) = 0$ where $\mathcal{N}(R)$ is the upper nil radical of the ring R.

Proposition 2.2. *If R is a commutative or artinian near-ring, then an $R-$ideal $P \lhd_R M$ is s-prime if and only if P is prime.*

Proposition 2.3. *Let M be an R-module. For an $R-$ideal $P \lhd_R M$, the following statements are equivalent:*

(1) P is s-prime.

(2) For every $a \in R$ and for every $m \in M$ if $x \in <a>$ and $x^n <m> \subseteq P$ for some $n \in \mathbb{N}$, then $m \in P$ or $<a> M \subseteq P$.

(3) P is a prime $R-$ideal and for every $a \in R$ such that $aM \nsubseteq P$ there exists $x \in <a>$ such that $x^n M \nsubseteq P$ for every $n \in \mathbb{N}$.

(4) $\mathcal{P} = (P:M)$ is an s-prime ideal and $(P:<m>) = \mathcal{P}$ for every $m \in M \setminus P$.

Proposition 2.4. *Let R be a ring and $\mathcal{P} \lhd R$, $\mathcal{P} \neq R$. The following are equivalent:*

(1) $\mathcal{P}$ is an s-prime ideal of R.

(2) There exists a s-prime R-module M such that $\mathcal{P} = (0:M)_R$.

Definition 2.6. Let R be a near-ring and M an R-module. A nonempty set $S \subseteq M\{0\}$ is called an s-system if, for each ideal A of R and for all submodules K, L of M, if $(K+L) \cap S \neq \emptyset$ and $(K+AM) \cap S \neq \emptyset$, then there exists $x \in A$ such that $(x^n L + K) \cap S \neq \emptyset$ for every $n \in N$.

Corollary 2.4. *Let M be an R-module. Then, the $R-$ideal P of M is s-prime if and only of $M \setminus P$ is an s-system.*

Proposition 2.5. *Let M be an R-module, P a proper $R-$ideal of M and let $S =: M \setminus P$. Then, the following statements are equivalent:*

(1) *P is s-prime.*
(2) *S is an s-system.*
(3) *For every $A \lhd R$ and for all $L \leq M$, if $L \cap S \neq \emptyset$ and $AM \cap S \neq \emptyset$, then there exists $a \in A$ such that $a^n N \cap S \neq \emptyset$ for every $n \in \mathbb{N}$.*
(4) *For every $a \in R$ and every $m \in M$, if $< m > \cap S \neq \emptyset$ and $aM \cap S \neq \emptyset$, then $a^n < m > \cap S \neq \emptyset$ for all $n \in \mathbb{N}$.*

Proposition 2.6. *Let M be an R-module $S \subseteq M$ an s-system and P an $R-$ideal of M maximal with respect to the property that $P \cap S = \emptyset$. Then P is an s-prime $R-$ideal.*

Definition 2.7. Let R be a near-ring and M and R-module. For an $R-$ideal N of M, if there is an $s-$prime $R-$ideal containing N, then we define

$$\mathcal{S}(N) =: \{s \in M : \text{every s-sytem containing } s \text{ meets } N\}$$

Theorem 2.10. *Let M be an R-module and N an $R-$ideal of M. Then, either $\mathcal{S}(N) = M$ or $\mathcal{S}(N)$ equals the intersection of all s-prime $R-$ideals of M containing N.*

In what follows, let R be any near-ring (need not have an identity). If M is an $R-$module, define $\mathcal{N}({}_RM) = \cap\{S \lhd_R M : S$ is an s-prime $R-$ ideal of $M\}$. $\mathcal{N}({}_RM)$ is the **upper nil radical** of the R-module M. Let $\mathcal{N}(R)$ be the upper nilradical of the near-ring R i.e. the intersection of all the s-prime ideals of R. Recall that $\mathcal{N}(R)$ is also equal to the sum of all the nil ideals of R.

For the near-ring R, consider the $R-$module ${}_RR$. We have the following:

Lemma 2.1. $\mathcal{N}({}_RR) \subseteq \mathcal{N}(R)$.

Let $R = \{\begin{bmatrix} x & y \\ 0 & 0 \end{bmatrix} : x, y \in Z_2\}$ and $M = {}_RR$. It is easy to check that (0) is an s-prime R-ideal.

Hence, $\mathcal{N}({}_RR) = 0$. Now, we have $(0 : R)_R$ is an s-prime ideal of R, $(0 : R)_R \neq (0)$, for if $b \neq 0$, $b \in Z_2$, then $\begin{bmatrix} 0 & b \\ 0 & 0 \end{bmatrix} R = 0$. Hence, $\mathcal{N}(R) \subseteq (0 : R)_R$. But since $(0 : R)_R(0 : R)_R = 0$ we have $(0 : R)_R \subseteq \mathcal{N}(R)$. Hence, $\mathcal{N}(R) = (0 : R)_R \neq 0$.

Lemma 2.2. *For any near-ring and any R-module M we have*

$$\mathcal{N}(R) \subseteq (\mathcal{N}({}_RM) : M)_R$$

Let $R = \mathbb{Z}$ and $M = \mathbb{Z}_{p^\infty} \oplus \mathbb{Z}$ for some prime number p. $\mathcal{N}_R(M) = \mathbb{Z}_{p^\infty}$ and $\mathcal{N}(R) = (0)$, i.e, $\mathcal{N}(R)M = (0)$.

Proposition 2.7. *For any near-ring R, $\mathcal{N}(R) = (\mathcal{N}({}_RR) : R)_R$.*

Remark 2.2. If R is any near-ring then $\mathcal{N}({}_RR) = \mathcal{N}(R)$ if and only if for $x \in R$, $xR \subseteq \mathcal{N}({}_RR)$ implies $x \in \mathcal{N}({}_RR)$.

Lemma 2.3. *For all R-modules M*

(1) $\mathcal{N}({}_RM) = \{x \in M : Rx \subseteq \mathcal{N}({}_RM)\}$.
(2) If $\mathcal{N}(R) = R$ *then* $\mathcal{N}({}_RM) = M$.

Proposition 2.8. *Let R be any ring. Then any of the following conditions implies* $\mathcal{N}(R) = \mathcal{N}({}_RR)$.

(1) R is commutative.
(2) $x \in xR$ *for all* $x \in R$, *e.g. if R has an identity or R is Von Neumann regular.*

Unfortunately, when directly extending various KA-radicals from rings to near-rings, we obtain a radical which is not only "bad" but is also "ugly" in the sense that we may lose one or both conditions (H_3) and (H_4) of the definition of a KA-radical. Many near-ring "radicals" are H-radicals but not necessarily KA-radicals. This leads to two basic paths one can take to obtain "nice" radicals (i.e., KA-radicals) for near-rings. First, one could add more properties to the radical in question so that it is a KA-radical on the class of near-rings and still coincides with its ring theoretic ancestor on the class of rings. The second path we can take (the one we will take in what follows) is to use the direct near-ring analogue of a ring radical but restrict the class of near-rings to which it is applied.

The majority of the so-called radicals of near-rings are defined not as radical classes but as mappings. Unfortunately these classes, as usual, are homomorphically closed but they are not necessarily closed under ideals i.e., the class is not a universal class and, therefore, the traditional theory of radicals cannot be applied here. To handle this situation we have the following:

H1 $(\rho(R)+I)/I \subseteq \rho(R/I)$ for all $I \lhd R$;
H2 $\rho(R/\rho(R)) = 0$ for all R;
H3 **complete** if $\rho(I) = I \lhd R$ implies $I \subseteq \rho(R)$;
H4 **idempotent** if $\rho(\rho(R)) = \rho(R)$.

Definition 2.8. Let σ be a mapping which assigns to the near-ring R an ideal $\sigma(R)$ and let $\mathcal{T}$ be a homomorphically closed class of near-rings. The mapping σ is called a $\mathcal{T}$-radical map if:

(a) σ satisfies (H1) and (H2);
(b) σ satisfies (H3) and (H4) for all $R \in \mathcal{T}$.

If in the above definition $\mathcal{T}$ is the class of all near-rings then the concepts $\mathcal{T}$-radical and KA-radical are the same.

From Groenewald[15] we have

Definition 2.9. A class $\mathcal{F}$ of near-rings is $\mathcal{T}$-special if:

R1. All near-rings from $\mathcal{F}$ are 2-prime;

R2. $R \in \mathcal{T} \cap \mathcal{F}$ and $A \triangleleft R$ implies $A \in \mathcal{F}$;
R3. $I \triangleleft J \triangleleft N$ and $J/I \in \mathcal{F}$ implies $I \triangleleft R$;
R4. If I is an essential ideal of R ($I \triangleleft \cdot R$) and $I \in \mathcal{F}$, then $R \in \mathcal{F}$ (i.e. $\mathcal{F}$ is closed under essential extensions).

The mapping σ is called a $\mathcal{T}$-*special radical* radical map if $\sigma(R) = \cap\{I : R/I \in \mathcal{F}\}$ where $\mathcal{F}$ is a $\mathcal{T}$-special class of near-rings. This definition extends the concept of a special radical for rings Divinsky[10] and Szasz[26] to near-rings.

Let R be a near-ring and $I \triangleleft R$. Let $r \in R$ and $m \in M$. *If M is an R/I-module, then with respect to $rm = (r+I)m$, M becomes an R-module and $I \subseteq (0 : M)_R$. If M is an R-module and $I \subseteq (0 : M)_R$, then M is an R/I-module with respect to $(r+I)m = rm$. In both cases, we have that $(0 : M)_{R/I} = (0 : M)_R/I$.*

Now let $\mathcal{T}$ be a nonempty class of zero-symmetric right near-rings which is closed under homomorphic images. For each near-ring R, let $\mathcal{M}_R$ be a class of R-modules (possibly empty). Let $\mathcal{M} = \cup\{\mathcal{M}_R : R \text{ is a near-ring}\}$. Then we introduce the notion of a $\mathcal{T}$-special class of near-ring modules:

Definition 2.10. A class $\mathcal{M} = \cup\{\mathcal{M}_R : R \text{ is a near-ring}\}$ of near-ring modules is called a $\mathcal{T}$-**special class** if it satisfies the following conditions:

M1 *If $M \in \mathcal{M}_R$ and $I \triangleleft R$ with $IM = 0$, then $M \in \mathcal{M}_{R/I}$.*
M2 *If $I \triangleleft R$* and *$M \in \mathcal{M}_{R/I}$, then $M \in \mathcal{M}_R$.*
M3 *If $M \in \mathcal{M}_R$ and $I \triangleleft R \in \mathcal{T}$ with $IM \neq 0$, then $M \in \mathcal{M}_I$.*
M4 *If $M \in \mathcal{M}_R$, then $RM \neq 0$ and $R/(0 : M)_R$ is a 2-prime near-ring.*
M5 *If $I \triangleleft R \in \mathcal{T}$ and $M \in \mathcal{M}_I$, then there exists an R-module $N \in \mathcal{M}_R$ such that $(0 : N)_I \subseteq (0 : M)_I$.*
M6 *If $K \triangleleft I \triangleleft R \in \mathcal{T}$ and there exists a faithful I/K-module $M \in \mathcal{M}_{I/K}$, then $K \triangleleft R$.*

There were numerous relationships between a near-ring and its modules. In particular, prime R-ideals of the R-module M led to prime ideals of R and, under certain conditions, the converses are also true. It is, therefore, natural to assume that there is a relationship between special radicals of near-rings and special radicals of their modules. In the two theorems that follow, we show the construction of a special class of near-rings from a special class of near-ring modules and the reversal of the process.

Theorem 2.11. *Let $\mathcal{M} = \cup\{\mathcal{M}_R : R \text{ is a near-ring}\}$ be a $\mathcal{T}$-special class of near-ring modules. Then*

$\mathcal{F} = \{R : \text{there exists } M \in \mathcal{M}_R \text{ with } (0 : M)_R = 0\} \cup \{0\}$ is a $\mathcal{T}$-special class of near-rings.

Theorem 2.12. *Let $\mathcal{F}$ be a $\mathcal{T}$-special class of near-rings and for the near-ring R, let $\mathcal{M}_R = \{M : M \text{ is an } R\text{-module, } RM \neq 0 \text{ and } R/(0 : M)_R \in \mathcal{F}\}$. If $\mathcal{M} = \cup\{\mathcal{M}_R : R \text{ is a near-ring}\}$ then $\mathcal{M}$ is a $\mathcal{T}$-special class of near-ring modules.*

Proposition 2.9. *Let $\mathcal{M}$ be a $\mathcal{T}$-special class of near-ring modules and suppose $I \lhd R \in \mathcal{R}_0$, where $\mathcal{R}_0$ denotes the class of zero-symmetric near-rings. Let $\mathcal{F}$ be the corresponding $\mathcal{T}$-special class of near-rings. Then $R/I \in \mathcal{F}$ if and only if $I = (0:M)_R$ for some $M \in \mathcal{M}_R$.*

Let $\mathcal{K}$ be a $\mathcal{T}$-special class of modules, let $\mathcal{M}_K$ be the class of near-rings defined by $\mathcal{M}_K := \{R :$ there exists $M \in \mathcal{K}_R$ with $(0:M)_R = 0\}$. Then, $\mathcal{M}_K$ is a $\mathcal{T}$-special class of near-rings and if $\mathcal{R}$ is the corresponding radical then, $\mathcal{R}(R) = \cap\{(0 : M)_R : M \in \mathcal{K}_R\}$ for each near-ring R. Conversely, if $\mathcal{F}$ is a $\mathcal{T}$-special class of near-rings, let $\mathcal{M}_R := \{M$ is an R-module, $RM \neq 0$ and $R/(0:M)_R \in \mathcal{F}\}$ for each near-ring R. Then $\mathcal{M} := \cup\{\mathcal{M}_R\}$ is a $\mathcal{T}$-special class of modules and $r(M) = \cap\{S \leq M : M/S \in \mathcal{M}\}$.

In 1958 Andrunakievich[1] proved the following lemma for associative rings:

"If I is an ideal of a ring R and K is an ideal of I, then $< K >_R^3 \subseteq K$ where $< K >_R$ is the ideal of R generated by K."

This result has had far reaching consequences through its application to radical theory of associative rings. Because in general the Andrunakievich Lemma is not satisfied for near-rings, the notion of $\mathcal{A}$-near-rings were introduced Birkenmeier *et al.* in Ref. 6. To get the best possible results about the prime radicals which are not KA-radicals, we shall make use of the concept of an $\mathcal{A}$-near-ring. An ideal I of a near-ring R is called an *$\mathcal{A}$-ideal* if for each ideal K of the near-ring I there is some $n \geq 1$, perhaps depending on K, such that $(\langle K\rangle_R)^n \subseteq K$. R is called an *$\mathcal{A}$-near-ring* if every ideal of R is an $\mathcal{A}$-ideal.

The class $\mathcal{A}$ is wide and varied, including all distributively generated near-rings and all near-rings which are neither nilpotent nor strongly regular. These and many other examples and the basic properties of $\mathcal{A}$-near-rings are given in Ref. 6.

Proposition 2.10. *Let R be any near-ring and $\mathcal{M}_R := \{M : M$ is a prime R-module$\}$. If $\mathcal{M} = \cup\mathcal{M}_R$, then $\mathcal{M}$ is an $\mathcal{A}-$special class of R-modules.*

Remark 2.3. If $\mathcal{M}_p$ denotes the $\mathcal{A}-$special class of prime near-ring modules, then the $\mathcal{A}-$special radical induced by $\mathcal{M}_p$ on a near-ring R is given by:

$$\begin{aligned}\wp(R) &= \cap\{(0:M)_R : M \text{ is a prime } R-\text{module}\}\\ &= \cap\{I \lhd R : I \text{ a 2-prime ideal}\}.\end{aligned}$$

Proposition 2.11. *Let R be any $\mathcal{A}-$near-ring and $\mathcal{M}_R := \{M{:}M$ is an s-prime R-module$\}$. If $\mathcal{M}_s = \cup\mathcal{M}_R$, then $\mathcal{M}_s$ is a $\mathcal{A}-$special class of near-ring modules.*

Proposition 2.12. *If $\mathcal{M}_s$ is a $\mathcal{A}-$special class of near-ring modules, then the $\mathcal{A}-$special radical induced by $\mathcal{M}_s$ on a near-ring R is given by:*

$$\begin{aligned}\mathcal{N}(R) &= \cap\{(0:M)_R : M \textit{ is an s-prime } R-\textit{module}\}\\ &= \cap\{I \lhd R : I \textit{ an s-prime ideal}\}.\end{aligned}$$

Bibliography

1. V. A. Andrunakievič, *Radicals in associative rings I, Math Sbornik*, **44**(1958), 179–212 (in Russian); English translation in Amer. Math. Soc. Transl., **52**(1966), 95–128.
2. G. F. Birkenmeier, Andrunakievich's lemma in near-rings, *Contributions to General Algebra 9, Proceedings of the workshop on General Algebra, Linz 1994 (H. Kaiser, W. Müller, G. Pilz (eds.)), Verlag Holder-Pichler-Tempsky*, (1995), 1–12.

3. G. F. Birkenmeier, H. Heatherly and E. Lee, Prime ideals in near-rings, *Results in Mathematics,* **24**(1993), 27–48.
4. G. F. Birkenmeier, H. Heatherly and E. Lee, Prime ideals and prime radicals in near-rings, *Monatshefte für Mathematik* **117**(1994), 179–197.
5. G. F. Birkenmeier, H. E. Heatherly and E. K. Lee, Near-rings in which every prime factor is integral, *Pure Math. Appl.* **59**(1994), 257–279.
6. G. F. Birkenmeier, H. Heatherly and E. Lee, Special radicals for near-rings, *Tamkang Journal of Mathematics* **27**(1996), 281–288.
7. G. F. Birkenmeier, N. J. Groenewald and W. A. Olivier, On class pairs and radicals, *Comm. Algebra* **29**(11)(2001), 5307–5328.
8. N. J. Divinsky and A. Sulinski, Radical pairs, *Can. J. Math.* **29**(5)(1997), 1086–1091.
9. G. L. Booth, N. J. Groenewald and S. Veldsman, A Kurosh-Amitsur prime radical for near-rings, *Comm. Algebra* **18**(1990), 3111–3122.
10. N. Divinsky, *Rings and Radicals*, Univ. Toronto Press, Ontario, 1965.
11. G. Ferrero and C. Ferrero-Cotti, *Nearrings, Some Developments Linked to Semigroups and Groups*, Kluwer Academic Publishers, Dordrecht, The Netherlands, 2002.
12. N. J. Groenewald, The completely prime radical in near-rings, *Acta Math. Hungar.* **51**(1988), 301–305.
13. N. J. Groenewald, Different prime ideals in near-rings, *Comm. Algebra* **19**(10)(1991), 2667–2675.
14. N. J. Groenewald, The almost nilpotent radical for near-rings, *Near-rings and Near-fields (Stellenbosch, 1997)* Kluwer Acad. Publ. (2001), 84–93.
15. N. J. Groenewald, *Strongly prime near-rings*, Proc. Edinb. Math. Soc. **31**(1988), 337–343.
16. M. Holcombe, A hereditary radical for near-rings, *Studia Sci. Math. Hungar* **17**(1982), 453–456.
17. S. U. Hwang, Y. C. Jeol and Y. Lee, *Structure and topological conditions of NI rings*, Journal of Algebra, **302** (2006), 186–199.
18. K. Kaarli, Special radicals of distributively generated near-rings (in Russian), *Tartu Riikl. Ül. Toimetised* **610**(1982), 53–68.
19. K. Kaarli and T. Kriss, Prime radical of near-rings, *Tartu Riikl. Ül. Toimetised* **764**(1987), 23–29.
20. R. Miltz and S. Veldsman, Radicals and subdirect decompositions of Ω-groups, *J. Austral. Math. Soc. (Series A)* **48**(1990), 171–198.
21. G. Pilz, *Near-Rings* (revised edition), North-Holland, Amsterdam, New York, Oxford, 1983.
22. D. Ramakotaiah and G. Koteswara Rao, IFP near-rings, *Journal of Australian Math. Soc.* (Series A) **27**(1979), 365–370.
23. R. L. Snider, Lattices of radicals, *Pacific J. Math.*, **40**(1972), 207–220.
24. S. Veldsman, An overnilpotent radical theory for near-rings, *J. Algebra* **144**(1991), 248–265.
25. A. P. J. Van der Walt, Prime ideals and nil radicals in near-rings, *Arch. Math.* **15**(1964), 408–414.
26. F. A. Szasz, *Radicals of Rings*, John Wiley & Sons, New York, 1981.

Topics in group theory

B. R. Shankar

Department of Mathematical and Computational Sciences,
National Institute of Technology Karnataka,
Surathkal, Karnataka, India
E-mail: brs@nitk.ac.in

Keywords: Cyclic groups, random walk, generators, torsion group, Grigorchuk group.

1. Introduction

Most introductory texts on algebra begin with Group theory. One possible reason is that only one binary operation is needed to define a group. Yet, the sheer variety of groups that exist may not be that obvious to a beginner. In this short article we intend to highlight some aspects of group theory that are not commonly encountered. We give a detailed account of Polya's recurrence theorem and a very brief introduction to the first Grigorchuk group.

A **Group** is defined as a nonempty set G together with a binary operation $*$, which is associative, has identity and inverses, denoted by $(G, *)$. We denote $x * y = xy$. If $xy = yx, \forall\ x,\ y \in G$, then G is said to be **commutative** or **abelian.** Otherwise it is nonabelian. A group is said to be **cyclic** if all its elements can be written as powers of a single element. The cyclic groups form an important and perhaps the simplest class of groups. Indeed the structure theorem for finite abelian groups states that all finite abelian groups are direct products (or sums) of finite cyclic groups. Compare this with the fact that the every positive integer is a unique product of prime numbers. Thus, finite cyclic groups are the building blocks of all finite abelian groups. All cyclic groups of a given order are isomorphic. If G is cyclic of order n then G has exactly one subgroup of order d for each divisor d of n. The only infinite cyclic group is isomorphic to the additive group of integers, $\mathbb{Z}$. For each $n \geq 1$, the set of all nth roots of unity, denoted by $G = \{\omega^k : 0 \leq k \leq n-1\}$ where $\omega = e^{\frac{2\pi i}{n}}$ is a cyclic group of order n. Inspite of this, it is surprising that when computational issues are considered, there can be wide differences in the effort needed to solve equations. We illustrate it as follows:

Let G be a cyclic group of order n, generated by a, say $G = \langle a \rangle$. Consider the equation $a^x = b$ in G. Given any two of a, x, and b, one can always find the third. In general, given x it is easy to find b. However given b, the problem of finding x is known as the **discrete log problem.** Of late it has gained enormous importance due to its applications in cryptography, especially digital signatures.

(i) If $G = \mathbb{Z}_n$, the additive group of integers modulo n, the problem is easy via Euclid's Algorithm.

(ii) If $G = \mathbb{Z}_p^*$, the multiplicative group of nonzero integers modulo a large prime p, the problem is computationally hard.

(iii) If G is a cyclic group via an elliptic curve over a large finite field, the problem is much harder.

2. Polya's Recurrence Theorem

Consider the **simple random walk** on the lattice $\mathbb{Z}$ of integers.

At time zero the walker is at the origin and moves one step to the left or to the right with equal probability after each unit of time. The question of **recurrence for** $\mathbb{Z}$ is to ask whether the walker has a 100% chance of visiting the origin again infinitely many times. The answer is yes.

Proof: Note that the number of steps in any path must be even to return to the origin. The number of all paths of length $2n$ is 2^{2n}. Among these, the number of paths ending at the origin is $\binom{2n}{n}$ because they involve a choice of n steps to the right and n steps to the left. Hence the probability u_{2n} of being at the origin after $2n$ steps is

$$u_{2n} = \frac{1}{2^{2n}}\binom{2n}{n}$$

By Stirling's formula for approximating $k!$ by $k^k e^{-k}\sqrt{2\pi k}$ we get

$$u_{2n} \sim \frac{1}{\sqrt{\pi n}}$$

As $\sum_1^\infty \frac{1}{\sqrt{n}} = \infty$, we have

$$\sum_{k=1}^{\infty} u_k = \sum_{n=1}^{\infty} u_{2n} = \infty$$

This means that the simple random walk on $\mathbb{Z}$ is **recurrent,** i.e., the random walker has probability 1 returning to the origin infinitely often. This can be seen as follows: for $n = 1, 2, 3, \ldots$, let $f_n =$ the probability that the walker returns to O(origin) for the **first time** at step n; set $f_0 = 0$ (because no return at step 0). Decompose the event "walker is at O at time n, $n \geq 1$" into **first** returns; i.e., if walker is at O at time n, he may or may not be there for the first time, i.e., he may have visited O earlier too. Thus

$$u_1 = f_0 u_1 + f_1 u_0$$

$$u_2 = f_0 u_2 + f_1 u_1 + f_2 u_0$$

$$\vdots$$

$$u_n = f_0 u_n + f_1 u_{n-1} + \cdots + f_n u_0.$$

Also, $u_0 = 1$ since the walker is at O initially. In u_n, the term $f_k u_{n-k}$ represents the probability that the walker returns to O for the first time after k steps and then returns to O in

$n-k$ more steps. Denote

$$U(s)=\sum_{m=0}^{\infty}u_m s^m \text{ and } F(s)=\sum_{m=0}^{\infty}f_m s^m.$$

Then,

$$\begin{aligned}U(s)F(s)&=f_0u_0+(f_0u_1+f_1u_0)s+\cdots\\&=0+u_1s+u_2s^2+\cdots\\&=U(s)-1\end{aligned}$$

Thus $U(s)[1-F(s)]=1$.
Now $\sum_0^N u_n \le \lim_{s\to 1} U(s)$ and so if $\sum_0^\infty u_n=\infty$, then $\lim_{s\to 1} U(s)=\infty$ and $f=\sum_0^\infty f_n=\lim_{s\to 1} F(s)=1$.
Hence if $\sum_0^\infty u_n=\infty$, then the probability of returning to O is 1. Otherwise there is a positive probability of not returning to O.
Consider the analogous problem on the lattice $\mathbb{Z}^2$ of the Euclidean plane. The walk is again **recurrent**.

Sketch of proof: The number of all paths of length $2n$ is now 4^{2n}. Among these, the number of paths that return to the origin after k steps North, k steps South, $n-k$ steps East, and $n-k$ steps West, is the multinomial coefficient

$$\binom{2n}{k,\ k,\ n-k,\ n-k}=\frac{(2n)!}{k!k!(n-k)!(n-k)!}.$$

Hence the probability of being at the origin after $2n$ steps is

$$u_{2n}=\frac{1}{4^{2n}}\sum_{k=0}^{n}\frac{(2n)!}{k!k!(n-k)!(n-k)!}=\frac{1}{4^{2n}}\frac{(2n)!}{n!n!}\sum_{k=0}^{n}\binom{n}{k}\binom{n}{n-k}$$

Again using Stirling's approximation we get $u_{2n}\sim\frac{1}{\pi n}$ and

$$\sum_{k=1}^{\infty}u_k=\sum_{n=1}^{\infty}u_{2n}=\infty$$

Thus the walk is again recurrent.
The situation is different in higher dimensions: The simple random walk on $\mathbb{Z}^3$ in Euclidean space $\mathbb{R}^3$ is **transient**. In this case one can show that

$$\sum_{k=1}^{\infty}u_k=\sum_{n=1}^{\infty}u_{2n}\le K\sum_{n=1}^{\infty}\frac{1}{n^{\frac{3}{2}}}<\infty$$

for a suitable constant K. The above discussion leads to

Theorem 2.1. *(Polya, 1921) The simple random walk on $\mathbb{Z}^d$ is*

$$\begin{cases}\textit{recurrent}, & \textit{if } d=1 \textit{ or } d=2\\ \textit{transient}, & \textit{if } d\ge 3.\end{cases}$$

This has been summarized as follows:

- A drunk man will find his way home, but a drunk bird may get lost for ever.

  \- Shizuo Kakutani

- All roads leads to Rome except the cosmic paths.

A **probability measure** on a group G is a function $p : G \to [0,1]$ such that $\sum_{g \in G} p(g) = 1$. It is **symmetric** if $p(g^{-1}) = p(g)\ \ \forall g \in G$. A probability p on G defines a **left invariant random walk,** with the probability that a walker at some point $g_1 \in G$ will go in one step to a point $g_2 \in G$ being $p(g_1^{-1}g_2)$.
The following result due to Varopoulos is very deep:

Theorem 2.2. *(Varopoulos, 1986.) Let G be a finitely generated group and p be a symmetric probability measure on G with finite support which generates G. If the random walk defined by G and p is recurrent, then*

either G is a finite group,
or G has a subgroup of finite index isomorphic to $\mathbb{Z}$
or G has a subgroup of finite index isomorphic to $\mathbb{Z}^2$.

3. Burnside's Problem

Elements of finite order in a group G form a subgroup, known as the **torsion subgroup** of G. If all elements of G are of finite order then G is said to be **torsion group or periodic group.**
Q: If G is a torsion group, must G itself be finite?
Answer is **No.**
Consider the group of **all** roots of unity. Every element has finite order but the group is not even finitely generated. Hence infinite. The **Burnside problem,** was posed by **William Burnside** in 1902 and is one of the oldest and most influential questions in group theory.

Figure 1: William Burnside (1852–1927).

The **general Burnside problem** asks: "If G is a finitely generated periodic group, is G necessarily a finite group?" Answer is **Yes** under the **additional assumption** that G is a subgroup of $GL(n,K)$ for some positive integer n and some field K. In general the answer is **No.** This was first shown by Golod and Shafarevich (1964). Many counterexamples are known now. In 1981, the Russian mathematician Rostislav Ivanovich Grigorchuk gave a counterexample, now known as the **first Grigorchuk group**. It turns out to have several interesting properties. In 2015 Rostislav Grigorchuk was awarded the AMS Leroy P. Steele Prize for Seminal Contribution to Research. We give a short list of some well-known examples of groups:

Figure 2: Rostislav Ivanovich Grigorchuk (1953–).

(1) The group $\mathbb{Z}^n$ is finitely generated for any integer n.
(2) The multiplicative group $\mathbb{Q}^*$ of $\mathbb{Q}$ is not finitely generated.
(3) The additive group $\mathbb{Q}$ is not finitely generated.
(4) If $n \geq 2$, the group $SL(n, \mathbb{Q})$ is not finitely generated.
(5) The four matrices s_1, s_2, s_3 and s_4 generate $GL(n, \mathbb{Z})$ for any integer $n \geq 2$.

$$s_1 = \begin{pmatrix} 0\,0\,0 \cdots 0\,1 \\ 1\,0\,0 \cdots 0\,0 \\ 0\,1\,0 \cdots 0\,0 \\ \vdots\,\vdots\,\vdots\,\ddots\,\vdots\,\vdots \\ 0\,0\,0 \cdots 0\,0 \\ 0\,0\,0 \cdots 1\,0 \end{pmatrix} \quad s_2 = \begin{pmatrix} 0\,1\,0 \cdots 0\,0 \\ 1\,0\,0 \cdots 0\,0 \\ 0\,0\,1 \cdots 0\,0 \\ \vdots\,\vdots\,\vdots\,\ddots\,\vdots\,\vdots \\ 0\,0\,0 \cdots 1\,0 \\ 0\,0\,0 \cdots 0\,1 \end{pmatrix}$$

$$s_3 = \begin{pmatrix} 1\,1\,0 \cdots 0\,0 \\ 0\,1\,0 \cdots 0\,0 \\ 0\,0\,1 \cdots 0\,0 \\ \vdots\,\vdots\,\vdots\,\ddots\,\vdots\,\vdots \\ 0\,0\,0 \cdots 1\,0 \\ 0\,0\,0 \cdots 0\,1 \end{pmatrix} \quad s_4 = \begin{pmatrix} -1\,0\,0 \cdots 0\,0 \\ 0\;\;1\,0 \cdots 0\,0 \\ 0\;\;0\,1 \cdots 0\,0 \\ \vdots\;\;\vdots\,\vdots\,\ddots\,\vdots\,\vdots \\ 0\;\;0\,0 \cdots 1\,0 \\ 0\;\;0\,0 \cdots 0\,1 \end{pmatrix}$$

Hua and Reiner have shown that s_1, s_3 and s_4 are enough to generate $GL(n, \mathbb{Z})$. Also s_1 and s_3 generate $GL(n, \mathbb{Z})$ when n is even and $SL(n, \mathbb{Z})$ when n is odd. (Coxeter and Moser.)

We now give a beautiful example (1983) due to N. Gupta and S. Sidki which shows that a finitely generated group, all whose elements have finite $p-$ power order (for a fixed prime p), can be infinite.

Let p be a fixed odd prime and let X be the set of all finite strings of symbols from the alphabet $\{0, 1, ..., p-1\}$. The empty string is of length 0. For $r \geq 0$, denote by 0^r the string of length r consisting of r zeros. Addition/subtraction of symbols from the alphabet is done modulo p. Define two permutations t and z on X as follows: They fix the empty string and on nonempty strings are defined by

(1) t changes the first symbol i to $(i+1)$ and leaves the rest of the string unchanged.
(2) For a string $0^r ijw$ with $i \neq 0$ and $r \geq 0$,

$$(0^r ijw)^z = 0^r i(j+i)w.$$

Thus, z only changes the symbol j which follows the first nonzero symbol i (if any) to $(j+i)$.

Let G be the group of permutations of X generated by t and z. Both t and z leave the lengths of strings invariant and hence all orbits of G are finite.

Theorem 3.1. *The group G is infinite and all its elements have finite, p-power order.*

We now describe the **first Grigorchuk group,** denoted by Γ and list a **few** of its **many remarkable properties**:

Let T denote the rooted binary tree and $\mathrm{Aut}(T)$ denote the group of all automorphisms of T.

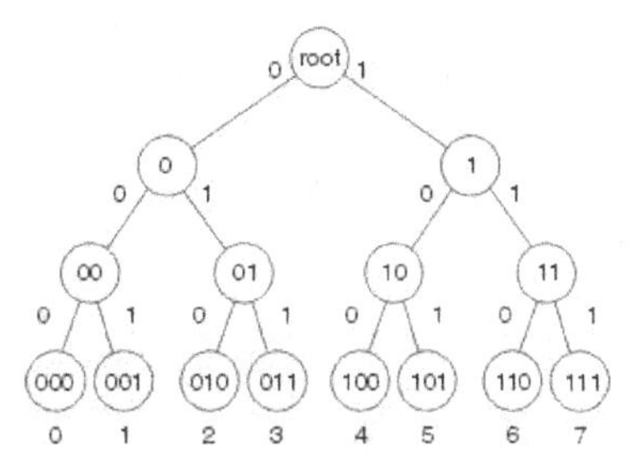

Figure 3: Part of a binary tree.

(1) $\mathrm{Aut}(T)$ is is uncountable.
(2) Γ is a subgroup of $\mathrm{Aut}(T)$ **generated by four elements**.
(3) Γ is **residually finite.** (A group G is residually finite for every nontrivial element $g \in G$ there exists a homomorphism from G to a finite group that maps g to a nontrivial element.)
(4) Γ is **Hopfian.** (A group G is Hopfian if any surjective homomorphism G to G is also injective.)
(5) Γ is **a 2- group**, i.e., for every g in Γ there exists an integer $n \geq 0$ such that $g^{2^n} = 1$.
(6) Γ is **just infinite**, i.e., Γ is infinite but every proper quotient group of Γ is **finite**. (Example: the additive group of integers $\mathbb{Z}$, all proper subgroups are of the form $n\mathbb{Z}$ and the quotient groups $\mathbb{Z}/n\mathbb{Z}$ are all finite.)
(7) The **word problem** is solvable in Γ.
(8) Γ is finitely generated but **not finitely presented**.

(9) Γ has **intermediate growth** i.e., faster than polynomial but slower than exponential.
(This gives a negative answer to a question of John Milnor 1968.)

Let $T = (V, E)$ be a tree with vertex set V all finite sequences of $\{0,1\}$. Denote vertex as $(j_1, j_2, ..., j_k)$. Two vertices are connected by an edge in E if their lengths as sequences differ by one, and the shorter sequence is obtained from the longer one by deleting its last term. The empty set is also a vertex and considered as the root of the tree.

Let $L(k)$ denote all sequences of length k, called a **level** of T. Note that

$$V = \bigcup_{k=0}^{\infty} L(k).$$

Automorphisms of the binary tree fix the root and permute subtrees that begin on the same level. All adjacent vertices remain adjacent, so we can think of an automorphism as a series of **twists** of branches of the tree.

Consider an automorphism $g \in \mathrm{Aut}(T)$. It permutes vertices of each level, but allowed permutations of a given level depend on the permutations of the previous levels (since adjacent vertices remain adjacent.) Thus we may write g as sequence of permutations $\{x_i\}$, where x_1 is the permutation of $L(1)$, x_2 is the permutation of the vertices of $L(2)$ after x_1 and so on. This enables one to prove that the group $\mathrm{Aut}(T)$ is uncountable.

Let $T(k)$ be the subtree of T spanned by all vertices with length at most k. Define $\mathrm{Aut}(T(k))$ analogously to $\mathrm{Aut}(T)$. Note that $\mathrm{Aut}(T(k))$ is finite for any given k.

Let

$$St(k) = \{g \in \mathrm{Aut}(T) : \forall x \in T(k), g(x) = x\}$$

be the set of all automorphisms in $\mathrm{Aut}(T)$ that fix $T(k)$.

Theorem 3.2. *$St(k)$ is a normal subgroup of finite index in $Aut(T)$.*

Proof. Let $\phi : \mathrm{Aut}(T) \longrightarrow \mathrm{Aut}(T(k))$ be the function taking an automorphism $g \in \mathrm{Aut}(T)$ to the automorphism it performs on $T(k)$. This is a homomorphism with kernel $St(k)$, thus $St(k)$ is normal in $\mathrm{Aut}(T)$. By the first isomorphism theorem, $St(k)$ has finite index in $\mathrm{Aut}(T)$. □

Let T_j be the subtree of all vertices whose sequence begins with j. Define the homomorphism $\delta_j : T \longrightarrow T_j$ that takes the vertex x to the vertex jx by concatenation of the sequences j and x. Given two automorphisms g_0 and g_1, we can define $g = (g_0, g_1)$ which acts on g_i as $\delta_i g_i \delta_i^{-1}$.

Definition 3.1. If $j \in \{0,1\}$ define $\overline{j}$ by $\overline{0} = 1$ and $\overline{1} = 0$. Let $a : V \longrightarrow V$ be defined by $a(j_1, j_2, ..., j_k) = (\overline{j_1}, j_2, ..., j_k)$. Let e be the identity automorphism. Define b, c, d recursively as follows. All fix the root and $L(1)$ and

$$b = (a, c) \quad c = (a, d) \quad d = (e, b).$$

As an example we compute
$b(1,0,1,1) = \delta_1(c(\delta_1^{-1}(1,0,1,1))) = \delta_1(c(0,1,1)) = \delta_1(\delta_0(a(\delta_0^{-1}(0,1,1)))) = \delta_1(\delta_0(a(1,1))) = \delta_1(\delta_0(0,1)) = \delta_1(0,0,1) = (1,0,0,1)$

Definition 3.2. The Grigorchuk group Γ is defined as the group of automorphisms generated by a, b, c and d, i.e.,

$$\Gamma = \langle a, b, c, d \rangle.$$

It can be shown that

(i) $a^2 = b^2 = c^2 = d^2 = e$
(ii) $bc = cb = d, \quad cd = dc = b, \quad db = bd = c$
(iii) Γ is infinite
(iv) Γ is a 2- group. i.e., every element has order equals a power of 2.

Conclusion

We conclude with some general remarks and open problems. If it is known that a group G with m generators and exponent n is finite, can one conclude that the order of G is bounded by some constant depending only on m and n? Equivalently, are there only finitely many finite groups with m generators of exponent n, up to isomorphism? The answer is **yes,** was proved by **Efim Zelmanov,** awarded Fields Medal in 1994.

Figure 4: Efim Zelmanov (1955–).

For which positive integers m, n is the free Burnside group $B(m,n)$ finite?
The full solution to Burnside's problem in this form is not known. Burnside considered some easy cases in his original paper:

(i) $B(1,n)$ is the cyclic group of order n.
(ii) $B(m,2)$ is the direct product of m copies of the cyclic group of order 2 and hence finite.
(iii) The following additional results are known (Burnside, Sanov, M. Hall): $B(m,3)$, $B(m,4)$, and $B(m,6)$ are finite for all m.

The particular case of $B(2,5)$ remains open: as of 2005 it was not known whether this group is finite. A list of references for further reading is given. The book[1] is excellent and has a detailed exposition of the first Grigorchuk group. The article[2] by B. K. Sahoo and

B. Sury is very readable and expository, with a detailed proof of the example of N. Gupta and S. Sidki[3]. Kostrikin's monograph on the Burnside problem[4] is a classic.

Bibliography

1. P. de La Harpe, *Topics in geometric group theory*. University of Chicago Press, 2000.
2. B. K. Sahoo and B. Sury, "What is the Burnside problem?" *Resonance*, vol. 10, no. 7, pp. 34–48, 2005.
3. N. Gupta and S. Sidki, "On the burnside problem for periodic groups," *Mathematische Zeitschrift*, vol. 182, no. 3, pp. 385–388, 1983.
4. A. I. Kostrikin, *Around Burnside*. Springer-Verlag, Berlin, 1990.
5. K. WADDLE, "The Grigorchuk group," 2008.
6. M. Vaughan-Lee, *The restricted Burnside problem*. Oxford University Press, 1993.

On the structure of composition rings

Stefan Veldsman

Dept. Mathematics, Nelson Mandela Metropolitan University,
Port Elizabeth, South Africa
E-mail: stefan.veldsman@nmmu.ac.za

Composition p-rings and composition t-rings are two types of composition rings introduced to facilitate the study of composition rings; in particular those that have foundation zero.

1. Introduction

Let G be a group. The algebraic home of the set $M(G) = \{f \mid f : G \to G\}$ of all self-maps on G with pointwise addition and composition, as well as the set of polynomials $G[x] = \{g_0 + n_1x + g_1 + n_2x + ... + n_kx + g_k \mid k \geq 0, 0 \neq n_i \in \mathbb{Z}$ and $g_i \neq 0$ for $i = 1,2,3,...,k-1\}$ with respect to the usual addition and composition is the variety of near-rings. When G is replaced by a ring R, then both $M(R)$ and $R[x]$ have one more operation namely pointwise multiplication. In addition to the near-ring structure with respect to composition, there is also a ring structure with respect to pointwise multiplication. The natural home of these algebraic structures is the variety of composition rings.

A composition ring C is a quadruple $(C,+,\cdot,\circ)$ where $C_1 := (C,+,\cdot)$ is a ring, $C_2 := (C,+,\circ)$ is a near-ring and $ab \circ c = a \circ c \cdot b \circ c$ for all $a,b,c \in C$. Unless indicated otherwise by brackets, juxtaposition (which represents multiplication) has a higher priority than composition (in the order of executing the operations) and composition has a higher priority than multiplication when the latter is denoted by "$\cdot$". For example, $ab \circ c = a \circ c \cdot b \circ c$ means $(ab) \circ c = (a \circ c) \cdot (b \circ c)$.

The *foundation of* C, denoted by F, is the constant part of the near-ring C_2, i.e. $F = Found\ (C) = \{c \in C | c \circ 0 = c\}$. F is a subcomposition ring of C, but usually when we refer to the foundation, we mean the ring $F = (F,+,\cdot)$. C_0 denotes the 0-symmetric part of of the near-ring C_2, i.e. $C_0 = \{c \in C | c \circ 0 = 0\}$. C_0 is an ideal of the ring C_1 and a left ideal of the near-ring C_2.

The near-ring C_2 may have an identity, called the *composition identity* and usually denoted by x. Often it is required that x is a *commuting composition identity*, i.e. for all $c \in C, c \circ x = c = x \circ c$ and $xc = cx$. The ring C_1 may also have an identity i (i.e. $ci = c = ic$ for all $c \in C$). In case $i \in F$, as is often the case, it will be called a *constant multiplicative identity*.

The standard examples of composition rings are the polynomials $R[x]$ and the 0-symmetric polynomials $R_0[x]$; the formal power series $R_N[[x]]$ where N is a nil ideal of R (in particular the special case $R_0[[x]]$ with $N = 0$); the transformations $M(R) = \{f \mid f$ is a function $f : R \to R\}$ and $M_0(R) = \{f \mid f$ is a function $f : R \to R$ with $f(0) = 0\}$ and the polynomial functions $\overline{R[x]}$ and $\overline{R_0[x]}$. R will always be a commutative ring with identity unless mentioned otherwise.

The study of these composition rings has inevitably led to the relationships between the properties of the composition ring and those of the underlying base ring R. In the case where there are constants around (eg in $R[x]$ and $M(R)$), the base ring is easily identified as all those elements c for which $c \circ 0 = c$ (i.e. the foundation). In a general composition ring C one thus work with the foundation $F = C \circ 0$. But for $R_0[x]$ and $R_N[[x]]$ things are problematic. For these two cases, a different approach is required and one is led to consider $D(c) \circ 0$ where $D(c)$ denotes the formal derivation of c. Formally this is problematic, for example in $R_0[x]$, $D(ax) = a \notin R_0[x]$ which takes one outside the considered composition ring. Moreover, for the composition ring $M_0(R)$ there may be no realistic derivation in sight to enable us to capture R.

This problem has been addressed earlier in Ref. 7. It seems that a slightly different approach to one of the types of composition rings considered there may capture the essence of these types better and this problem will be revisited here. We will recall and survey many of the earlier results and as an application investigate the relationship between maximal ideals in the composition ring and its base ring. The proofs of the statements can be found in Refs. 5–8.

The motivation for the approach using the base is from the diagram:

$$\begin{array}{ccccccc} R_0[x] & \hookrightarrow & R[x] & & M(R) & \hookleftarrow & M_0(R) \\ \uparrow & & \nwarrow & & \nearrow & & \uparrow \\ (Rx,+,\circ) & \cong & & (R,+,\cdot) & & \cong & (\overline{R},+,\circ) \end{array}$$

where $\overline{R} = \{\overline{a} \mid a \in R\}$ and $\overline{a}$ is the map $\overline{a} : R \to R$, $\overline{a}(t) = at$ for all $t \in R$. We then recall from Ref. 7:

Definition 1.1. Let C be a composition ring. The base B of C, written as $Base(C)$, is defined as $B = Base(C) = \{v \in C \mid \text{for all } a,b \in C,\ v \circ (a+b) = v \circ a + v \circ b \text{ and } v \circ ab = v \circ a.b\}$.

A number of elementary properties from Ref. 7 are recalled below together with a few new ones. Recall that for a subset S of C, $(0:S)_{C_1} = \{c \in C \mid cS = 0\}$ and $(0:S)_{C_2} = \{c \in C \mid c \circ S = 0\}$.

Proposition 1.1. *Let C be a composition ring with foundation $Found(C) = F$.*

(1) $(B,+,\circ)$ is a subring of the near-ring $(C_0,+,\circ)$.

(2) If x is a commuting composition identity in C, then:

(i) $x \in B$.

(ii) $(v \circ c)x = vc$ for all $v \in B$ and $c \in C$.

(iii) $\psi : (F,+,\cdot) \to (Fx,+,\circ)$ defined by $\psi(n) := nx$ is a surjective (ring) homomorphism with $\ker\psi = (0:x)_{C_1} \cap F$ and $Fx \lhd (B,+,\circ)$. ψ is an isomorphism $\Leftrightarrow$ for all $n \in F, nx = 0$ implies $n = 0$. This will be the case, for example, if the ring F is semiprime or if it has an identity.

(3) Let x be a commuting composition identity in C. Then the following are equivalent:

(i) $(B,+,\circ) = (Fx,+,\circ)$.

(ii) There is an $i \in F$ such that $x = ix$.

(iii) C has a constant multiplicative identity.
If any one of these conditions hold, then $(B,+,\circ) = (Fx,+,\circ) \cong (F,+,\cdot)$.

An element s of C is a *semi-constant* if for any $f \in C_0$,
$$s \circ f \circ s = \begin{cases} s \text{ if } f \circ s \neq 0 \\ 0 \text{ if } f \circ s = 0 \end{cases}.$$
A semi-constant is necessarily 0-symmetric and if C has a commuting composition identity, then $s \circ s = s$. Also note that for any $u \in C_0$, $u \circ s$ is a semi-constant whenever s is a semi-constant. We will discuss semi-constants in much more detail in Section 3 below.

Proposition 1.2. *Let C be a composition ring with a commuting composition identity x and a multiplicative identity i. Then:*
(1) For all $c \in C, i \circ c = 0$ implies $c = 0$.
(2) For all $v \in B, v = (v \circ i)x$. Moreover, $vx = 0 \Rightarrow v \circ i = 0 \Rightarrow v = 0$.
(3) For all $c,d \in C$, $(i - i \circ d).c \circ d = 0$ and $i \circ c.c = c = c.i \circ c$.
(4) For $s \in C$, let $S = C \circ s$. Then $(0 : S)_{C_1} = 0$ if and only if $i \in S$.
(5) i is a constant multiplicative identity if and only if $(0 : F)_{C_1} = 0$.
(6) i is a 0-symmetric multiplicative identity if and only if $C = C_0$ (i.e. $F = 0$).
(7) If C has a non-zero semi-constant s, then i is a semi-constant if and only if $F = 0$ and $(0 : S)_{C_1} = 0$ where $S = C \circ s$.

It is appropriate to say something here about the two two's in a composition ring. Let C be a composition ring with commuting composition identity x and a multiplicative identity i. The ring of integers $\mathbb{Z}$ acts on C in the usual way with $(n+m)c = nc + mc$, $n(c+d) = nc + nd$, $(nm)c = n(mc)$, $n(cd) = (nc)d = c(nd)$ and $n(c \circ d) = (nc) \circ d$ for all $n,m \in \mathbb{Z}$ and $c,d \in C$. In particular, $1c = c$, $(-1)c = -c$ and $0c = 0$.

Let $2^* := i + i = 2i$ and $2^{**} := x + x = 2x = 2ix = 2^*x$. Now 2^* and 2^{**} are elements in C and $2^{**} \circ c = 2c = 2^*c$ while $2^{**}c = 2(xc)$ and $2^* \circ c = 2(i \circ c)$ for any $c \in C$. Then clearly $charC = 2$ iff $2^* = 0$ iff $2^{**} = 0$. The natural home of 2^* is $(C,+,\cdot)$ and that of 2^{**} is $(C,+,\circ)$, so when we talk about their inverses, it will be on home territory. If i is a constant multiplicative identity, then $(F,+,\cdot) \cong (Fx,+,\circ)$ and so $2^* \in F$ is invertible in $(F,+,\cdot)$ if and only if $2^{**} = 2^*x$ is invertible in $(Fx,+,\circ)$. But this is true in general:

Proposition 1.3. *Let C be a composition ring with commuting composition identity x and multiplicative identity i. Then $2^{**} = x + x$ is invertible in $(C,+,\circ)$ if and only if $2^* = i + i$ is invertible in $(C,+,\cdot)$.*

Next we recall many examples of composition rings from Ref. 7.

Example 1.1. If not mentioned otherwise, R below denotes a commutative ring with identity.

(1) Each of $R[x]$, $R_0[x]$ and $R_N[[x]]$ has base $(Rx,+,\circ) \cong (R,+,\cdot)$; all three have commuting composition identity x and only the first has a (constant) multiplicative identity. If

the ring R has no identity, then these composition rings do not have composition identity nor multiplication identity. Their foundations are $R, 0$ and N respectively.

(2) Both $M(R)$ and $M_0(R)$ have commuting composition identity $x = 1_R$ (the identity map on R) and base $(\overline{R}, +, \circ) \cong (R, +, \cdot)$. Both have multiplication identities, the first the constant function $\widehat{1}$ defined by $\widehat{1}(r) = 1$ for all $r \in R$ and the second the "nearly constant" function $\widetilde{1}$ defined by $\widetilde{1}(r) = 1$ for $r \neq 0$ and $\widetilde{1}(0) = 0$. If the ring R does not have an identity, each of these composition rings will have a composition identity but no multiplicative identity.

(3) Let R and S be rings with $\alpha : S \to R$ a fixed function. Let $C = M(R, S, \alpha)$ be the sandwich composition ring determined by R, S and α. This means C consists of all the functions from R to S, the addition and multiplication are component wise and the composition, here denoted by $*$, is defined by: $f * g = f \circ \alpha \circ g$. Then $B = Base(C) = \{f \in C \mid f \circ \alpha \in Ends(S, +)\}$.

(4) For rings R without identity, the base of $M_0(R)$, or $M(R)$, need not be isomorphic to R. Let $R = 2\mathbb{Z}$ and let $C = M_0(R)$. In this case, $Base(M_0(R)) = \{\overline{a} | a \in \mathbb{Z}\} \cong \mathbb{Z} \ncong R$.

(5) Let R be a commutative ring with $S \neq R$ a subring of R with identity $1_S \in S$. Let $C = \{f \in M(R) | f(R) \subseteq S\}$. Then C is a subcomposition ring of $M(R)$. Note that C does not have a composition identity. However, it has many composition left identities: Let $s_0 \in S$ and let $x(t) = \begin{cases} t \text{ if } t \in S \\ s_0 \text{ if } t \in R \backslash S. \end{cases}$

Then x is a left identity. Here we have $Found$ $(C) = S$ and $B = Base(C) = \{a_g | a \in S$ *and* $g \in C\}$ where $a_g : R \longrightarrow R$ is defined by

$$a_g(t) = \begin{cases} at \text{ if } t \in S \\ g(t) \text{ if } t \in R \backslash S. \end{cases}$$

(6) Let $C = \overline{R[x]}$ where $\overline{R[x]}$ denotes all the polynomial functions. Then C is a subcomposition ring of $M(R)$ and $B = Base$ $(C) = \{rx | r \in R\}$. Note that here we may have $B \cap C_0 x \neq 0$. For example, if $R = \mathbb{Z}_4$, then $0 \neq 2x = 2x^2 \in B \cap C_0 x$.

(7) Let $C = (R, +, \cdot, \circ)$ where $(R, +, \cdot)$ is a ring and $(R, +, \circ)$ is any near-ring.
(a) If $(R, +, \cdot)$ is a zero ring (i.e. $R^2 = 0$) then $Base(C) = \{c \in C \mid c$ is left distributive$\}$. If $(R, +, \circ)$ is a constant near-ring, then $Base(C) = 0$.
(b) If $(R, +, \circ)$ is a zero ring, then $Base(C) = (C, +, \circ)$.

(8) Let $(R[x, y], +, \cdot)$ be the ring of all polynomials in the two commuting indeterminates x and y. As is well-known, we can define a near-ring multiplication $\circ$ on $(R[x, y], +)$ by: $f(x, y) \circ g(x, y) := f(g(x, y), g(x, y))$. Then $C = (R[x, y), +, \cdot, \circ)$ is a composition ring with constant multiplicative identity 1. But C has no composition identity. Both x and y are composition left identities; in fact so are $c + x - c \circ x$ for all $c \in C$. These are all in $B = Base(C)$ and also $0 \neq (0 : C)_{C_2} \subseteq B$. It can be shown that $B = \{v \in C \mid v = \sum_{i=1}^{k} \sum_{j=0}^{i} c_{ij} x^{i-j} y^j$ for some $k \geq 1$ and $c_{ij} \in R$ for all i, j where $\sum_{j=0}^{i} c_{ij} = 0$ for all $i = 2, 3, ..., k\}$.

(9) With $(R[x, y], +, \cdot)$ as in (8) above, define $\circ$ by $f(x, y) \circ g(x, y) = f(g(x, x), g(y, y))$. Then $C = (R[x, y], +, \cdot, \circ)$ is a composition ring with right identities x and y and $Found(C) = R$. In this case, it can be shown that $B = Base(C) = 0$.

(10) Let $(R[x,y],+,\cdot)$ be as in (8) above. On the direct sum of two copies of this ring with itself, define $\circ$ by: $(f(x,y),\ g(x,y)) \circ (h(x,y),\ k(x,y)) := (f(h(x,y),\ k(x,y)),\ g(h(x,y),\ k(x,y)))$. Then $C := (R[x,y] \times R[x,y],+,\cdot,\circ)$ is a composition ring with constant multiplicative identity $(1,1)$ and commuting composition identity (x,y). Here $Found(C) = R \times R$ and $Base(C) = (R \times R)(x,y) = Rx \times Ry$.

Recall, an abstact affine near-ring is a near-rings N with $(N,+)$ abelian, $N_c \lhd N$ and $N/N_c \cong N_0$ is a ring. The base in a composition ring always leads to an abstact affine near-ring: Let C be a composition ring with $F = Found(C) \neq 0$ and $B = Base(C) \neq 0$. Let $S = F + B$. Then $(S,+,\circ)$ is a subnear-ring of $(C,+,\circ)$ and it can be verified that S is an abstract affine near-ring.

The question now is, to what extent can we use the base of a composition ring in the place of the foundation; especially when the foundation is 0? Let us first note that the foundation F has two useful connections to the composition ring.

Firstly, there always is a surjective ring homomorphism $\beta : (C,+,\cdot) \to (F,+,\cdot)$ defined by $\beta(c) = c \circ 0$ for all $c \in C$. Two particular cases are $\beta : R[x] \to R$ given by $\beta(f(x)) = f(0)$ and $\beta : M(R) \to R$ given by $\beta(f) = f(0)$. Of course, when C has foundation $F = 0$, these homomorphisms are not interesting.

The second useful fact is that the ring F is a (left) C-module in the canonical sense. In terms of representation theory, this means there is a composition ring homomorphism $\psi : (C,+,\cdot,\circ) \to (M(F),+,\cdot,\circ)$ given by $\psi(c) = \psi_c$ where $\psi_c : F \to F$ is defined by $\psi_c(n) = c \circ n$ for all $n \in F$. Once again, when $F = 0$, this is of no use.

If $F = C \circ 0 = 0$, one can mimmick the above by replacing 0 with another element $e \in C$ for which $C \circ e \neq 0$. Then $(C \circ e,+,\cdot)$ is a subring of $(C,+,\cdot)$, $\beta : (C,+,\cdot) \to (C \circ e,+,\cdot)$ defined by $\beta(c) = c \circ e$ is a surjective ring homomorphism and $\psi : (C,+,\cdot,\circ) \to (M(C \circ e),+,\cdot,\circ)$ with $\psi(c) = \psi_c$ and $\psi_c : C \circ e \to C \circ e$ defined by $\psi_c(d \circ e) = c \circ d \circ e$ is a composition ring homomorphism. So we have what we want, but we may have no idea what the ring $(C \circ e,+,\cdot)$ is. For example, in $C = R_0[x]$ take e as simple as possible, but not simpler; i.e. we can take $e = x^2$ but not $e = x$. Then $C \circ e = \{f(x) \circ x^2 \mid f(x) \in R_0[x]\} = \{f(x^2) \mid f(x) \in R_0[x]\}$ which is not useful at all. We would rather like to work with the base of C in the place of $C \circ e$. However, for this we have to impose some restrictions on the composition ring and we will distinguish between two types of composition rings.

2. Composition p-rings

A composition ring C is called a *composition p-ring* [7] if it has a commuting composition identity x and $(C_0,+) = (B,+) \oplus (C_0x,+)$. Then $(C,+) = (F,+) \oplus (C_0,+) = (F,+) \oplus (B,+) \oplus (C_0x,+)$ and by decomposing C_0 again, we get $C = F \oplus B \oplus Bx \oplus C_0x^2 = F \oplus B \oplus Bx \oplus Bx^2 \oplus C_0x^3 = \dots$. If C has a constant multiplication identity, then $B = Fx$. In this case, $C_0 = B + C_0x = Fx + C_0x = (F + C_0)x = Cx$. The first result here shows that a multiplicative identity in a composition p-ring must always be constant.

Proposition 2.1. *Let C be a composition p-ring with foundation F and base B. The following four conditions are equivalent:*

(1) C has a multiplication identity.

(2) C has a constant multiplication identity.
(3) There is an $i \in F$ with $x = ix$.
(4) $B = Fx$.

Corollary 2.1. *A 0-symmetric composition p-ring does not have a multiplicative identity.*

(1) $R[x]$, $R_0[x]$ and $R_N[[x]]$ are composition p-rings. For the corresponding composition rings of polynomial functions, things are not straightforward. If R is an infinite field, then $R[x] \cong \overline{R[x]}$ which is a composition p-ring and $\overline{R[x]} \subsetneq M(R)$. But when R is a finite field, $\overline{R[x]} = M(R)$ which is not a composition p-ring (in this case $B \cap C_0 x = B \neq 0$). In fact, in general we have: Let $C = \overline{R[x]}, R$ a commutative ring with $1 \in R$. Then C is not a composition p-ring if and only if there exists an $f = f_1 x + f_2 x^2 + \cdots + f_k x^k$ in $R[x]$ with $f_1 \neq 0$ and $f(t) = 0$ for all $t \in R$. As examples of such elements, one may consider $2x + 2x^2$ in $\overline{\mathbb{Z}_4[x]}$ or $3x + 3x^2$ in $\overline{\mathbb{Z}_6[x]}$.

(2) Let $I \lhd R, R$ a commutative ring with $1 \in R$. Then $C := \{c \in R[x] | c \circ 0 \in I\}$ is a subcomposition ring of $R[x]$ which contains x and it is a composition p-ring.

(3) On $R[x,y] \times R[x,y]$ define $\circ$ by: $(f(x,y), g(x,y)) \circ (h(x,y), k(x,y)) = (f(h(x,y),y), g(x,k(x,y)))$. Then $C = (R[x,y] \times R[x,y], +, \cdot, \circ)$ is a composition ring with commuting composition identity (x,y) and constant multiplicative identity $(1,1)$. Here $F = Found(C) = \{(f(x,y), g(x,y)) \in C \mid f(x,y)$ is a polynomial in y and $g(x,y)$ is a polynomial in x, i.e. $f(x,y) = f(r,y)$ and $g(x,y) = g(x,r)$ for all $r \in R\}$. Moreover, $B = Base(C) = \{(f(x,y)x, g(x,y)y) \mid (f(x,y), g(x,y)) \in F\}$. Now $C_0 = \{(f(x,y), g(x,y)) \mid$ every term of $f(x,y)$ contains at least one x, and every term of $g(x,y)$ contains at least one $y\}$ and $(C_0, +) = (B, +) \oplus (C_0(x,y), +)$. Hence C is a composition p-ring.

Let C be a composition p-ring with foundation $F = 0$ but $B \neq 0$. Then $C = B + C_0 x$ and we can define $\beta : (C, +, \circ) \to (B, +, \circ)$ as the projection of C onto B. Then β is a surjective homomorphism with $\ker \beta = C_0 x$ and $(\frac{C}{C_0 x}, +, \circ)$ is isomorphic to the ring $(B, +, \circ)$. But there is another possibility inspired by the following: If $C = R_0[x]$, then $B = Rx$. Choose any $0 \neq r \in R$ fixed and define $\beta : (R_0[x], +, \cdot) \to (B, +, \circ)$ by $\beta(f(x)) = f(r)x$. In this case we have a surjective ring homomorphism with $\ker \beta = \{f(x) \in R_0[x] \mid f(r) = 0\}$. To do this in general is not straightforward, since we do not have $r \in R$ with which to work. However, we can work with $rx \in Rx = B$, the base of $R_0[x]$.

Suppose C is a 0-symmetric composition p-ring with $(C, +) = \bigoplus_{i=0}^{\infty} (Bx^i, +)$. Choose $0 \neq b \in B$ fixed and define $\beta : (C, +, \cdot) \to (B, +, \circ)$ by $\beta(c) = c * b$ where $*$ is defined as follows:

For $c = v_0 + v_1 x + \ldots + v_k x^k \in \bigoplus_{i=0}^{\infty} (Bx^i, +), c * b = (v_0 + v_1 x + \ldots + v_k x^k) * b = v_0 \circ b + v_1 \circ b^{(2)} + \ldots + v_k \circ b^{(k+1)}$ where $b^{(n)} = b \circ b \circ \ldots \circ b$, n times. The motivation is that one should think of $*$ as composition $\circ$ and $vx^n \circ b = v \circ b.(x^n \circ b) = v \circ b.b \circ \ldots \circ b = (v \circ b) \circ (b \circ \ldots \circ b) = v \circ b^{(n+1)}$ by remembering that the "multiplication" in the ring B is $\circ$ and therefore the "." in $v \circ b.(x \circ b.x \circ b. \ldots .x \circ b)$ changes to "$\circ$". For the next result, it is worthwhile to first note the following:

(1) If $(B,+,\circ)$ is commutative, then for all $u,v\in B$, $uv=(u\circ v)x=(v\circ u)x=vu$.

(2) Since $B^2=Bx$, for every $u,v\in B$ and $i,j\geq 1$, there are $w_1,w_2\in B$ such that $vx^i.ux^j=w_1x^{i+j+1}\in Bx^{i+j+1}$ and $vx^i\circ ux^j=(v\circ u)u^ix^{(i+1)j}=w_2x^ix^{(i+1)j}=w_2x^{i+(i+1)j}\in Bx^{i+(i+1)j}$. Hence $\bigoplus_{i=0}^{\infty}Bx^i$ is closed under multiplication as well as composition.

(3) In view of the above, if $(B,+,\circ)$ is commutative, then $C=\bigoplus_{i=0}^{\infty}Bx^i$ has commutative multiplication.

(4) For $v\in B$ and $c\in C$, we know $vc=(v\circ c)x$. An inductive argument then gives $v_1v_2...v_kc=(v_1\circ v_2\circ...\circ v_k\circ c)x^k$ for $v_i\in B,c\in C,k\geq 1$.

(5) In a commutative ring the multinomial identity holds, i.e. $(a_1+a_2+...+a_m)^n=\sum\binom{n}{k_1,k_2,...,k_m}a_1^{k_1}a_2^{k_2}...a_m^{k_m}$ where the sum is over all $k_1+k_2+...+k_m=n$, $0\leq k_i\leq n$ and $\binom{n}{k_1,k_2,...,k_m}=\frac{n!}{k_1!k_2!...,k_m!}$. We will use this w.r.t. both the multiplication in C as well as the composition in B.

We then have, after lengthy and tedious verifications:

Proposition 2.2. *Let C be a 0-symmetric composition p-ring with $(C,+)=\bigoplus_{i=0}^{\infty}(Bx^i,+)$ and $(B,+,\circ)$ a commutative ring. Choose $0\neq b\in B$ fixed and define $\beta:(C,+,\cdot)\to(B,+,\circ)$ by $\beta(c)=c*b$. Then β is a surjective ring homomorphism. Moreover, define $\psi:C\to M(B)$ by $\psi(c)=\psi_c$ where $\psi_c:B\to B$ is the map $\psi_c(b)=c*b$. Then ψ is a composition ring homomorphism.*

In our motivating example $C=R_0[x]$, the above gives $\psi_f(rx)=f(r)x$ as desired. It may be worthwhile to generalize the definition of a composition p-ring to include composition rings with one-sided composition identities; for example many of those with some composition defined on the ring $R[x,y]$. Next we look at our second type of composition ring.

3. Composition t-rings

As motivation we have $M_0(R)$ in mind. Here we have *Base* $M_0(R)=\overline{R}=\{\overline{a}\mid a\in R\}$ where $\overline{a}:R\to R$, $\overline{a}(t)=at$ for all $t\in R$. This composition ring has no constant maps, but we have the "0-symmetric constant" maps

$$\widetilde{a}:R\to R,\ \widetilde{a}(t)=\begin{cases}a \text{ if } t\neq 0\\ 0 \text{ if } t=0\end{cases}.$$

Let us call these maps for the time being semi-constant.

Note that here the composition identity $x=1_R$ is the identity mapping on R and $\overline{a}=\widetilde{a}1_R=\widetilde{a}x$, i.e. $B=\{\text{semi-constants}\}x$; very much in line with polynomials where we have $B=Fx=Rx=\{\text{constants}\}x$. So, our task at hand is to isolate these semi-constants in a general composition ring.

A first attempt could be to say that $e \in C = M_0(R)$ is a semi-constant if for all $f \in C$, $e \circ f = \begin{cases} e \text{ if } f \neq 0 \\ 0 \text{ if } f = 0 \end{cases}$; but this won't do. One could have $f \neq 0$, but $f(r) = 0$ for some $r \neq 0$ with $0 \neq e(r) = (e \circ f)(r) = 0$.

Semi-constants were already defined in Ref. 7, but with hindsight this earlier approach needs to be refined. In the original definition of a semi-constant, both the multiplication and the composition of the composition ring were used. This is not desirable, because semi-constants can be found in near-rings which are not necessarily composition rings. One should rather define a semi-constant in a near-ring only in terms of the two near-ring operations (addition and near-ring multiplication). With this in mind, we recall from Section 1 above:

Definition 3.1. $s \in C$ is a *semi-constant* if for all $f \in C_0$,

$$s \circ f \circ s = \begin{cases} s \text{ if } f \circ s \neq 0 \\ 0 \text{ if } f \circ s = 0 \end{cases}.$$

A semi-constant is necessarily 0-symmetric. If s is a semi-constant, then so is $u \circ s$ for any $u \in C_0$. Indeed, let $f \in C_0$ such that $f \circ u \circ s \neq 0$. Then $(u \circ s) \circ f \circ (u \circ s) = u \circ (s \circ f \circ u \circ s) = u \circ s$ since $f \circ u \in C_0$ and $f \circ u \circ s \neq 0$. A semi-constant in a composition ring with a composition identity is idempotent with respect to composition. In $M(R)$ and $M_0(R)$ the semi-constants are exactly the maps $\widetilde{a}$. In $R[x]$ and $R_0[x]$ there are no non-zero semi-constants. The semi-constants are used to define the second type of composition ring we want to consider:

Definition 3.2. A composition ring C is called a *composition t-ring* if:

(i) C has a commuting composition identity x.

(ii) C has an unital semi-constant e; i.e. a semi-constant e for which $x = ex$.

(iii) For all semi-constants s in C and for all $w \in C_0$,

$$s \circ w.w = sw \text{ and } w.s \circ w = ws.$$

Note that this definition of a composition t-ring is different to that given in the earlier paper[7]. If C is a composition t-ring with unital semi-constant e, then $e \circ e = ee = e \neq 0$. The last equality follows from the more general property $eu = u = ue$ for all $u \in C_0$. Indeed, $u = x \circ u = ex \circ u = e \circ u.u = eu$ by the definition of a composition t-ring. Using $x = xe$ the equality $u = ue$ can be obtained. Some of the main properties of these composition rings are given in:

Proposition 3.1. *Let C be a composition t-ring. Then:*

(i) $S := C_0 \circ e$ is the class of all semi-constants in C.

(ii) For any two semi-constants s and t with $t \neq 0$, $s \circ t = s$.

(iii) $S = C_0 \circ s$ for any semi-constant $s \neq 0$.

(iv) For any semi-constant $s \neq 0$ and $f \in C_0, s \circ f = 0 \Rightarrow f = 0$.

(v) If B_0 is the base of the composition ring $(C_0, +, \cdot, \circ)$, i.e. $B_0 = \{v \in C_0 \mid$ for all $a, b \in C_0$, $v \circ (a+b) = v \circ a + v \circ b$ and $v \circ ab = v \circ a.b\}$, then $B \subseteq B_0 = Sx = (C_0 \circ e)x$.

The semi-constants give a nice analogy with the role constants play in a near-ring:

$x = ex$	
$F \neq 0$	$F = 0$
e constant	e semi-constant
$B = Fx$	$B = Sx$
$F = C \circ 0 = C \circ n$ for all $n \in F$	$S = C \circ e = C \circ s$ for all $s \in S$

We have seen above that e is a multiplication identity for $(C_0, +, .)$. Next we investigate the existence and properties of a multiplication identity in a composition t-ring.

Proposition 3.2. *Let C be a composition t-ring and suppose C has a multiplication identity i. Then the following are equivalent:*
(1) $C = C_0$.
(2) i is 0-symmetric.
(3) $i = e$.
(4) i is a semi-constant.

We note that $C = M(R)$ is a composition t-ring with composition identity $x = 1_R$, constant multiplication identity $i = \hat{1}$ and unital semi-constant $e = \tilde{1}$ which is a multiplication identity for C_0. Having a constant multiplication identity is not exceptional; in fact we have:

Proposition 3.3. *Let C be a composition t-ring with foundation F and suppose C has a multiplication identity i. If $F \neq 0$, then i is a constant multiplication identity.*

Before we give the main result, we fix some notation. For any $z \in C$, let $D_z := \{c \in C \mid c \circ z = 0\}$. Then $D_z \lhd (C, +, \cdot)$ and $D_z \lhd_l (C, +, \circ)$. If $z \in C_0$, then $D_z \subseteq C_0$ (if $c \circ z = 0$, then $c \circ 0 = c \circ z \circ 0 = 0 \circ 0 = 0$). The main result is then:

Theorem 3.1. *Let C be a composition t-ring. Then $(C_0, +) = (B_0, +) \oplus (D_e, +)$ where $B_0 = Sx = (C_0 \circ e)x$ is the base of the composition ring C_0 and $D_e = \{c \in C \mid c \circ e = 0\}$. Moreover, if $C = C_0$, then there is a ring homomorphism $\beta : (C, +, \cdot) \to (B, +, \circ)$ defined by $\beta(c) = (c \circ e)x$ and a representation $\psi : C \to M(B)$ given by $\psi(c) = \psi_c$ where $\psi_c : B \to B$ is the map $\psi_c(sx) = (c \circ s)x$ for all $sx \in Sx = B$. Moreover, $\ker \psi = (0 : S)_{C_2}$.*

In the decomposition $C_0 = B_0 + D_e = (C_0 \circ e)x + D_e$, the C_0 in the definition of B_0 does not decompose any further, since $B_0 = (C_0 \circ e)x = ((B_0 + D_e) \circ e)x = B_0 \circ e.x + D_e \circ e.x = B_0 + 0 = B_0$.

If C is a composition t-ring, then so is C_0; hence any composition t-ring is of the form $C = F + C_0 = F + B_0 + D_e$ where B_0 is the base of C_0.

In general composition p-rings and composition t-rings are different. In fact, a non-zero composition ring C cannot be a composition p-ring and a composition t-ring. Indeed, if C is a composition p-ring, then $C = F + B + C_0x$ where $C_0 = B + C_0x$ and $B \cap C_0x = 0$. If C is also a composition t-ring, then $C = F + B_0 + D_e$ where B_0 is the base of C_0. We know $B_0 = Sx \subseteq C_0x$; so we have $B \subseteq B_0 = Sx \subseteq C_0x$. Thus $B = 0$ and so $x \in B = 0$; a contradiction.

The above captures exactly what we were looking for. In the 0-symmetric composition t-ring $C = M_0(R)$ we have: $e = \widetilde{1}$, $S = \{\widetilde{a} \mid a \in R\}$, $B = Sx = \{\widetilde{a}1_R \mid a \in R\} = \{\overline{a} \mid a \in R\}$, $\beta(f) = f(1)x$ and $\psi_f(\widetilde{a}x) = \widetilde{f(a)}x$.

Note also:

(1) Let R be a commutative ring with $1 \in R$ and let C be a subcomposition ring of $M(R)$ such that $x = 1_R \in C$. Then C is a composition t-ring iff $\widetilde{1} \in C$.

(2) For a finite field R, $\overline{R[x]} = M(R)$ is a composition t-ring but for other rings R this need not be the case. If R is a commutative ring with identity which contains a non-zero nilpotent element b, then $\overline{R[x]}$ is not a composition t-ring. Indeed, suppose $b^n = 0$ but $b^{n-1} \neq 0$. If $\overline{R[x]}$ is a composition t-ring, it contains a semi-constant e with $x = ex$, say $e = e_1x + e_2x^2 + \ldots + e_kx^k$. For any $t \in R$, $t = e(t)t = e_1t + e_2t^2 + \ldots + e_kt^k$ and so $t^{n-1} = e_1t^n + e_2t^{n+1} + \ldots + e_kt^{n+k-1}$ for all t. For $t = b$, we get $0 \neq b^{n-1} = 0$; a contradiction.

The foundations have been laid; one can now generalize the results for the composition rings $R[x]$, $R_0[[x]]$ and $R_N[[x]]$ to arbitrary composition rings C with non-zero foundation (there is already fairly extensive results on these) and when the foundation is 0, to those that have non-zero base.

4. Ideals in composition rings

We should point out that Kautschitsch and Mlitz[4] have given a general method to describe the maximal ideals of $R_0[x], R[x]$ and $R_N[[x]]$. This was done within the framework of composition subrings of $R[[x]]$. The results presented here will be more general and amongst others, will serve as a guideline on how to study polynomials and power series under our more general setting.

As mentioned earlier, the foundation F of C is always an C-module in the cannonical sense with respect to the module action $(c,n) := c \circ n$. A subset J of F is an *C-ideal* of F if J is an ideal of the ring F and $c \circ (n+j) - c \circ n \in J$ for all $c \in C, n \in F$ and $j \in J$. F is *C-simple* if it has no non-trivial proper C-ideals. Clearly, if F is simple, then it is C-simple. The composition rings C for which the notions "C-ideal" and "ideal" of F coincide, are called *compatible*. More information on these and also as background for what we present here, can be found in Refs. 5–7. For a non-empty subset S of C, $\langle S \rangle$ will denote the ideal in C generated by S.

Proposition 4.1.[5] *Let $J \lhd F, F = Found\ (C), C$ a composition ring. Then the following are equivalent:*
(1) J is an C-ideal of F.
(2) $(J : F)_{C_2} \lhd C$.
(3) $J = \langle J \rangle \cap F$.

Proposition 4.2.[5] *Let $I \lhd C$. Then there is a unique C-ideal J of F (namely $J = F \cap I$) such that $\langle J \rangle \subseteq I \subseteq (J : F)_{C_2}$.*

Proposition 4.3.[5] *Let C be a composition ring with Found$(C) = F \neq 0$. Then $(0:F)_{C_2}$ is a maximal ideal of C if and only if F is C-simple and $(0:F)_{C_2} + \langle F \rangle = C$.*

Corollary 4.1.[5] *Let C be a composition ring with Found$(C) = F \neq 0$ and $\langle F \rangle = C$. Then:*
(1) F is C-simple if and only if $(0:F)_{C_2}$ is the unique maximal ideal of C.
(2) C is simple if and only if F is C-simple and $(0:F)_{C_2} = 0$.

Further consequences of the above are:

Corollary 4.2.[5] *(1) Let C be a composition ring with foundation F such that $\langle F \rangle = C$. Then $(0:F)_{C_2} = 0$ and F simple implies C is simple.*

(2) Suppose C has a constant multiplication identity. If $(0:F)_{c_2} = 0$ and F is simple, then any composition subring D of C with Found $(D) = F$ is simple. Conversely, if every subcomposition ring D of C with Found$(D) = F$ is simple, then $(0:R)_{c_2} = 0$ and F is simple.

Let $\psi : C \to M(F)$ be the composition ring homomorphism defined by $\psi(c) = \psi_c : F \to F$ $(F = Found(C)), \psi_c(n) = c \circ n$ for all $n \in F, c \in C$. $Ker\psi = (0:F)_{C_2}$ and if $Ker\psi = 0$, then C can be regarded as a subcomposition ring of $M(F)$. In particular, if C is simple and $F \neq 0$, then C is a subcomposition ring of $M(F)$. As a special case of (2) in the previous corollary, we get the result of Adler[1]:

Proposition 4.4.[1] *Let R be a ring. If every subcomposition ring C of $M(R)$ with Found $(C) = R$ is simple, then R is simple. Conversely, if R is a simple ring with identity, then every subcomposition ring C of $M(R)$ with Found$(C) = R$ is simple.*

If C is infinite and $0 \neq Found\ (C) = F$ is finite, then C is not simple. Indeed, $(0:F)_{C_2}$ is non-zero (otherwise $\psi : C \to M(F)$ defined by $\psi(c) = \psi_c : F \to F$, $\psi_c(n) = c \circ n$, is an injective mapping from the infinite set C into the finite set $M(F)$) and $(0:F)_{C_2} \neq C$ (since $F \neq 0$). From Ref. 8 we know $FB \subseteq B$ and for all $n \in F, v \in B$ and $c \in C$, $nc = cn, (v \circ c)x = vc$ and $v \circ nx = nx \circ v = nv$. Futhermore:

Proposition 4.5.[7] *Let C be a composition ring with commuting composition identity x, Found$(C) = F$ and Base$(C) = B$.*
(1) If $J \lhd F$, then $Jx \lhd (Fx, +, \circ)$.
(2) If J is an C-ideal of F, then $B \circ J + J \circ B \subseteq J$ and $Jx \lhd (B, +, \circ)$
(3) Let $K \lhd B$. Let $J = \{n \in F | nx \in K \cap F\}$. Then $Jx = K \cap Fx$ and $J \lhd F$. If $B = Fx$, then $K \lhd B$ if and only if $K = Jx$ for some $J \lhd F$.

We now investigate the ideals of the near-ring $(C, +, \circ)$ in a composition ring C. As a special case of our results, Theorems 17.7 and 17.8 in Clay[2] will follow. It is interesting to note that essentially the same proofs work for the more general composition p-rings. We start with

Lemma 4.1. *Let C be a composition ring with commuting composition identity x. If $I \lhd (C, +, \circ)$, then $I \cap F \lhd (F, +, \cdot)$.*

Theorem 4.1. *Let C be a composition ring with commuting composition identity x, $F = Found(C)$ a field with at least three elements and $(0:F)_{C_2} = 0$. Then the near-ring $(C,+,\circ)$ is simple.*

For an infinite field F and $f(x) \in F[x]$, $f(a) = 0$ for all $a \in F$ is only possible when $f(x) = 0$. Hence we have Theorem 17.7 from Clay[2]:

Corollary 4.3.[2] *For an infinite field F, the near-ring $(F[x],+,\circ)$ is simple.*

Care should be taken when formulating results in terms of the base when the foundation is 0. For example, let $C = F_0[x]$ where F is a field. Then C has foundation 0 and base $(B,+,\circ) = (Fx,+,\circ) \cong (F,+,\cdot)$ with $(0:B)_{C_2} = 0$. But $(C,+,\circ)$ is not simple $(0 \neq Cx \lhd (C,+,\circ))$. Rather, what is valid is: Let C be a composition ring with commuting composition identity x, $F = Found(C)$ and $B = Base(C)$ a field with at least three elements. Suppose $(0:F)_{C_2} = 0$ and $Fx \neq 0$. Then $(C,+,\circ)$ is simple. This follows from the theorem since Fx is a nonzero ideal of the field B (and so $(B,+,\circ) = (Fx,+,\circ) \cong (F,+,\cdot)$).

Theorem 4.2. *Let C be a composition ring with commuting composition identity x and commutative multiplication. If $2^{**} = x+x$ is invertible in $(C,+,\circ)$, then every ideal of the near-ring $(C,+,\circ)$ is an ideal of C.*

As another corollary we have Theorem 17.18 from Clay[2]:

Corollary 4.4.[2] *If F is a finite field with characteristic > 2, then every ideal of the near-ring $(F[x],+,\circ)$ is an ideal of C.*

We note that Clay[2], 17.19 has shown that if the finite field F has characteristic 2, then $F[x]$ has near-ring ideals that are not ideals of the composition ring.

Lastly we discuss the maximal ideals of the near-ring $(C,+,\circ)$ for a composition p-ring C. We have $C = F + B + C_0x$ where F is the foundation and B is the base. We know $Fx \lhd (B,+,\circ)$ and we suppose $Fx \neq B$, i.e. C does not have a constant multiplicative identity. In other words, $x \in B$ (always), but $x \notin Fx$. From Section 2 we know $\beta : (C,+,\circ) \to (B/Fx,+,\circ)$ is a surjective near-ring homomorphism (just the projection onto B followed by the cannonical mapping) with $\ker\beta = F + C_0x + Fx = F + (C_0 + F)x = F + Cx \lhd (C,+,\circ)$. At times we will suppose that the composition ring C satisfies:

(I) $v + ux \in B + C_0x$ $(v \in B, u \in C_0)$ is invertible in $(C,+,\circ)$ if and only if v is invertible in $(B,+,\circ)$.

Properties required below are listed in:

Proposition 4.6. *Let C be a composition p-ring with $Fx \neq B$.*

(1) Let $J \lhd B$ such that $J + Fx$ is a maximal ideal of B. Then $\frac{J+Fx}{Fx}$ is a maximal ideal of B/Fx and $F + J + Cx = \beta^{-1}(\frac{J+Fx}{Fx})$ is a maximal ideal of $(C,+,\circ)$.

(2) Let M be a maximal ideal of $(C,+,\circ)$ with $F + Cx = \ker\beta \subseteq M$. If $\beta(M) = \frac{J}{Fx}$, then J is a maximal ideal of B and $Fx \subseteq J$.

(3) Let $I \lhd (C,+,\circ)$. Then:

(i) $\beta^{-1}(\beta(I)) = I$ if and only if $F + Cx = \ker\beta \subseteq I$.

(ii) Suppose $\beta(I) = \beta(C)$. *Then* $I = C$ *if and only if* $\ker\beta \subseteq I$.

(iii) Suppose $Fx \subseteq I$ *and* C *fulfills condition* (I). *Then* $\beta(I) = \beta(C)$ *if and only if* $I = C$.

Theorem 4.3. *Let* C *be a composition p-ring which fulfills condition* (I) *and has foundation* $F = 0$. *Let* $M \lhd (C,+,\circ)$. *Then* M *is a maximal ideal of* $(C,+,\circ)$ *if and only if* $M = J + C_0x$ *where* J *is a maximal ideal of* $(B,+,\circ)$.

Corollary 4.5. *Let* C *be a composition p-ring which fulfills condition* (I), *has foundation* $F = 0$ *and the base* B *is a field. Then* $(C,+,\circ)$ *has a unique maximal ideal* C_0x $(= Cx)$.

Theorem 4.4. *Let* C *be a composition p-ring which fulfills condition* (I). *Let* $M \lhd (C,+,\circ)$ *with* $Fx \subseteq M$. *Then* M *is a maximal ideal in* $(C,+,\circ)$ *if and only if* $M = F + J + Cx$ *where* $J \lhd B$ *with* $J + Fx$ *a maximal ideal of* $(B,+,\circ)$.

We note that Kautschitsch[3] has shown that the requirement $Fx \subseteq M$ in the composition rings $F_N[[x]]$ of formal power series is not necessary. In this case this inclusion $Fx \subseteq M$ is a consequence of the ideal M being maximal.

Bibliography

1. I. ADLER. Composition Rings. *Duke J. Math.* **29** (1962), 607–623.
2. J. R. CLAY. *Nearrings: Geneses and Applications.* Oxford Science Publications, New York, 1992.
3. H. KAUTSCHITSCH. Maximal ideals in the nearring of formal power series. *Proc. Conf. Nearrings and near-fields*, San Benedetto del Trente (Italy) 1981.
4. H. KAUTSCHITSCH and R. MLITZ. Maximal ideals in composition-rings of formal power series. *Contr. to General Algebra* **6** Verlag Holder - Pichler - Tempsky, Wien, 1988.
5. Q. N. PETERSEN and S. VELDSMAN. Composition near-rings. *Nearrings, Nearfields and K-loops* (Proc. Conf. Nearrings and Nearfields, Hamburg 1995, Editors: G. Saad and M. J. Thomsen), Kluwer Academic Publishers, The Netherlands, 1997, 357–372.
6. Q .N. PETERSEN and S. VELDSMAN. Composition near-rings. *Nearrings, Nearfields and K-loops* (Proc. Conf. Nearrings and Nearfields, Hamburg 1995, Editors: G. Saad and M. J. Thomsen), Kluwer Academic Publishers, The Netherlands, 1997, 357–372.
7. S. VELDSMAN. On the radical of composition near-rings. *Proc. Conf. Near-rings and Nearfields*, (Stellenbosch, July 1997). Editors Y. Fong *et al.*, Kluwer Acad. Publ., Netherlands, 2001, 198–201.
8. S. VELDSMAN. Polynomial and transformation composition rings. *Contr. Algebra and Geometry.* **41** (2000), 489–511.
9. S. VELDSMAN. Composition *p*-rings and composition *t*-rings revisited. Manuscript.

Centers and generalized centers of near-rings

Mark Farag

*Department of Mathematics,
Fairleigh Dickinson University,
Teaneck, NJ, 07666, USA
Email: mfarag@fdu.edu*

Kent M. Neuerburg

*Department of Mathematics,
Southeastern Louisiana University,
Hammond, LA 70402, USA
Email: kneuerburg@southeastern.edu*

This chapter introduces background, history, and recent results on centers and generalized centers of near-rings.

Keywords: near-ring; center; generalized center.

1. Introduction

For many algebraic structures, the center — the subset of elements which commute with all other elements under a given operation — exhibits the same structure as the larger set: the center of a semigroup is a subsemigroup, the center of a group is a subgroup, and the multiplicative center of ring is a subring. However, the analogous result is not, in general, true in the case of near-rings.

Let $(N, +, \circ)$ be a right near-ring. We refer the reader to the books of Pilz[14] and Clay[9] for general definitions and background on near-rings. Further, we adhere to the convention of writing $a \circ b$ as ab. The *center* of N is $C(N) = \{a \in N \mid an = na \text{ for all } n \in N\}$. When nonempty, $(C(N), \circ)$ is a subsemigroup of $(N, \circ)$. However, analysis of the centers of near-rings is complicated by the fact that $C(N)$ need not be a subnear-ring of N. In fact, $C(N)$ need not even be additively closed; i.e., $(C(N), +)$ need not be an additive subgroup of $(N, +)$.

With some notable exceptions in the area of near-fields (see, in particular, Wähling[16]), there had been little investigation of centers of near-rings until 2004, when Aichinger and Farag[2] gave a systematic analysis of several classes of near-rings, N, in which $C(N)$ is a subnear-ring of N. Subsequent work has continued along these lines of studying the centers of specific classes of near-rings.

In this chapter, we summarize many of the most significant results to date on centers and the related concept of generalized centers of near-rings. We begin in the next section by discussing a series of results on centers of near-rings. In Section 3, we treat the related notion of generalized centers of near-rings. Then we cover more recent results on centers and generalized centers of "Malone-like" near-rings in Section 4. Finally, in Section 5, we end with some suggestions for possible future research along these lines.

2. Centers of near-rings

Since, for a given near-ring N, the center $C(N)$ is not necessarily additively closed, the question of when $C(N)$ is a subnear-ring of N is of particular interest to near-ring theorists. This question is, in general, a nontrivial one and remains unanswered for many classes of near-rings. In this section, we discuss a series of results on centers of near-rings.

We begin by noting a few general facts that are useful in the study of centers of near-rings.

- For a near-ring N, the condition "$C(N)$ is additively closed in N" is weaker than the condition "$C(N)$ is a subnear-ring of N", as observed by Aichinger and Farag[2]. In fact, examples exist in which $C(N)$ is additively closed, but not a subnear-ring (see, for example Birkenmeier, Heatherly, and Pilz[5]). However, if N is finite the two conditions above are equivalent.
- If $C(N)$ is a subnear-ring of N, then N is zero-symmetric, as noted by Aichinger and Farag[2].
- If N is a near-ring with identity and $C(N)$ is additively closed, then $(N,+)$ is abelian (see Cannon, Farag, and Kabza[6]).

2.1. *Near-rings of functions*

We begin by considering the centralizer near-ring $M_0(G)$.

Definition 2.1. Let $(G,+)$ be a group (written additively but not necessarily abelian). Then $M_0(G) = \{f : G \to G \,|\, f(0) = 0\}$, the set of zero-preserving functions from G to G.

It is well-known[14] that $(M_0(G),+,\circ)$ is a right near-ring under pointwise addition and function composition. Aichinger and Farag[2] gave the following complete characterization of those near-rings of the form $N = M_0(G)$ for which $C(N)$ is a subnear-ring.

Theorem 2.1. *For a group G, let $N = (M_0(G),+,\circ)$. Then $(C(N),+,\circ)$ is a subnear-ring of N if and only if $exp(G) = 2$.*

The following result of Cannon, Farag, and Kabza[6] builds upon the preceding theorem.

Theorem 2.2. *Let G be a non-trivial finite group and $N = (M_0(G),+,\circ)$. The following are equivalent:*
1. $C(N)$ is a subnear-ring of N.
2. $G \cong \mathbb{Z}_2 \times \cdots \times \mathbb{Z}_2$.
3. $C(N) = GC(N) = \{0, id\}$, where id is the identity function on G.
4. $C(N)$ is a subring of N.

Definition 2.2. Let G be a non-trivial finite group and A a group of automorphisms of G. The set $N = M_A(G) = \{f : G \to G \,|\, f(0) = 0 \text{ and } f \circ a = a \circ f \text{ for all } a \in A\}$, along with the operations of pointwise addition and function composition, is the *centralizer near-ring determined by A and G.*

Cannon, Farag, and Kabza[6] showed that centralizer near-rings whose centers are subnear-rings must have a very particular kind of underlying group.

Theorem 2.3. *If $C(N)$ is a subnear-ring of $N = (M_A(G), +, \circ)$ then every non-identity element of G has prime order.*

Note that the converse is not true. Also, unlike in the case of $M_0(G)$, here $C(N)$ can be a subnear-ring of $N = M_A(G)$ even if the non-identity elements of G have different prime orders. For example, let $N = M_I(S_3)$ where $I = Inn(S_3)$, the set of all inner automorphisms of S_3. In this case, one can show that $N \cong \mathbb{Z}_6$, a ring. Hence, $C(N)$ is a subnear-ring of N.

2.2. *von Neumann regular near-rings*

Recall that a near-ring N is *von Neumann regular* if and only if for all $a \in N$ there is $b \in N$ such that $a = aba$ (see, for example, Pilz[14]). Aichinger and Farag[2] proved the following results concerning centers of such near-rings.

Theorem 2.4. *Let N be a von Neumann regular near-ring with identity. If $C(N)$ is a subnear-ring of N, then $C(N)$ is also von Neumann regular (and hence a subdirect product of fields).*

Corollary 2.1. *Let N be a von Neumann regular near-ring with identity that has no central zero-divisors. If $C(N)$ is a subnear-ring, then $C(N)$ is a field.*

2.3. *Matrix near-rings*

Using arrays with entries from a near-ring and the standard "row-by-column" array multiplication does not, in general, yield an associative structure. Therefore, in 1986 Meldrum and van der Walt[13] defined the $k \times k$ matrix near-ring over a near-ring as follows.

Let N be a near-ring and $k \in \mathbb{Z}^+$. Let ι_j be the jth coordinate injection map and π_j be the jth coordinate projection map. For $n \in N$, let λ_n be the left multiplication map; $\lambda_n : N \to N$ by $\lambda_n(m) = nm$. Then $f_{ij}^n = \iota_i \lambda_n \pi_j$ for $1 \leq i, j \leq k$ are functions in $M(N^k)$, the set of all mappings on N^k, which is a near-ring under pointwise addition and composition of functions. These functions form the basis of the following definition.

Definition 2.3. The $k \times k$ matrix near-ring over N is defined as $\mathbb{M}_k(N) = < \{f_{ij}^n = \iota_i \lambda_n \pi_j \,|\, 1 \leq i, j \leq k, n \in N\} >$, the subnear-ring of $M(N^k)$ generated by $\{f_{ij}^n = \iota_i \lambda_n \pi_j \,|\, 1 \leq i, j \leq k, n \in N\}$.

As established by Meldrum and van der Walt[13], $\mathbb{M}_k(N)$ is isomorphic to the standard ring of $k \times k$ matrices over N in case N is a ring.

The center of a matrix near-ring with identity and conditions under which this center forms a subnear-ring were determined by Aichinger and Farag[2].

Theorem 2.5. *Let N be a near-ring with identity. Then $C(\mathbb{M}_k(N)) = \{f_{11}^c + f_{22}^c + \cdots + f_{kk}^c \,|\, c \in C(N)\}$. Further, $C(\mathbb{M}_k(N))$ is a subnear-ring of $\mathbb{M}_k(N)$ if and only if $C(N)$ is a subnear-ring of N.*

2.4. *Polynomial near-rings*

Analogous to the case of matrix near-rings, using standard multiplication of polynomials with coefficents in a near-ring does not, in general, yield an associative structure. Therefore in 1997 Bagley[4] defined the polynomial near-ring over a zero-symmetic near-ring with identity as follows.

Let N be a zero-symmetric near-ring with identity. Let K denote the set of nonnegative integers and define $x : N^K \to N^K$ to be the mapping on the set of all sequences of elements of N defined via:

$$x(n_0, n_1, n_2, \ldots) = (0, n_0, n_1, n_2, \ldots).$$

For $a \in N$, let $L_a : N^K \to N^K$ denote the left multiplication map:

$$L_a(n_0, n_1, n_2, \ldots) = (an_0, an_1, an_2, \ldots);$$

the elementwise right multiplication map R_a is defined similarly.

Definition 2.4. The polynomial near-ring $N[x]$ in (one commuting indeterminant) x with coefficents in N is the subnear-ring of $M_N(N^K) = \{f \in M(N^K) \,|\, f(R_a(v)) = R_a(f(v))$ for all $v \in N^K$ and all $a \in N\}$ generated by $\{L_a \,|\, a \in N\} \cup \{x\}$.

Bagley[4] also showed that $N[x]$ is isomorphic to the usual polynomial ring over N in case N is a ring.

For the next result, we will require the following.

Definition 2.5. Given a near-ring N, the *distributor ideal* of N, denoted $D(N)$, is the ideal of N generated by the set $\{a(b+c) - ac - ab \,|\, a, b, c \in N\}$.

Notation 2.1. Given a near-ring N, we will denote by $\mathcal{U}(N)$ the set of multiplicative units of N.

Farag[10] proved, *inter alia*, the following result.

Theorem 2.6. *Let N be an abelian zero-symmetric near-ring with identity satisfying $N = D(N)$ and $C(N) \setminus \{0\} \subseteq \mathcal{U}(N)$. Then $C(N[x]) = \{cx^k \,|\, c \in C(N), k \in K\}$.*

2.5. *Near-rings on* $\mathbb{Z}_p \times \mathbb{Z}_p$

Let p be a prime. From a result of Maxson[12], we know that any near-ring with identity having $\mathbb{Z}_p$ as its additive group is necessarily a ring. Therefore Aichinger and Farag[2] analyzed near-rings on the additive group $N = \mathbb{Z}_p \times \mathbb{Z}_p$ and developed an exhaustive list of ten cases in which $C(N)$ is additively closed. Recall that for finite near-rings, additively closed is equivalent to being a subnear-ring.

Theorem 2.7. *If $C(N)$ is additively closed then N is isomorphic to one of the following.*

- *the ring $\mathbb{Z}_{p^2}$*
- *the field with p^2 elements*
- *a Dickson non-field with p^2 elements*
- *the exceptional near-field with 11^2 elements and solvable multiplicative group*
- *the exceptional near-field with 23^2 elements*
- *the exceptional near-field with 59^2 elements*
- *the square of the field with p elements*
- *the ring $\mathbb{Z}_p[x]/(x^2)$*
- *one of two near-rings $< \mathbb{Z}_p \times \mathbb{Z}_p, +, \circ >$ each with a specific multiplication.*

3. The generalized center of a near-ring

Since it can be a nontrivial task to characterize the exact elements of $C(N)$ and, even when such a characterization exists, $C(N)$ need not a subnear-ring of N, Farag looked at subnear-rings of N containing the subnear-ring generated by $C(N)$. In 2001, he defined a new class of centralizer near-rings [10].

We first recall the definition of the set of distributive elements of a near-ring.

Definition 3.1. Let N be a right near-ring. Then the set of *distributive elements* of N is $N_d = \{a \in N \mid a(b+c) = ab + ac \text{ for all } b, c \in N\}$.

Definition 3.2. Let N be a near-ring and $\emptyset \neq S \subseteq N_d$. Then $M_S(N) = \{a \in N \mid as = sa \text{ for all } s \in S\}$.

The set $M_S(N)$ is of interest since it is always a subnear-ring containing $C(N)$.

Theorem 3.1. *For any near-ring N, $M_S(N)$ is a subnear-ring of N containing $< C(N) >$.*

For $S = N_d$ Farag defined $M_S(N)$ to be the *generalized center* of N and denoted it by $GC(N)$.

For any zero-symmetric near-ring N, it is easy to see that

$$\{0\} \subseteq C(N) \subseteq GC(N) \subseteq N.$$

We note that, in the middle inclusion, equality is possible but not required. For example, if N is distributive (i.e., $N = N_d$), then we clearly have $C(N) = GC(N)$. There are many examples in which $C(N) \subsetneq GC(N)$, as we shall see in the sequel.

Importantly, we observe that if N is a ring then $C(N) = GC(N)$, so $GC(N)$ is a generalization of the center that both coincides with the usual definition of center when N is a ring and is always a subnear-ring of N.

Generalized centers have been investigated and compared with centers for several classes of near-rings. Unless otherwise noted, the remaining results in this section appear in a paper of Cannon, Farag, and Kabza [6].

3.1. *Distributively generated near-rings*

Definition 3.3. A near-ring N is said to be *distributively generated* if the additive group of N is generated by the distributive elements.

The following results relate centers and generalized centers in such near-rings.

Theorem 3.2. *Let N be a distributively generated near-ring. Then $C(N) = GC(N)$ if and only of $GC(N) \subseteq N_d$.*

Theorem 3.3. *Let N be a distributively generated near-ring with identity. The following are equivalent.*
1. $C(N)$ is a subnear-ring of N.
2. N is a ring.
3. $C(N) = GC(N)$.
4. $GC(N) \subseteq N_d$.

3.2. *Matrix near-rings*

In the case of matrix near-rings we have the following result relating the center and generalized center.

Theorem 3.4. *Let N be a near-ring with identity. If $C(N) = GC(N)$ then $C(\mathbb{M}_2(N)) = GC(\mathbb{M}_2(N))$.*

3.3. *Polynomial near-rings*

Farag[10] found sufficient conditions on an abelian zero-symmetric nonring with identity N which guarantee that the subnear-ring of $N[x]$ generated by $C(N[x])$ is properly contained in $GC(N[x])$.

Theorem 3.5. *Let N be an abelian zero-symmetric near-ring with identity satisfying $N = D(N)$ and $\langle C(N)\rangle \subsetneq N_d$. Then $\langle C(N)\rangle \subsetneq GC(N)$ implies that $\langle C(N[x])\rangle \subsetneq GC(N[x])$.*

3.4. *Near-rings of polynomials*

Let R be a commutative ring with identity and N be the set of polynomials over R with zero constant term. Then, with the usual addition and substitution of polynomials, $N = (R_0[x], +, \circ)$ is a zero-symmetric near-ring with identity, x. The following two results help to characterize when the center of a near-ring of polynomials is a subnear-ring.

Theorem 3.6. *Let R be a commutative ring with identity and let $N = (R_0[x], +, \circ)$ be the near-ring of polynomials over R with zero constant term. If $C(N)$ is a subnear-ring of N, then $char(R) = 2$.*

Theorem 3.7. *Let R be a commutative ring with $char(R) = 2$ with identity. Then $N = (R_0[x], +, \circ)$ satisfies $C(N) \subseteq GC(N) \subseteq \{\sum_{j=0}^{m} b_j x^{2^j} \mid 0 \leq m \in \mathbb{Z}, b_j = b_j^2 \text{ for all } b_j\} \subseteq N_d$.*

A student of Cannon, Matt Arbo, continued this investigation in his Honors Thesis[3]. Let R be a commutative ring with identity and $char(R) = 2$. Let $N = (R_0[x], +, \circ)$ be the corresponding near-ring of polynomials with zero constant term. Finally, let $S = \{bx^{2^n} \mid n \geq 0, b^2 = b, \text{ and } ba^{2^n} = ab \text{ for all } a \in R\}$. Arbo showed that $S \subseteq N_d$. Further, he established the following result.

Theorem 3.8. *If N and S are as described above, then*

$$C(N) \subseteq GC(N) = < S > = < C(N) > .$$

Arbo concluded his thesis with the next theorem.

Theorem 3.9. *Let R be a commutative ring of characteristic 2 with identity and let $N = (R_0[x], +, \circ)$. The following are equivalent*

- *$C(N)$ is a subnear-ring of N.*
- *$C(N) = GC(N)$.*
- *$C(N) = \{bx \mid b^2 = b \text{ and } b \in C(R)\}$.*
- *For $b = b^2 \in R$ and $n \geq 0$, if $ba^{2^n} = ab$ for all $a \in R$, then $b = 0$ or $n = 0$.*

4. Recent results on centers and generalized centers: Malone-like near-rings

In this section we provide some more recent results on centers and generalized centers of near-rings defined by certain multiplications on groups. We note that almost all near-rings herein do not possess a two-sided multiplicative identity element.

4.1. *Malone trivial near-rings*

Definition 4.1. Let G be a nontrivial group and suppose $S \subseteq G^*$. Define a multiplication on G via:

$$a \cdot b = \begin{cases} a \text{ if } b \in S \\ 0 \text{ if } b \notin S \end{cases} .$$

Then $N = (G, +, \cdot)$ is a right, zero-symmetric near-ring[11], which we call a *Malone trivial near-ring*. For such near-rings, Cannon, Farag, Kabza, and Neuerburg[7] have established the following basic results and examples.

Lemma 4.1. *A Malone trivial near-ring N as defined above has a two-sided multiplicative identity, 1, if and only if $N = \{0, 1\}$ and $S = \{1\}$, i.e., $N \cong \mathbb{Z}_2$.*

Proposition 4.1. *Let N be a Malone trivial near-ring defined as above. If $S = \emptyset$, then $C(N) = N$; if $S \neq \emptyset$, then $C(N) = \{0\}$ except when $N \cong \mathbb{Z}_2$.*

So only the extreme cases $C(N) = N$ and $C(N) = \{0\}$ are possible for Malone trivial near-rings. The same is true for $\langle C(N) \rangle$, the subnear-ring of N generated by $C(N)$. However, the generalized center admits more interesting results.

Theorem 4.1. *Let N be a Malone trivial near-ring.*
1. If $S = \emptyset$, then $C(N) = GC(N) = N$.
2. If $S \neq \emptyset$ and $|N| = 2$, then $\{0\} \neq C(N) = GC(N) = N$.
3. If $S \neq \emptyset$, $|N| > 2$, and $N_d = \{0\}$, then $\{0\} = C(N) \subsetneq GC(N) = N$.
4. If $S \neq \emptyset$, $|N| > 2$, $N_d \cap S = \emptyset$, and $N_d \neq \{0\}$, then $\{0\} = C(N) \subsetneq N \setminus S = GC(N) \subsetneq N$.
5. If $S \neq \emptyset$, $|N| > 2$, $N_d \cap S \neq \emptyset$, and $|N_d| = 2$, then $\{0\} = C(N) \subsetneq N_d = GC(N) \subsetneq N$.
6. If $S \neq \emptyset$, $|N| > 2$, $N_d \cap S \neq \emptyset$, and $|N_d| > 2$, then $\{0\} = C(N) = GC(N) \subsetneq N$.

We indicate examples for a few of the cases from the preceding theorem, some of which were found using the SONATA package[1] for GAP[15].

Example 4.1. Case 4: Let $G = \mathbb{Z}_8$ and $S = \{1,3,5,7\}$. Then $N_d = \{0,4\}$, so that $N_d \cap S = \emptyset$. It follows that $\{0\} = C(N) \subsetneq \{0,2,4,6\} = N \setminus S = GC(N) \subsetneq N$.

Example 4.2. Case 5: Let $G = \mathbb{Z}_6$ and $S = \{1,3,5\}$. Then $N_d = \{0,3\}$, so that $N_d \cap S \neq \emptyset$, and $|N_d| = 2$. It follows that $\{0\} = C(N) \subsetneq \{0,3\} = N_d = GC(N) \subsetneq N$.

Example 4.3. Case 6: Let $G = S_3$ and $S = \{(12),(13),(23)\}$. Then $N_d = \{0,(12),(13),(23)\}$, so that $N_d \cap S \neq \emptyset$, and $|N_d| > 2$. It follows that $\{0\} = C(N) = GC(N) \subsetneq N$.

4.2. *Complemented Malone near-rings*

In this section, we introduce another "Malone-like" multiplication on groups. Its definition, and all the remaining definitions and results in Section 4, can be found in a paper of Cannon, Farag, Kabza, and Neuerburg[7].

Definition 4.2. Let $(G,+)$ be a nontrivial abelian group and suppose $\emptyset \neq S \subseteq G^*$ such that for all $x \in S$, $-x \notin S$. Define a multiplication on G by

$$a \cdot b = \begin{cases} a \text{ if } b \in S \\ -a \text{ if } -b \in S \\ 0 \text{ if } b \notin S \text{ and } -b \notin S \end{cases} .$$

One may readily verify the following.

Proposition 4.2. *Given G as above, $N = (G,+,\cdot)$ is a zero-symmetric right near-ring with $|N| \geq 3$.*

We refer to the near-ring $(N,+,\cdot)$ from the preceding Proposition as a *complemented Malone near-ring* because of the similarity of its multiplication to that of ordinary Malone trivial near-rings and the additional condition that negatives of elements of S must be in the complement of S.

We have the following.

Lemma 4.2. *Let N be a complemented Malone near-ring as above. The following are equivalent:*

1. $C(N) \neq \{0\}$.
2. N *has a multiplicative left identity.*
3. $|N| = 3$.
4. $N \cong \mathbb{Z}_3$.

This Lemma is used to classify centers and generalized centers of complemented Malone near-rings.

Theorem 4.2. *Let* N *be a complemented Malone near-ring as above.*

1. If $|N| = 3$, *then* $N \cong \mathbb{Z}_3$.
2. If $|N| > 3$ *and* $N_d = \{0\}$, *then* $\{0\} = C(N) \subsetneq GC(N) = N$.
3. If $|N| > 3$, $N_d \neq \{0\}$, *and* $N_d \cap S = \emptyset$, *then* $\{0\} = C(N) \subsetneq N \setminus (S \cup (-S)) = GC(N) \subsetneq N$.
4. If $|N| > 3$, $|N_d| = 3$, *and* $N_d \cap S \neq \emptyset$, *then* $\{0\} = C(N) \subsetneq N_d = GC(N) = \{0, y, -y\} \subsetneq N$ *for some* $y \neq -y$.
5. If $|N| > 3$, $|N_d| > 3$, *and* $N_d \cap S \neq \emptyset$, *then* $\{0\} = C(N) = GC(N) \subsetneq N$.

In all cases, $C(N)$ is a subnear-ring of N.

4.3. *TS near-rings*

Definition 4.3. Let $(G, +)$ be a finite group of even order, not necessarily abelian. Suppose there exists $\emptyset \neq T \subseteq G^*$ such that $G \backslash T$ is a (normal) subgroup of G of index 2. Further suppose there is $\emptyset \neq S \subseteq T$ with $S = S_1 \dot{\cup} S_2 \dot{\cup} \cdots \dot{\cup} S_n$, a partition of S, and that there are distinct elements $q_1, q_2, \ldots, q_n$ of order 2 in $G \backslash (T \cup \{0\})$.

Define a multiplication on G by

$$a \cdot b = \begin{cases} q_1 & \text{if } a \in T, b \in S_1 \\ q_2 & \text{if } a \in T, b \in S_2 \\ \vdots & \\ q_n & \text{if } a \in T, b \in S_n \\ 0 & \text{otherwise} \end{cases}.$$

Proposition 4.3. $N = (G, +, \cdot)$ *with* G *as above is a right, zero-symmetric near-ring without multiplicative identity.*

We call the near-ring N above a TS *near-ring*.

Theorem 4.3. *Let* N *be a TS near-ring with* G, S, *and* T *as described previously.*

1. If $n = 1$ *and* $S = T$, *then* $C(N) = N_d = GC(N) = N$, *making* N *a commutative near-ring.*
2. If $n = 1$ *and* $S \subsetneq T$, *then* $N \setminus T = C(N) = N_d \subsetneq GC(N) = N$.
3. If $n \geq 2$, *then* $N \setminus T = C(N) = N_d \subsetneq GC(N) = N$.

In all cases, $C(N)$ is a subnear-ring of N.

4.4. *TSI near-rings*

Definition 4.4. Let $(G,+)$ be a group of even order, not necessarily abelian. Suppose there exists $\emptyset \neq T \subseteq G^*$ such that $G \backslash T$ is a (normal) subgroup of G of index 2. Let $\emptyset \neq I \subseteq T$ and $\emptyset \neq S \subseteq G^* \backslash I$ such that $T = I \cup (S \cap T)$. Partition S into $S = S_1 \dot{\cup} S_2 \dot{\cup} \cdots \dot{\cup} S_n$ such that for each $1 \leq i \leq n$, $S_i \subseteq S \cap T$ or $S_i \subseteq S \backslash T$. Furthermore, choose distinct $q_i \in S_i$ such that $2q_i = 0$ for each $1 \leq i \leq n$. Define a multiplication on G by

$$ab = \begin{cases} a & \text{if } b \in I \\ q_1 & \text{if } a \in T, b \in S_1 \\ \vdots & \\ q_n & \text{if } a \in T, b \in S_n \\ 0 & \text{otherwise} \end{cases}.$$

Proposition 4.4. *$N = (G,+,\cdot)$ as above is a right zero-symmetric near-ring, and N has a two-sided identity, 1, if and only if $I = \{1\}$, $S = \{q_1, q_2, \ldots, q_n\}$, and $N \setminus (S \cup T) = \{0\}$.*

We call the near-ring N above a *TSI near-ring*.

Theorem 4.4. *Let N be a TSI near-ring with $S \cap T = \emptyset$.*
1. $C(N) = Q \cup \{0\}$, which is a subnear-ring of N if and only if $Q \cup \{0\}$ is a subgroup of $G \backslash T$;
2. If $N_d = Q \cup \{0\}$, then $GC(N) = N$. If $N_d \neq Q \cup \{0\}$, then $GC(N) = N \backslash T$.

Theorem 4.5. *Let N be a TSI near-ring with $S \subsetneq T$.*
1. If $N_d \neq \{0\}$, then $S = S_1$, $Q = \{q_1\}$, $N_d = \{q_1, 0\} = C(N)$, and $GC(N) = N$.
2. If $N_d = \{0\}$, then $C(N) = \{0\}$ and $GC(N) = N$.

Theorem 4.6. *Let N be a TSI near-ring such that $S \cap T \neq \emptyset$ with $S \not\subseteq T$ and $N_d \neq \{0\}$.*
1. If $N_d = \{0, i\}$ for some $i \in I$, then $GC(N) = Q \cup \{0, i\}$. Furthermore, if $I = \{i\}$, $S = Q$, and $N \backslash (S \cup T) = \{0\}$, then $C(N) = \{0, i\}$; otherwise $C(N) = \{0\}$.
2. If $N_d = \{0, s\}$ for some $s \in (S_j \cap T) \backslash Q$, then $GC(N) = S_j \cup (N \backslash (S \cup T))$ and $C(N) = \{0\}$.
3. If $N_d = \{0, q_j\}$ for some $q_j \in S_j \cap T \cap Q$, then $GC(N) = I \cup S_j \cup (N \backslash (S \cup T))$ and $C(N) = \{0\}$.

The center $C(N)$ is a subnear-ring of N if and only if N does not have a two-sided multiplicative identity or N has a two-sided multiplicative identity of additive order two.

5. Future research

Many avenues of future research on centers and generalized centers of near-rings remain open. We end by indicating a few of these.

- There are many classes of near-rings for which we still do not know when $C(N)$ is a subnear-ring, when $GC(N) = C(N)$, or when $GC(N) = < C(N) >$. Study centralizer near-rings using particular groups, refine results on matrix near-rings, consider planar near-rings.

- For a near-ring N, $GC(N)$ is only one possible generalized center. Develop and study other possible generalizations.
- We have presented only some of the possible Malone-like multiplications to generate near-rings; others have been defined and studied by Cannon and his students [8]. These and other similarly definied near-rings provide many possibilities for future study.

Bibliography

1. E. Aichinger, F. Binder, J. Ecker, R. Eggetsberger, P. Mayr and C. Nöbauer, *SONATA - system of near-rings and their applications, Package for the group theory system GAP4*, Division of Algebra, Johannes Kepler University (Linz, Austria, 1999).
2. E. Aichinger and M. Farag, On when the multiplicative center of a near-ring is a subnear-ring, *Aequationes Math.* **68** (2004), no. 1–2, pp. 46–59.
3. M. Arbo, Centers and generalized centers of near-rings of polynomials, Honors Thesis, (Southeastern Louisiana University, USA, 2007).
4. S. W. Bagley, Polynomial near-rings: polynomials with coefficients from a near-ring, in *Near-rings, Nearfields, and K-Loops*, eds. G. Saad and M. J. Thomsen, Mathematics and Its Applications, Vol. 426 (Springer Netherlands, 1997), pp. 179–190.
5. G. Birkenmeier, H. Heatherly and G. Pilz, Homomorphisms on groups, I. Distributive ad d.g. near-rings, *Comm. Algebra* **25** (1997), no. 1, pp. 185–211.
6. G. A. Cannon, M. Farag and L. Kabza, Centers and generalized centers of near-rings, *Comm. Algebra* **35** (2007), no. 2, pp. 443–453.
7. G. A. Cannon, M. Farag, L. Kabza and K. M. Neuerburg, Centers and generalized centers of near-rings without identity defined via Malone-like multiplications, to appear, Math. Pannonica.
8. G. A. Cannon, V. Glorioso, B. B. Hall and T. Triche, Centers and generalized centers of near-rings without identity, to appear, Missouri J. of Math. Sci.
9. J. R. Clay, *Nearrings: Geneses and Applications*, Oxford University Press, London, 1992.
10. M. Farag, A new generalization of the center of a near-ring with applications to polynomial near-rings, *Comm. Algebra* **29** (2001), no. 6, pp. 2377–2387.
11. J. J. Malone, Near-rings with trivial multiplications, *Amer. Math. Monthly* **74** (1967), no. 9, pp. 1111–1112.
12. C. J. Maxson, On finite near-rings with identity, *Amer. Math. Monthly* **74** (1967), no. 10, pp. 1228–1230.
13. J.D.P. Meldrum and A.P.J. van der Walt, Matrix near-rings, *Arch. Math.* **47** (1986), pp. 312–219.
14. G.F. Pilz, *Near-rings*, 2nd ed., North-Holland Publishing Company, Amsterdam, New York, Oxford, 1983.
15. M. Schönert *et al.*, *GAP - Groups, Algorithms and Programming*, Lehrstuhl D für Mathematik, RWTH, (Aachen, 1994).
16. H. Wähling, *Theorie der Fastkörper*, Thales-Verlag, Essen, 1987.

Some semigroup theoretic aspects of nearrings

A. R. Rajan

Emeritus Professor, Department of Mathematics, University of Kerala, Kariavattom, Thiruvananthapuram 695 581

E-mail: arrunivker@yahoo.com

1. Introduction

A semigroup is a pair $(S,\cdot)$ where S is a nonempty set and $\cdot$ is an associative binary operation on S. A nearring $(S,+,\cdot)$ is a system with two binary operations $+$ and $\cdot$ such that $(S,+)$ is a group and $(S,\cdot)$ is a smigroup such that $\cdot$ is distributive over $+$ either on the left or on the right.

We consider nearrings $(S,+,\cdot)$ with right distribution property. That is,

$$(a+b)c = ac+bc$$

for all $a,b,c \in S$. We call them right nearrings.

We consider the multiplicative semigroup $(S,\cdot)$ of a nearring $(S,+,\cdot)$ and explore the properties of this semigroup. The objective is to determine how far this semigroup depends on the nearring.

It is well-known that any group $(S,+)$ can be made into a nearring by defining a multiplication making it a left zero semigroup. That is, $a\cdot b = a$ for all $a,b \in S$. The following converse question is more interesting. Given any semigroup $(S,\cdot)$ does there exist a group structure $(S,+)$ so that $(S,+,\cdot)$ is a nearring.

The answer is negative in general. For example the semigroup $(S,\cdot)$ where

$$S = \{(1,1),(1,2),(2,1),(2,2)\}$$

with product defined by $(i,j)(r,s) = (i,s)$ for $i,j,r,s \in \{1,2\}$ is a semigroup. We now show that this semigroup does not admit an addition making $(S,+,\cdot)$ a near ring.

The details are as follows. Suppose there is an addition so that $(S,+)$ is a group and $(S,+,\cdot)$ is a near ring. Let $(1,1)$ be the zero for the group $(S,+)$. In a right nearring S we have $0.x = 0$ for all $x \in S$. So $0\cdot(1,2) = 0$ and since $0 = (1,1)$ we have

$$0\cdot(1,2) = (1,1)(1,2) = (1,2).$$

This gives $(1,1) = (1,2)$ which is not true.

One of the fundamental questions here is to determine all semigroups $(S,\cdot)$ which admits a nearring structure $(S,+,\cdot)$ for some additive group $(S,+)$.

The semigroup given in the above example belongs to one of the interesting classes of semigroups considered in semigroup theory which are called rectangular bands. In general a rectangular band is semigroup on a set of the form $S = I\times\Lambda$ for nonempty sets I and Λ

and the product is defined by

$$(i,j)(r,s) = (i,s)$$

for all $(i,j),(r,s) \in I \times \Lambda$. We have the following result connecting rectangular bands and nearrings.

Theorem 1.1. *Let $(S,\cdot)$ be a rectangular band where $S = I \times \Lambda$ where I and Λ are nonempty sets and Λ contains more than one element. Then $(S,\cdot)$ can not be the multiplicative semigroup of a nearring $(S,+,\cdot)$.*

In this theorem if Λ is singleton then $(S,\cdot)$ is a left zero semigroup and it can be seen that any additive group $(S,+)$ makes $(S,+,\cdot)$ a nearring.

2. Categories

Categories play an important role in structure theory of semigroups. One class of semigroups for which category related structure theory is much developed is the class of regular semigroups. Here regularity means that every element x in the semigroup has a generalised inverse x' such that $xx'x = x$. This will enable us to use some of the well developed structure theory of regular semigroups to study nearrings. One such structure theory was developed by K. S. S. Nambooripad[3] known as theory of cross connections which uses categories in the description of the structure. The categories used here are called normal categories.

A category $\mathcal{C}$ is a structure consisting of a class $v\mathcal{C}$ called the class of objects of the category $\mathcal{C}$ and corresponding to every pair a,b of objects of $\mathcal{C}$ a set $\mathcal{C}(a,b)$ called the set of morphisms from a to b. When $f \in \mathcal{C}(a,b)$, we often write $f : a \to b$. Moreover for $a,b,c,d \in v\mathcal{C},\ \mathcal{C}(a,b) \cap \mathcal{C}(c,d) = \emptyset$ unless $a = c$ and $b = d$. A composition of morphisms is defined between composible pairs of morphisms. A morphism $f : a \to b$ is said to be composible with $g : c \to d$ if and only if $b = c$. The composition is associative whenever defined and admits left and right identity for each morphism. That is if $f : a \to b$ then f has a left identity denoted by $1_a : a \to a$ and a unique right identity is $1_b : b \to b$ so that

$$1_a f = f \text{ and } f 1_b = f.$$

A morphism $f : a \to b$ is said to be invertible if there is a $g : b \to a$ such that $fg = 1_a$ and $gf = 1_b$. Invertible morphisms are also called isomorphisms. A right cancellative morphism is called a monomorphism and a left cancellative morphism is called an epimorphism.

Two special categories we often use are groupoids and preorders. A groupoid is a category in which every morphism is an isomorphism. And a preorder is a category in which for every $a,b \in v\mathcal{C}$ the morphism set $\mathcal{C}(a,b)$ contains at most one element.

A strict preorder is a category $\mathcal{C}$ in which $\mathcal{C}(a,b) \cup \mathcal{C}(b,a)$ contains at most one element. A strict preorder $\mathcal{C}$ is essentially a partially ordered set. In a strict preorder $\mathcal{C}$ we define a partial order $\leq$ on $v\mathcal{C}$ as follows. For $a,b \in v\mathcal{C}$,

$$a \leq b \text{ if } \mathcal{C}(a,b) \neq \emptyset.$$

Then $\leq$ is a partial order on $v\mathcal{C}$. Conversely if $(X, \leq)$ is a partially ordered set then we can define a category $\mathcal{C}$ whose set of objects is X and for $x, y \in X$

$$\mathcal{C}(x,y) = \begin{cases} (x,y) \text{ if } x \leq y \\ \text{undefined otherwise.} \end{cases}$$

Then $\mathcal{C}$ is a strict preorder.

Now we proceed to associate some categories with regular semigroups. The first one is the category $\mathcal{L}(S)$ of principal left ideals of a semigroup S.

Let S be a regular semigroup. Then we define a category $\mathcal{L}(S)$ as follows. The objects of $\mathcal{L}(S)$ are principal left ideals of S. Since S is a regular semigroup every principal left ideal of S is generated by an idempotent[1]. Here an idempotent is an element e such that $e^2 = e$. So

$$v\mathcal{L}(S) = \{Se : e \in E(S)\}$$

where $E(S)$ is the set of idempotents of S. It may be observed that Se can be same as Se' for some $e \neq e'$.

The morphisms in $\mathcal{L}(S)$ are defined as follows. For objects Se, Sf in $\mathcal{L}(S)$ a morphism from Se to Sf is a right translation ρ_u induced by an element $u \in eSf$. That is ρ_u is the mapping given by $x\rho_u = xu$ for every $x \in Se$.

It may be noted that since $u \in eSf$ we have $eu = u$ so that ρ_u maps e to u. We denote ρ_u by $\rho(e,u,f) : Se \to Sf$. Further $\rho(e,u,f) = \rho(e',u',f')$ whenever $Se = Se'$, $Sf = Sf'$ and $u' = e'u$.

This category has some interesting factorization properties. Also we can identify certain clusters of morphisms described as cones in this category. Such categories are called normal categories[3].

Some classes of regular semigroups can be determined in terms of the associated normal categories. A wider application of normal categories is in determining the structure of regular semigroups by relations derived on translations of ideals. These relations are known as cross connections (see Ref. 3).

Now we proceed to give an abstract description of normal categories. This involves several concepts such as category with subobjects, normal factorization, normal cones etc.

Definition 2.1. A category with subobjects is a pair $(\mathcal{C}, \mathcal{P})$ where $\mathcal{C}$ is a category and $\mathcal{P}$ is a subcategory of $\mathcal{C}$ satisfying the following.

(1) $v\mathcal{C} = v\mathcal{P}$.
(2) $\mathcal{P}$ is a strict preorder.
(3) Every $f \in \mathcal{P}$ is a monomorphism in $\mathcal{C}$.
(4) If $f, g \in \mathcal{P}$ and if $f = hg$ for some $h \in \mathcal{C}$, then $h \in \mathcal{P}$.

In this case a morphism $j : a \to b$ in $\mathcal{P}$ is said to be an inclusion from a to b. We denote the inclusion from a to b by $j(a,b)$. Also we say $a \leq b$ in this case.

A category with normal factorization is defined as follows.

Definition 2.2. A category with normal factorization is a category with subobjects $(\mathcal{C}, \mathcal{P})$ satisfying the following.

N1 For each $j : a \to b$ in $\mathcal{P}$ there is a morphism $q : b \to a$ in $\mathcal{C}$ such that $jq = 1_a$. In this case q is said to be a retraction.

N2 Every morphism $f : a \to b$ in $\mathcal{C}$ has a factorization in the form

$$f = quj$$

where q is a retraction; u is an isomorphism and j is an inclusion. Such a factorization of f is called a *normal factorization.*

Remark 2.1. The normal factorization of a morphism f as $f = quj$ is not in general unique. But if $f = quj = q_1u_1j_1$ then it can be seen that $qu = q_1u_1$ and $j = j_1$. We denote this qu by f^0 and call it the epimorphic part of f. It may be noted that in this case f^0 is an epimorphism.

The next component in the description of normal category is normal cone which is defined as follows.

Definition 2.3. A normal cone in a category with subobjects $(\mathcal{C}, \mathcal{P})$ is a pair (γ, d) where $d \in v\mathcal{C}$ and γ is a mapping $\gamma : v\mathcal{C} \to \mathcal{C}$ such that for each $a \in v\mathcal{C}$, $\gamma(a)$ is a morphism from a to d satisfying the following.

(1) For every inclusion $j(a,b) : a \to b$, $\gamma(a) = j(a,b)\gamma(b)$.

(2) There exists $c \in v\mathcal{C}$ such that $\gamma(c) : c \to d$ is an isomorphism.

The cone (γ, d) is often referred to as a cone γ with vertex d. Now we define normal categories.

Definition 2.4. A normal category is a category with subobjects $(\mathcal{C}, \mathcal{P})$ satisfying the following.

N1 Every inclusion $j : a \to b$ in $\mathcal{C}$ has a right inverse, that is a morphism $q : b \to a$ in $\mathcal{C}$ such that $jq = 1_a$.

N2 Every morphism $f : a \to b$ in $\mathcal{C}$ has a normal factorization.

N3 For each $c \in v\mathcal{C}$ there is a normal cone ε with vertex c such that $\varepsilon(c) = 1_c$.

Now we can see that for any regular semigroup S, $\mathcal{L}(S)$ is a normal category. Also the normal cones in a normal category form a regular semigroup, with product as follows. Let γ, σ be normal cones in $\mathcal{C}$ with vertices c and d respectively. Then

$$(\gamma * \sigma)(a) = \gamma(a) \cdot (\sigma(c))^0$$

where $\cdot$ denotes the composition in $\mathcal{C}$. Then $\gamma * \sigma$ is a normal cone and so this defines a semigroup structure on the set $T\mathcal{C}$ of all normal cones in $\mathcal{C}$. It can be seen that this semigroup $T\mathcal{C}$ is a regular semigroup.

It may be noted that a normal cone σ is an idempotent in $T\mathcal{C}$ if and only if $\sigma(c) = 1_c$ where c is the vertex of σ.

The following theorem shows that every normal category arises as the normal category $\mathcal{L}(S)$ of a regular semigroup S upto isomorphism. Here isomorphism of normal categories are category isomorphisms which preserve the inclusions.

Theorem 2.1. *Let $(\mathcal{C}, \mathcal{P})$ be a normal category and $T\mathcal{C}$ be the semigroup of normal cones in $\mathcal{C}$. Then the category of principal left ideals of the regular semigroup $T\mathcal{C}$ is isomorphic to $\mathcal{C}$ as normal categories.*

The normal categories arising from regular rings have been studied in Ref. 7 and with suitable addtional structure on normal categories these categories have been characterised. The present objective is to modify the results for regular rings to the case of nearings.

Towards this we consider regular nearrings which are nearrings in which the multiplicative semigroup is a regular semigroup. The following properties of ideals of regular rings have been used significantly in describing the nature of the normal ctegory of a regular ring.

Theorem 2.2.[2] *Let S be a regular ring and $e, f, g, \ldots$ denote idempotents in S. Then the following hold.*

(1) The sum (join) of two principal left ideals is again a principal left ideal and more specifically
$Se + Sf = S(e+g)$
for any idempotent g such that $Sg = S(f - fe)$.
(2) The meet (intersection) of two principal left ideals is again a principal left ideal.
(3) The set of all principal left ideals is a relatively complemented lattice.

As a consequence we see that the resulting normal category $\mathcal{L}(S)$ of a regular ring S has the following properties.

Theorem 2.3.[7] *Let $\mathcal{L}(S)$ be the category of principal left ideals of a a regular ring S. Then*

(1) each morphism set $[Se, Sf]$ is an abelian group with the natural addition induced from the abelian group $(S, +)$.
(2) $v\mathcal{L}(S)$ is a relatively complemented lattice.
(3) every subset of $v\mathcal{L}(S)$ which has an upper bound has a grearest element.

The above properties in a normal category $\mathcal{C}$ enables one to define an addition on the set $T\mathcal{C}$ of normal cones so that $(T\mathcal{C}\, +, \cdot)$ is a regular ring[7].

3. Normal Categories from Nearrings

With the properties of the normal category of regular rings as the guiding relations we consider nearrings which give similar properties on principal left ideals. Several special conditions have to be imposed on the nearrings so that the resulting normal categories have the above properties. Hence the class of nearrings we consider here will be much restricted.

One such restriction we impose is to have partial distributivity on the left. For each subset X of a right nearring S we define

$$D(X) = \{a \in S : a(x+y) = ax + ay \text{ for all } x, y \in X\}.$$

We say that X is a distributing set if $D(X) = S$. For example if the nearring S is zero symmetric then $\{0\}$ is a distributing set.

We consider regular nearrings $(S,+,\cdot)$ in which for all idempotents e,f the set eSf is a distributing set. Further we restrict to nearrings $(S,+,\cdot)$ in which $(S,\cdot)$ is an inverse semigroup. Note that inverse semigroups are semigroups with the property that for every element $a \in S$ there is a unique element $a' \in S$ such that $aa'a = a$ and $a'aa' = a'$. In this case the set of idempotents is a commutative subsemigroup.

The following theorem gives some properties of such nearrings.

Theorem 3.1. *Let $(S,+,\cdot)$ be a right nearring which is $0-$ symmetric and satisfying the following properties.*

(1) $(S,\cdot)$ is an inverse semigroup
(2) For any idempotents e,f the set eSf is a distributing set.

Then the following hold.

(i) $Se \vee Sf = S(e+g)$ where $g = f - fe$ is an idempotent.
(ii) $v\mathcal{L}(S)$ is a lattice.

Proof. First we observe that $g = f - fe$ is an idempotent. Since S is an inverse semigroup $fe = ef$ and so $f, fe \in fSf$. Since fSf is a distributing set we have

$$(f-fe)(f-fe) = f - fef - fe + fe = f - fe.$$

To prove (i) we show that

$$Se \subseteq S(e+g) \text{ and } Sf \subseteq S(e+g)$$

and if for some idempotent k, $Se \subseteq Sk$ and $Sf \subseteq Sk$ then $S(e+g) \subseteq Sk$.

Now to see that $Se \subseteq S(e+g)$ first observe that $ge = (f-fe)e = 0$. Now choose an idempotent h such that $S(e+g) = Sh$. Then $e+g = xh$ for some $x \in S$. Then

$$e = (e+g)e = xhe = xeh \in Sh = S(e+g).$$

So $Se \subseteq S(e+g)$. Further since $S(e+g) = Sh$ is a group with respect to $+$ and $e, e+g \in Sh$ we have $g \in Sh$ so that $g+e \in Sh$. Let $g+e = yh$ for some $y \in S$. Now

$$\begin{aligned} f &= g + ef = (g+e)f \text{ (since } ef = fe \text{ and } g = gf) \\ &= yhf = yfh \in Sh = S(e+g). \end{aligned}$$

Thus $Se \subseteq S(e+g)$ and $Sf \subseteq S(e+g)$.

Now let k be an idempotent such that $Se \subseteq Sk$ and $Sf \subseteq Sk$. Then $ek = e$ and $fk = f$ and so $gk = (f-fe)k = f - fe = g$. Therefore

$$e + g = ek + gk \in Sk.$$

Thus $Se \vee Sf = S(e+g)$. Statement (ii) follows by defining join as in (i) here and meet by

$$Se \wedge Sf = S(ef).$$

□

Remark 3.1. From the steps in the proof we see that

$$Se \vee Sf = S(e+g) = S(g+e) \text{ for } g = f - fe \text{ and}$$

$$Se \vee Sf = S(f+k) = S(k+f) \text{ for } k = e - ef.$$

Now we have the following theorem on the category $\mathcal{L}(S)$ associated with such a nearring.

Theorem 3.2. *Let $(S,+,\cdot)$ be a nearring such that $(S,\cdot)$ is an inverse semigroup and for any idempotents e,f; eSf is a distributing set. Then $\mathcal{L}(S)$ is a normal category in which each morphism set $[Se,\ Sf]$ is an additive group. Further $T\mathcal{L}(S)$ is a nearring which contains S as a subnearring.*

Proof. From the definition of morphisms in $\mathcal{L}(S)$ every morphism is of the form $\rho(e,u,f)$ with $e \in eSf$ which maps $x \in Se$ to $xu \in Sf$. Consider $\rho(e,u_1,f),\ \rho(e,u_2,f) \in [Se,Sf]$ where $u_1,u_2 \in eSf$. We define

$$\rho(e,u_1,f) + \rho(e,u_2,f) = \rho(e,u_1+u_2,f).$$

Since $u_1,u_2 \in eSf$ and eSf is a distributing set we have $u_1+u_2 \in eSf$ and so the addition is well defined. Now $\rho(e,0,f)$ is the additive identity and $\rho(e,-u,f)$ is the inverse of $\rho(e,u,f)$. Thus $[Se,Sf]$ is an additive group.

Now we define a left nearring structure on the set $T\mathcal{L}(S)$ of all normal cones in $\mathcal{L}(S)$ in this case. From the results on normal categories (cf Ref. 3) the multiplicative semigroup structure on $T\mathcal{L}(S)$ is known.

We define an addition in $T\mathcal{L}(S)$ as follows. Let τ,σ be normal cones in $\mathcal{L}(S)$ with verices Sf and Sg respectively. Let $Sk = Sf \vee Sg$ where k is an idempotent. Since S is an inverse semigroup, we see from Ref. 6 that there is only one isomorphism component for a normal cone in $\mathcal{L}(S)$. Let $\tau(Se_1) = \rho(e_1,u_1,f)$ and $\sigma(Se_2) = \rho(e_2,u_2,g)$ be the isomorphism components of τ and σ respectively. Let $Sk' = S(u_1+u_2)$ for an idempotent k'. For every $Se \in v\mathcal{L}(S)$ define

$$(\tau+\sigma)(Se) = (\tau(Se)j(Sf,Sk) + \sigma(Se)j(Sg,Sk))q(Sk,Sk')$$

where j stands for inclusion and q stands for retraction with the given domain and codomain which in this case is unique by Ref. 6.

It is easy to see that when $Se' \subseteq Se$ we have

$$(\tau+\sigma)(Se') = j(Se',Se)(\tau+\sigma)(Se).$$

It follows that $\tau+\sigma$ is a normal cone with vertex Sk'. The normal cone 0 which has all components zero is the additive identity for $T\mathcal{L}(S)$. For any normal cone τ with vertex Sf the cone $-\tau$ defined by

$$-\tau(Se) = \rho(e,-u,f)$$

is a normal cone which is the additive inverse of τ where $\tau(Se) = \rho(e,u,f)$. Thus $T\mathcal{L}(S)$ is a group with respect to $+$ defined above. From the right distributivity of the nearring S it follows that $(T\mathcal{L}(S),+,\cdot)$ is a right nearring.

Now to see that S is isomorphic to a subnearring of $T\mathcal{L}(S)$ we consider a normal cone ρ^a corresponding to each $a \in S$ defined a follows.

$$\rho^a(Se) = \rho(e, ea, f)$$

where f is the unique idempotent such that $Sa = Sf$. Then the the map $a \mapsto \rho^a$ is an isomorphism of S into $T\mathcal{L}(S)$. Therefore S is isomorphic to a subnearring of $T\mathcal{L}(S)$. □

Bibliography

1. Clifford, A. H. and Preston, G. B., Algebraic Theory of Semigroups Vol. I, *Amer. Math. Soc.* 1961.
2. Goodearl, Von Neumann, Regular Rings, Pitman Publishing Ltd., London 1979.
3. Nambooripad, K. S. S., Theory of cross connections, Pub. No. 38, Centre for Mathematical Sciences, Trivandrum, 1984.
4. Nambooripad, K. S. S., Rajan, A. R. and Krishnan, E., Structure theory of regular semigroups, (to be published).
5. Pilz, G., Near rings, North Holland, 1983.
6. Rajan, A. R., Normal categories of inverse semigroups, *East West J. Math.* Vol. 16 (2), 2014, 122–130.
7. Rajan, A. R. and Sunny Luckose, Normal categories of regular rings, *Indian J. Pure and Applied Math.* 2010,

On ideal theory for lattices

S. Parameshwara Bhatta

Department of Mathematics, Mangalore University, Mangalagangothri, Mangalore - 574199, Karnataka
E-mail: s_p_bhatta@yahoo.co.in

Historically ideal theory for lattices was developed by Hashimoto [10]. He established that there is a one to one correspondence between ideals and congruence relations of a lattice L under which the ideal corresponding to a congruence relation is a whole congruence class under it if and only if L is a generalized Boolean algebra. His proof involved topological ideas which were later simplified, using lattice theoretic ideas, by Gratzer and Schmidt [8]. Also Gratzer [6] introduced the notion of standard elements and ideals in lattices which were extensively studied by Gratzer and Schmidt [9]. It was shown that standard ideals of lattices play a role somewhat similar to that of normal subgroups of groups or ideals of rings. Later, Fried and Schmidt [4] extended the notion of standard ideals of lattices to convex sublattices. Generalizations of some of these results to trellises (or also called weakly associative lattices) may be found in Bhatta and Ramananda [13] and Shashirekha [14].

Keywords: Lattices, Complete lattices, Algebraic lattices and ideals of lattices.

1. Introduction

For basic definitions and results we refer Gratzer [5], Crawley and Dilworth [2], and Szasz [16].

A *lattice* is a partially ordered set (or briefly, a poset) $\langle L;\leq\rangle$ in which every pair of elements has a greatest lower bound and a least upper bound; or 'equivalently' a *lattice* is an 'algebra' $\langle L;\wedge,\vee\rangle$, with two binary operations $\wedge(meet)$ and $\vee(join)$ on L, which are both commutative, associative and satisfy the 'absorption' identities: $a\wedge(a\vee b)=a$; $a\vee(a\wedge b)=a$.

By a *complement* of an element x in a bounded lattice L (i.e., a lattice with the minimum element 0 and the maximum element 1) we mean an element y of L with $x\wedge y=0$ and $x\vee y=1$. A bounded lattice L is said to be complemented if each element of L has a complement. A lattice L is said to be relatively complemented if every interval $[x,y]$ of L is complemented. A complete lattice is a poset in which every subset has a meet (i.e., a greatest lower bound) and a join (i.e., a least upper bound) or equivalently, a complete lattice is a lattice in which every subset has a meet and a join.

An ideal I of a lattice L is a 'nonempty' subset of L satisfying the following two conditions:

1) If $a,b\in I$, then $a\vee b\in I$,
2) If $a\in I$ and $x\in L$, then $a\wedge x\in I$.

Remark 1.1. In the above definition, condition (2) can be replaced by the following:
2') If $a\in I$ and $x\in L$ with $x\leq a$, then $x\in I$ (Or equivalently, if $a\in I$, then $(a]\in I$, where $(a]=\{x\in L: x\leq a\}$).

The ideal generated by a subset of a lattice and its description (Gratzer[5])

If H is a 'nonempty' subset of a lattice L, then the *ideal generated by* H, denoted by $(H]$, is the smallest ideal of L containing H or equivalently, it is the intersection of all ideals of L containing H. Also, $(H] = \{x \in L : x \leq h_1 \vee \ldots \vee h_n$ for some $n \geq 1$ and $h_i \in H$ for $1 \leq i \leq n\}$. In particular, if $H = \{a\}$, where $a \in L$, then $(H] = (a]$, called the *principal ideal generated by a.*

It is known that every ideal of a lattice L is *principal* if and only if the lattice satisfies the ascending chain condition (i.e., there is no infinite 'ascending' chain of elements of L of the form $a_1 < a_2 < a_3 \ldots$) or equivalently every nonempty subset of L has a maximal element.

An element of a complete lattice L is said to be *compact* (Crawley and Dilworth[2] and Gratzer[5]) if whenever $a \leq \bigvee S$, where S is a subset of L, there exists a finite subset T of S with $a \leq \bigvee T$. A complete lattice L is said to be *algebraic (or compactly generated)*, if every element of L is the join of compact elements of L. It is known that every element of a complete lattice L is compact if and only if it satisfies the ascending chain condition. In particular, it follows that any lattice which is bounded below and satisfying the ascending chain condition is algebraic.

Congruence relations in lattices (Gratzer[5])

A congruence relation Θ on a lattice L is an equivalence relation satisfying the following 'substitution property':
For $a,b,c,d \in L, a \equiv b(\Theta)$ and $c \equiv d(\Theta)$ imply $ac \equiv bd(\Theta)$ and $ac \equiv bd(\Theta)$. Or equivalently, for $a,b,c \in L, a \equiv b(\Theta)$ implies $ac \equiv bc(\Theta)$ and $ac \equiv bc(\Theta)$.

Remark 1.2. Simplest examples of congruences on a lattice L are the equality relation ω and the universal relation ι , where, for $a,b \in L, a \equiv b(\omega)$ iff $a = b$ (and $a \equiv b(\iota)$ for every $a,b \in L$). A lattice L is said to be simple if its only congruences are these two 'trivial' ones.

Quotient lattices of lattices (Gratzer[5])

A congruence relation on a lattice L, being an equivalence relation on L, induces a partition of L, into convex sublattices, given by, $L/\Theta = \{[a]\Theta : a \in L\}$, where $[a]\Theta = \{x \in L : x \equiv a(\Theta)\}$ is the congruence class of L modulo Θ containing a. L/Θ is itself a lattice, called the Quotient lattice of L modulo Θ, with respect to the binary operations $\wedge$ and $\vee$ on it given by the equations , $[a]\Theta \vee [b]\Theta = [a \wedge b]\Theta$ and $[a]\Theta \vee [b]\Theta = [a \wedge b]\Theta$.

For lattices L_1 and L_2, a map $f : L_1 \to L_2$ is called a homomorphism if $f(a \vee b) = f(a) \vee f(b)$ and $f(a \wedge b) = f(a) \wedge f(b)$ for all $a,b \in L$.

The homomorphism theorem for lattices (Gratzer[5])

Any homomorphic image of a lattice L is isomorphic to a suitable quotient lattice of L. More precisely, if $f : L_1 \to L_2$ is a homomorphism of a lattice L_1 onto a lattice L_2, then the relation on L_1 given by, $a \equiv b(\Theta)$ if and only if f(a) = f(b), is a congruence relation and the

quotient lattice L_1/Θ is isomorphic to L_2.
An isomorphism $g: L_1/\Theta \to L_2$ is given by the equation $g([a]\Theta) = f(a)$.

The three kernel concepts for lattices(Gratzer[5])
1) Let $f: L_1 \to L_2$ be a homomorphism of a lattice L_1 into a lattice L_2. Assume that L_2 has the least element 0 and further the set $f^{-1}(\{0\}) = \{x \in L_1 : f(x) = 0\}$ is nonempty. Then $f^{-1}(\{0\})$ is an ideal of L_1, called the ideal kernel of the homomorphism *f*.
2) Let L be a lattice and be a congruence relation on L. If the quotient lattice L/Θ has the least element, say $[a]\Theta$, then $[a]\Theta$ as a subset of L is an ideal, called the *ideal kernel* of the congruence relation.
3) Let $f: L_1 \to L_2$ be a homomorphism of a lattice L_1 into a lattice L_2. Then the relation Θ on L defined by, for $x, y \in L, x \equiv y(\Theta)$ if and only if $f(x) = f(y)$, is a congruence relation, called the congruence *kernel of the homomorphism* f (usually denoted by *Ker(f)*).

Remark 1.3. The Ideal kernel of a homomorphism may not be defined for all homomorphisms of a given lattice. Also, the ideal kernel of a congruence relation may not be defined for all congruence relations of a lattice. However, the congruence kernel of a homomorphism is always defined for any homomorphism of a lattice. Moreover, the notions of a congruence relation, the quotient lattice and the homomorphism theorem, can be easily extended to "universal algebras" as in Gratzer[7].
In the case of lattices, the three kernel concepts are useful. Relationship between these three notions of kernels, when all of them exist together, is given below:

- If I is the ideal kernel of a homomorphism $f: L_1 \to L_2$, then I is the ideal kernel of Θ , where Θ is the congruence kernel of *f*.

Illustrations for kernel concepts

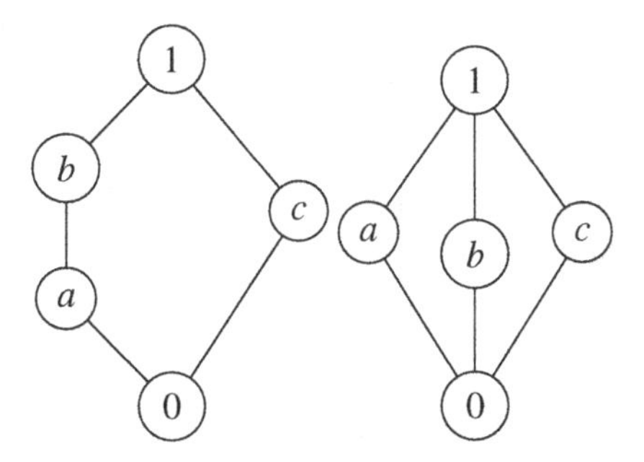

Figure 1: Lattices N_5 and M_5.

In the case of each of the lattices N_5 and M_5 (given by Figure 1), their only ideals are $(0], (a], (b], (c]$ and $(1]$. In N_5, the only ideals which are kernels of some homomorphisms/congruences are $(0], (b], (c]$ and $(1]$. In M_5, the only ideals which are kernels of some homomorphisms/congruences are $(0]$ and $(1]$. These two lattices are complemented.

M_5 is also a relatively complemented simple lattice whereas N_5 is neither relatively complemented nor simple.
A lattice satisfying the identity $x \wedge (y \vee z) = (x \wedge y) \vee (x \vee z)$, Or equivalently, its 'dual identity' $x \vee (y \wedge z) = (x \vee y) \wedge (x \vee z)$, is called a distributive lattice.
A lattice satisfying the condition, for $x, y, z \in L, x \wedge (y \vee z) = (x \wedge y) \vee z$, whenever $x \geq z$, is called a modular lattice.

Remark 1.4. Every distributive lattice is modular, but not conversely. It may be noted that N_5 and M_5 are the only 5-element lattices which are non-distributive and among them M_5 is modular and N_5 is not modular. Further, any lattice having at most 4 elements is always distributive. Dedekind's criterion asserts that a lattice is modular if and only if it has no sublattice isomorphic to N_5. Birkhoff's criterion asserts that a lattice is distributive if and only if it has no sublattice isomorphic to any of the lattices N_5 and M_5.

A complemented distributive lattice is called a *Boolean algebra*. The following results are well-known (Szasz [16]).

Theorem 1.1. *The following statements are equivalent for a lattice L.*

(1) L is a Boolean algebra.
(2) L is uniquely complemented and distributive.
(3) L is uniquely complemented and modular.
(4) L is uniquely complemented and relatively complemented.

A lattice L is said to be a *generalized Boolean algebra* if it is a relatively complemented distributive lattice with the least element 0.

A proper ideal P of a lattice L (i.e., an ideal $P \neq L$) is said to be a *prime ideal* if for $a, b \in L, a \wedge b \in P$ implies either $a \in P$ or $b \in P$.

Remark 1.5. A subset P of a lattice L is a prime ideal if and only if it is the ideal kernel of a homomorphism f of L onto the 2-element lattice $C_2 = \{0, 1\}$ with $0 < 1$. The following results are well-known:

- In a distributive lattice every 'maximal ideal' is a prime ideal but not conversely. In a relatively complemented lattice every prime ideal is maximal.
- Nachbin's Theorem (see Ref. 1) asserts that a distributive lattice is relatively complemented if and only if every prime ideal is maximal.

Moreover, the following theorem gives some useful characterizations of distributive lattices in terms of prime ideals (Balbes and Dwinger [1,5], Szasz [16]):

Theorem 1.2. *The following statements are equivalent for a lattice L: 1) L is distributive. 2) If I is an ideal and D is a dual ideal of L such that $I \cap D = \emptyset$, then there exists a prime ideal P of L such that $P \supseteq I$ and $P \cap D = \emptyset$ (Stone's theorem).*
3) If I is an ideal and $a \notin I$, then there exists a prime ideal P of L such that $a \notin P$ and $P \supseteq I$.
4) For any two distinct elements a and b of L, there exists a prime ideal P of L containing

one of these two elements but not containing the other.
5) Every ideal of L is the intersection of all prime ideals of L containing it.

Remark 1.6. Further, Hashimoto[10] proved that "every ideal of a lattice L is the kernel of at least one congruence relation on L if and only if L is distributive".

More about ideals and congruence relations in lattices.
The following theorem characterizes lattices in which there is a one to one correspondence between ideals and congruence relations under which the ideal corresponding to a congruence relation is its kernel (as in the case of Rings!).

Theorem 1.3. *(Hashimoto[10], Gratzer[5] and Szasz[16]):*
The following statements are equivalent for a lattice L :
1) Every ideal is the kernel of exactly one congruence relation and and every congruence relation of L has an ideal kernel (In other words, there is a one to one correspondence between ideals and congruences of L under which the ideal corresponding to a congruence relation is its kernel).
2) L is a generalized Boolean algebra.

Remark 1.7. It may be noted that Boolean algebras are Boolean rings in disguise (see Refs. 5 and 16). The above theorem asserts that the ideal lattice of a lattice 'coincides' with the congruence lattice of the lattice 'naturally' (in the sense of statement (1) of the theorem) only when the lattice is a generalized Boolean algebra. It is known that the congruence lattice of a lattice is a distributive algebraic lattice (see Refs. 5 and 2). It may also be noted that for a ring R, its congruence lattice and the ideal lattice 'coincide' naturally and it is a modular algebraic lattice.

Distributive, Standard and Neutral elements (and ideals) in lattices
The three types of elements (and respective ideals) were discovered by Ore, Gratzer, and Birkhoff, respectively.

Definition 1.1. (Gratzer[5]): Let L be a lattice and let $d \in L$.
1) The element d is said to be distributive if $d \wedge (x \vee y) = (d \wedge x) \vee (d \wedge y)$ for all $x, y \in L$.
2) The element d is said to be standard if $x \wedge (d \vee y) = (x \wedge d) \vee (x \wedge y)$ for all $x, y \in L$.
3) The element d is said to be neutral if $(d \wedge x) \vee (x \wedge y) \vee (y \wedge d) = (d \vee x) \wedge (x \vee y) \wedge (y \vee d)$ for all $x, y \in L$.

An ideal I of a lattice L is said to be distributive, standard or neutral, respectively, if I is distributive, standard or neutral as an element of the ideal lattice *I(L)* of *L*.

Remark 1.8. Using the notion of neutral elements in lattices, some structure theorems for lattices were obtained (Gratzer[5]):

- The direct decompositions of a bounded lattice L into two factors are in one to one correspondence with the complemented neutral elements.
- If L is a complemented modular lattice of finite length, then L is isomorphic to a direct product of simple lattices (Birkhoff-Menger Theorem).

The classical paper entitled "Standard Ideals in Lattices" by Gratzer and Schmidt[9], shows that the notion of a standard ideal in lattices corresponds to the notion of a normal subgroup in groups. It shows that many theorems of group theory may be "translated" (in one direction only!) to lattice theory using the following "dictionary":

subgroup $\rightarrow$ ideal
normal subgroup $\rightarrow$ standard ideal
quotient group $\rightarrow$ quotient lattice
group operation $\rightarrow$ join operation

Group theoretic results corresponding to isomorphism theorems, Zassenhaus lemma, Schreier's extension problem, and many others, have been successfully extended to lattices using Standard Ideals! Later E. Fried and E. T. Schmidt[4] extended the notion of standard ideals in lattices to convex sublattices called "standard sublattices" in lattices.

A trellise (or a weakly associative lattice) as a natural generalization of the lattice concept A reflexive and antisymmetric relation $\trianglelefteq$ on a set A is called a *pseudo-order* on A. A pseudo-ordered set or a psoset $\langle A;\trianglelefteq\rangle$ consists of a nonempty set A together with a pseudo-order $\trianglelefteq$ on A. The notion of an upper bound, a lower bound etc., can be defined for subsets of a psoset analogous to that for posets.
A *trellis* (or also called a weakly associative lattice) is a psoset $\langle A;\trianglelefteq\rangle$ in which any two elements have a greatest lower bound and a least upper bound. The notion of a trellis, introduced by Fried[3] and Skala[15] is a natural generalization of that of a lattice.
A trellis can also be regarded as an algebra $\langle A;\wedge\vee$ with two binary operations $\wedge$ and $\vee$ on A, which are both commutative and satisfy the 'absorption' identities:
$a\wedge(a\vee b)=a; a\vee(a\wedge b)=a$ and the identities:
$a\vee((a\wedge b)\vee(a\wedge c))=a=a\wedge((a\vee b)\wedge(a\vee c))$ for all $a,b,c\in A$ (the weak associativity).

Remark 1.9. Generalizations of some concepts and results of lattices to trellises may be found in Bhatta and Ramananda[13] and Shashirekha[14]. A characterization of neutral elements in lattices by the exclusion of 17 types of sublattices containing it can be found in S. P. Bhatta[11]. Also a characterization of standard elements in lattices by the exclusion of 19 types of sublattices containing it, along with certain additional conditions, can be found in Bhatta[12].

Bibliography

1. R. Balbes and Ph. Dwinger, *Distributive lattices*, University of Missouri Press, 1974.
2. P. Crawley and R. P. Dilworth, *Algebraic theory of lattices*, Prentice-Hall, Englewood Ciliffs, N.J., 1973.
3. E. Fried, *Tournaments and non-associative lattices*, Ann. Univ. Sci. Budapest, Sect. Math. Vol. 13, (1970), 151–164.
4. E. Fried and E. T. Schmidt, *Standard sublattices*, Algebra Universalis, Vol. 5, 1975, pp. 203–211.
5. G. Gratzer, *General Lattice Theory*, Second edition, Birkhauser Verlag, 2003.
6. G. Gratzer, *Standard ideals.* (Hungarian.), Magyar Tud. Akad. Mat. Fiz. Oszt. Kozl. Vol. 9, (1959), 81–97.

7. G. Gratzer, Universal Algebra, *University Series in Higher Mathematics*, Van Nostrand, Princeton, N.J., 1968.
8. G. Gratzer and E. T. Schmidt, *On ideal theory for lattices*, Acta Sci. Math. (Szeged), Vol. 19, (1958), 82–92.
9. G. Gratzer and E. T. Schmidt, Standard ideals in lattices, Acta Math. Sci. Hungar. Vol. 12, (1962), 17–86.
10. J. Hashimoto, *Ideal theory for lattices*, Math. Japon., Vol. 2, (1952), 149–186.
11. S. P. Bhatta, *A characterization of neutral elements by the exclusion of sublattices*, Discrete Mathematics, Vol. 309, (2009), 1691–1702.
12. S. P. Bhatta, *On the problem of characterizing standard elements by the exclusion of sublattices*, Order, Vol. 28, No. 3, (2011), 565–576.
13. S. P. Bhatta and H. S. Ramananda, *On ideals and congruence relations in trellises*, Acta Math. Univ. Comenianae, Vol. 79, No. 2, (2010), 209–216.
14. Shashirekha B. Rai, *Distributive, standard and neutral elements in trellises*, Acta Math. Univ. Comenianae, Vol. 77, No. 2, (2008), 167–174.
15. H. L. Skala, *Trellis Theory*, Algebra Universalis, Vol. 1, (1971), 218–233.
16. G. Szasz, *Introduction to Lattice Theory*, Academic Press, New York and London, 1963.

Hypernear-rings: Some developments linked to near-rings

B. Davvaz

Department of Mathematics, Yazd University,
Yazd, Iran
E-mail: davvaz@yazd.ac.ir

The concept of a hypernearring is a generalization of the concept of a nearring. The overall aim of this paper is to present an introduction to some of the results, methods and ideas about hypernearrings. We study:

(1) Basic concepts of algebraic hyperstructures and hypernearrings;

(2) Fundamental relation on hypernearrings;

(3) Hyper R-subgroups of a hypernearring;

(4) H_v-nearrings.

Keywords: Nearring; Algebraic hyperstructure; Hypernearring; Regular relation; Fundamental relation; Hyper R-subgroup; H_v-nearring.

1. Basic Definitions

A *hypergroupoid* $(H, \circ)$ is a non-empty set H together with a map $\circ : H \times H \to \mathcal{P}^*(H)$ called *(binary) hyperoperation*, where $\mathcal{P}^*(H)$ denotes the set of all non-empty subsets of H. The image of the pair (x, y) is denoted by $x \circ y$. If A, B are non-empty subsets of H and $x \in H$, then by $A \circ B$, $A \circ x$ and $x \circ B$ we mean

$$A \circ B = \bigcup_{\substack{a \in A \\ b \in B}} a \circ b, \ A \circ x = A \circ \{x\} \text{ and } x \circ B = \{x\} \circ B.$$

A hypergroupoid $(H, \circ)$ is called a *semihypergroup* if $(x \circ y) \circ z = x \circ (y \circ z)$, for all $x, y, z \in H$. This means that

$$\bigcup_{u \in x \circ y} u \circ z = \bigcup_{v \in y \circ z} x \circ v.$$

The associativity for semihypergroups can be applied for subsets, i.e., if $(H, \circ)$ is a semihypergroup, then for all non-empty subsets A, B, C of H, we have $(A \circ B) \circ C = A \circ (B \circ C)$. The element $a \in H$ is called *scalar* if $|a \circ x| = |x \circ a| = 1$, for all $x \in H$. An element e in a semihypergroup $(H, \circ)$ is called *scalar identity* if $x \circ e = e \circ x = \{x\}$, for all $x \in H$. An element e in a semihypergroup $(H, \circ)$ is called *identity* if $x \in e \circ x \cap x \circ e$, for all $x \in H$. An element $a' \in H$ is called an *inverse* of $a \in H$ if there exists an identity $e \in H$ such that $e \in a \circ a' \cap a' \circ a$. An element 0 in a semihypergroup $(H, \circ)$ is called *zero element* if $x \circ 0 = 0 \circ x = \{0\}$, for all $x \in H$.

Dasic[1] introduced the notion of hypernearrings. This concept was discussed further by several researchers, for example, Davvaz[2–4], Gontineac[7], and Kim *et al.*[5,8]. A *left hypernearring* is an algebraic hyperstructure $(R, +, \cdot)$ which satisfies the following axioms:

(1) $(R,+)$ is a *quasi canonical hypergroup* (not necessarily commutative), i.e., in $(R,+)$ the following conditions hold:

(a) $x+(y+z)=(x+y)+z$, for all $x,y,z \in R$;
(b) There is $0 \in R$ such that $x+0=0+x=x$, for all $x \in R$;
(c) For every $x \in R$ there exists one and only one $x' \in R$ such that $0 \in x+x'$ (we write $-x$ for x' and we call it the opposite of x);
(d) $z \in x+y$ implies $y \in -x+z$ and $x \in z-y$.

(2) With respect to the multiplication, $(R,\cdot)$ is a semigroup having absorbing element 0, i.e., $x \cdot 0 = 0$ for all $x \in R$. But, in general, $0x \neq 0$ for some $x \in R$.

(3) The multiplication is distributive with respect to the hyperoperation $+$ on the left side, i.e., $x \cdot (y+z) = x \cdot y + x \cdot z$, for all $x,y,z \in R$.

Similarly, we can define the notion of *right hypernearring*. Here, by a *hypernearring* we mean only a left hypernearring. Further, we write ab for $a \cdot b$ just for simplicity of notation. A hypernearring R is called *zero symmetric* if $0x = x0 = 0$ for all $x \in R$. Note that for all $x,y \in R$, we have $-(-x)=x, 0=-0, -(x+y)=-y-x$ and $x(-y)=-xy$. A subhypergroup $A \subseteq R$ is called *normal* if for all $x \in R$ we have $x+A-x \subseteq A$. A normal subhypergroup A is

(1) a *left hyperidel* of R if $x \cdot a \in A$ for all $x \in R$ and $a \in A$;
(2) a *right hyperideal* of R if $(x+A) \cdot y - x \cdot y \subseteq A$, for all x and $y \in R$;
(3) a *hyperideal* of R if $(x+A) \cdot y - x \cdot y \cup z \cdot A \subseteq A$, for all x,y and $z \in R$.

If I is a hyperideal of a hypernearring R, then we define the relation $x \equiv y (mod\, I)$ if and only if $(x-y) \cap I \neq \emptyset$. This is a congruence relation on R. The class $x+I$ is represented by x and we denote it with $C(x)$, see Dasic[1]. Moreover, $C(x)=C(y)$ if and only if $(x-y) \cap I \neq \emptyset$. We can define a hyperoperation $\oplus$ on R/I, the set of all equivalence classes, by $C(x) \oplus C(y) = \{C(z) \mid z \in C(x)+C(y)\}$ and an operation $\odot$ on R/I by $C(x) \odot C(y) = C(x \cdot y)$, for all $C(x), C(y) \in R/I$. In this way, we obtain a hypernearring $(R/I, \oplus, \odot)$, which is called the *factor hypernearring*.

2. Fundamental Relation

The aim of this section is to present a review to some of the results, methods and ideas about hypernearrings. We study the congruence relations on nearrings, and regular and strongly regular relations on hypernearrings. Fundamental relations are special kind of strongly regular relations and they are important in the theory of algebraic hyperstructures. Indeed, the main tools connecting the class of hypernearrings with the ordinary nearrings are the fundamental relations. So, by using the fundamental relations we make a connection between hypernearrings and nearrings. By using a certain type of equivalence relations, we can connect semihypergroups to semigroups, hypergroups to groups, hypernearrings to nearrings. These equivalence relations are called strong regular relations. More exactly, starting with a (semi)hypergroup and using a strong regular relation, we can construct a (semi)group structure on the quotient set. A natural question arises: Do they also exist regular relations? The answer is positive, regular relations provide us new (semi)hypergroup structures on the quotient sets. Let us define these notions. First, we do some notations. Let $(H,\circ)$ be a semihypergroup and ρ be an equivalence relation on H. If A and B are nonempty subsets of H, then

$$\begin{aligned} A\overline{\rho}B & \text{ means that } \forall a\in A, \exists b\in B \text{ such that } a\rho b \text{ and} \\ & \qquad \forall b'\in B, \exists a'\in A \text{ such that } a'\rho b'; \\ A\overline{\overline{\rho}}B & \text{ means that } \forall a\in A, \forall b\in B, \text{ we have } a\rho b. \end{aligned}$$

The equivalence relation ρ is called

(1) *regular on the right* (*on the left*) if for all x of H, from $a\rho b$, it follows that $(a\circ x)\overline{\rho}(b\circ x)$ $((x\circ a)\overline{\rho}(x\circ b)$ respectively);
(2) *strongly regular on the right* (*on the left*) if for all x of H, from $a\rho b$, it follows that $(a\circ x)\overline{\overline{\rho}}(b\circ x)$ $((x\circ a)\overline{\overline{\rho}}(x\circ b)$ respectively);
(3) ρ is called *regular* (*strongly regular*) if it is regular (strongly regular) on the right and on the left.

Let $(H,\circ)$ be a semihypergroup and ρ be an equivalence relation on H.

(1) If ρ is regular, then H/ρ is a semihypergroup, with respect to the following hyperoperation: $\overline{x}\otimes\overline{y}=\{\overline{z}\mid z\in x\circ y\}$;
(2) If the above hyperoperation is well defined on H/ρ, then ρ is regular.

If $(H,\circ)$ is a hypergroup and ρ is an equivalence relation on H, then ρ is regular if and only if $(H/\rho,\otimes)$ is a hypergroup. The fundamental relation β^* was introduced on (semi)hypergroups by Koskas[9], and studied by many authors. The fundamental relation β^* is defined on semihypergroups as the smallest equivalence relation so that the quotient is a group. Let H be a semihypergroup and U be the set of all finite products of elements of H and define the relation β on H as follows: $x\beta y$ if and only if $\{x,y\}\subseteq u$, for some $u\in U$. Freni[6] proved that for hypergroups we have $\beta^*=\beta$. The fundamental relation on hyperrings were obtained by Vougiouklis[10,11]. We analyze here the fundamental relation in the context of hypernearrings. Let R be a hypernearring. We define the relation γ as follows: $a\gamma b$ if and only if $\{a,b\}\subseteq u$ where u is a finite sum of finite products of elements of R. We denote the transitive closure of γ by γ^*. The equivalence relation γ^* is called the *fundamental equivalence relation* in R. We denote the equivalence class of the element a (also called the *fundamental class of a*) by $\gamma^*(a)$. Let $\mathcal{U}$ be the set of all finite sums of products of elements of R. We can rewrite the definition of γ^* on R as follows: $a\gamma^*b$ if and only if $\exists z_1,\dots,z_{n+1}\in R$ with $z_1=a, z_{n+1}=b$ and $u_1,\dots,u_n\in\mathcal{U}$ such that $\{z_i,z_{i+1}\}\subseteq u_i$ for $i\in\{1,\dots,n\}$. Let R be a hypernearring. Then, the relation γ^* is the smallest equivalence relation in R such that the quotient R/γ^* is a nearring. The both $\oplus$ and $\odot$ on R/γ^* are defined as follows:

$$\begin{aligned} \gamma^*(a)\oplus\gamma^*(b)&=\gamma^*(c), \text{ for all } c\in\gamma^*(a)+\gamma^*(b), \\ \gamma^*(a)\odot\gamma^*(b)&=\gamma^*(d), \text{ for all } d\in\gamma^*(a)\cdot\gamma^*(b). \end{aligned}$$

If $u = \sum_{j\in J}\left(\prod_{i\in I_j} x_i\right) \in \mathcal{U}$, then for all $z \in u$

$$\gamma^*(u) = \oplus\sum_{j\in J}\left(\odot\prod_{i\in I_j}\gamma^*(x_i)\right) = \gamma^*(z),$$

where $\oplus\sum$ and $\odot\prod$ denote the sum and the product of classes. In order to speak about canonical maps, we need the following notion. Let R_1 and R_2 be two hypernearrings. The map $f : R_1 \to R_2$ is called an *inclusion homomorphism* if for all $x,y \in R$, the following conditions hold: $f(x+y) \subseteq f(x)+f(y)$, $f(x\cdot y) \subseteq f(x)\cdot f(y)$ and $f(0) = 0$. Moreover, f is called a *strong homomorphism* if for all $x,y \in R$, we have $f(x+y) = f(x)+f(y)$, $f(x\cdot y) = f(x)\cdot f(y)$ and $f(0) = 0$. Let R be a hypernearring. We denote by β_+ the following binary relation: $x\beta_+y$ if and only if $\exists z_1,...,z_n \in R$ such that $\{x,y\} \subseteq z_1+...+z_n$. We denote the transitive closures of the relation β_+ by β_+^*, and we call β_+^* the *fundamental relation* with respect to the addition. For all $a \in R$ we denote the corresponding equivalence class of a by $\beta_+^*(a)$ and we have $\beta_+^*(a) \subseteq \gamma^*(a)$. Let us consider the following canonical maps

$$\varphi_+ : R \to R/\beta_+^*,\ \varphi_+(x) = \beta_+^*(x),$$
$$\varphi^* : R \to R/\gamma^*,\ \ \varphi^*(x) = \gamma^*(x).$$

We notice that the maps $\varphi_+ : (R,+) \to (R/\beta_+^*,\oplus)$, $\varphi^* : (R,+,\cdot) \to (R/\gamma^*,\oplus,\odot)$ are strong homomorphisms. We denote by ω_+, ω^* the kernels of φ_+, φ^*, respectively. If $\bar{0}$ is the zero element of R/β_+^* or R/γ^*, then $\omega_+ = ker\varphi_+ = \{x \in R \mid \varphi_+(x) = \bar{0}\}$ and $\omega^* = ker\varphi^* = \{x \in R \mid \varphi^*(x) = \bar{0}\}$. We have $\omega_+ \subseteq \omega^*$. Let R be a hypernearring. Then,

(1) $R\omega^* \subseteq \omega^*$, $\omega^*R \subseteq \omega^*$;
(2) If $(R,+)$ is a regular hypergroup, then ω^* is a hyperideal of R.

For every hypernearring we have $\gamma^* = \beta_+^*$. Let R be a hypernearring.

3. Hyper R-subgroups

Kim, Davvaz and Roh [8] introduced the notion of hyper R-subgroups of a hypernearring and investigated some properties of hypernearrings with respect to the hyper R-subgroups. A *two-sided hyper R-subgroup* of a hypernearring R is a subset H of R such that

(1) $(H,+)$ is a subhypergroup of $(R,+)$, i.e.,
  (1i) $a,b \in H$ implies $a+b \subseteq H$,
  (1ii) $a \in H$ implies $-a \in H$,
(2) $RH \subseteq H$,
(3) $HR \subseteq H$.

If H satisfies (1) and (2), then it is called a *left hyper R-subgroup* of R. If H satisfies (1) and (3), then it is called a *right hyper R-subgroup* of R. Let $(R,+,\cdot)$ be a hypernearring.

(1) The subset $R_0 = \{x \in R \mid 0x = 0\}$ of R is called a *zero-symmetric part* of R.
(2) The subset $R_c = \{x \in R \mid xy = y$, for all $y \in R\}$ is called *constant part* of R.

(3) If $R = R_0$ (respectively, $R = R_c$), we say that R is a *zero-symmetric* (respectively, *constant) hypernearring*.

If $0y = y$ for each y in a hypernearring R, then R is a constant hypernearring. Let $(R,+,\cdot)$ be a constant hypernearring. Then, R is the only right hyper R-subgroup of R. Let $(R,+,\cdot)$ be a constant hypernearring. If $(H,+)$ is a subhypergroup of $(R,+)$ then H is a left hyper R-subgroup of R. Let R be a hypernearring. Then, R_0 is a zero symmetric subhypernearring of R. Now let H be a subhypernearring of hypernearring R. Define the sets by $B_t = \{0x \mid x \in H\}$ and $B = \{0r \mid r \in R\}$. Let H be a sub-hypernearring of hypernearring R. Then, we have (1) B_t is a two-sided hyper H-subgroup of H, (2) B_t is a left hyper B-subgroup of B. If A is a normal subhypergroup of R, then $A + x = x + A$ and $(A+x)+(A+y) = A+x+y$ for all $x, y \in R$. Let R be a hypernearring and H be a sub-hypernearring. Then, $H = (R_0 \cap H) + B_t$ and $(R_0 \cap H) \cap B_t = \{0\}$. For an element x of a hypernearring R, the *(right) annihilator* of x is $Ann(x) = \{r \in R \mid xr = 0\}$. For a non-empty subset B of a hypernearring R, the annihilator of B is $Ann(B) = \cap\{Ann(x) \mid x \in B\}$. For any element x of a zero symmetric hypernearring R, $Ann(x)$ is a right R-subgroup of R[8]. If e is any element of a hypernearring R, then $eR = \{er \mid r \in R\}$ is a right hyper R-subgroup of R. An element e of a hypernearring R is an *idempotent* if $e^2 = e$. For a hypernearring R, if $e \in R_c$ then $e^2 = e$, so e is an idempotent[8]. Let e be an idempotent element of a zero symmetric hypernearring R. Then, $Ann(e) \cap eR = \{0\}$ and for all $r \in R$, there exists a unique element $a \in Ann(e)$ and there exists a unique element $b \in eR$ such that $r \in A + b$[8].

Let $f : R \to R'$ be a strong homomorphism of hypernearrings. Then, the following statements are true.

(1) If f is onto and M is a hyper R-subgroup of R, then $f(M)$ is a hyper R'-subgroup of R'.
(2) If N is a hyper R'-subgroup of R', then $f^{-1}(N)$ is a hyper R-subgroup of R.
(3) $f(R_0) \subseteq R'_0$.
(4) $f(R_c) \subseteq R'_c$.
(5) If f is an isomorphism, then so is f^{-1}.

Let $(R,+,\cdot)$ be a hypernearring.

(1) If K is a left hyperideal of R and L is a left hyper R-subgroup of R, then $L+K$ is a left hyper R-subgroup of R.
(2) If K is a right hyperideal of R and L is a right hyper R-subgroup of R, then $L+K$ is a right hyper R-subgroup of R.

Let R be a hypernearring, S be a sub-hypernearring of R and H a left (resp. *right, two-sided*) hyper R-subgroup of R. Then, $H \cap S$ is a left (respectively, *right, two-sided*) hyper R-subgroup of R. Let H be a normal hyper R-subgroup of hypernearring R. If we define a relation $x \sim y \pmod{H}$ if and only if $x - y \cap H \neq \emptyset$ for all $x, y \in H$, then this relation is a congruence on H. Let $\rho(x)$ be the equivalence class of the element $x \in H$ and define R/H as follows: $R/H = \{\rho(x) \mid x \in H\}$. Define the hyperoperation $\oplus$ and the multiplication $\odot$ on R/H by $\rho(a) \oplus \rho(b) = \{\rho(c) \mid c \in \rho(a) \oplus \rho(b)\}$ and $\rho(a) \odot \rho(b) = \rho(a \cdot b)$. Then,

$(R/H, \oplus, \odot)$ is a hypernearring, *factor hypernearring*. If H be a normal hyper R-subgroup of R, then $\rho(x) = H + x$. Let f be a homomorphism from R into R' with kernel K such that K is a normal hyper R-subgroup of R, then $R/H \cong Imf$[8]. Let R be a hypernearring and K a normal hyper R-subgroup of R. Then, the following statement are equivalent:

(1) K is the kernel of a hypernearring homomorphism.
(2) $(a+x)y - xy \subseteq K$ for all $x, y \in R$ and all $a \in K$.
(3) $-xy + (a+x)y \subseteq K$ for all $x, y \in R$ and all $a \in K$.

4. H_v-nearrings

H_v-structures first introduced by Vougiouklis[11] in the fourth AHA congress. The concept of H_v-structures constitute a generalization of the well-known algebraic hyperstructures (hypergroup, hyperring, hypermodule and so on). Actually some axioms concerning the above hyperstructures such as the associative law, the distributive law and so on are replaced by their corresponding weak axioms. Then, Davvaz[3] introduced the notion of H_v-nearrings. In this section, we review the concept of an H_v-near ring. According to Davvaz[3], an H_v-nearring is an algebraic structure $(R, +, \cdot)$ which satisfies the following axioms: (1) $(R, +)$ is a weak canonical hypergroup, i.e., (i) for every $x, y, z \in R$, $x + (y+z) \cap (x+y) + z \neq \emptyset$ (weak associative axiom), (ii) there exists $0 \in R$ such that $x + 0 = 0 + x = x$ for all $x \in R$, (iii)] for every $x \in R$ there exists one and only one $x' \in R$ such that $0 \in x + x' \cap x' + x$ (we write $-x$ for x'), (iv) $z \in x + y$ implies $y \in -x + z$ and $x \in z - y$; (2) Relating to the multiplication, $(R, \cdot)$ is a semigroup; (3) The multiplication is weak distributive with respect to the hyperoperation $+$ on the left side or on the right side, i.e., $x \cdot (y+z) \cap x \cdot y + x \cdot z \neq \emptyset$ or $(x+y) \cdot z \cap x \cdot z + y \cdot z \neq \emptyset$ for all $x, y, z \in R$. Of course, there are various modifiers for the various H_v-near rings. Note that for all $x, y \in R$ we have $-(-x) = x$, $0 = -0$, 0 is unique and $-(x+y) = -y - x$. For an H_v-near ring, if the right weak distributive law is valid, then $(R, +, \cdot)$ is called a *right H_v-near ring*. You guessed it, if the left weak distributive is valid, $(R, +, \cdot)$ is called a *left H_v-nearring*. If both the right and left weak distributive are valid, $(R, +, \cdot)$ is a *distributive H_v-near ring*. In this section, when considering H_v-near rings in general, we assume the H_v-near ring to be a left H_v-near ring. An H_v-near ring $(R, +, \cdot)$ for which $a \cdot 0 = 0 \cdot a = 0$ for each $a \in R$ is called a *zero-symmetric H_v-near ring*. Let $(A, +, \cdot)$ be a zero-symmetric near-ring and I be an ideal of A; for $a, b \in A$ we say a is congruent to b $mod(I)$, written $a \equiv b \ mod(I)$ if $a - b \in I$. The relation $a \equiv b \ mod(I)$ is an equivalence relation and it is denoted by $a\sigma b$ if and only if $a \equiv b \ mod(I)$. Let $\sigma^*(a)$ be the equivalence class of the element $a \in A$. Suppose that $A/\sigma = \{\sigma(x) | \ x \in A\}$. On A/σ we consider the hyperoperation $\oplus$ and the multiplication $\odot$ defined as follows: $\sigma(a) \oplus \sigma(b) = \{\sigma(c) | \ c \in \sigma(a) + \sigma(b)\}$ and $\sigma(a) \odot \sigma(b) = \sigma(a \cdot b)$. Then, $(A/\sigma, \oplus, \odot)$ is an H_v-near ring[3]. A subhypergroup $K \subseteq R$ is called *normal* if for all $x \in R$ holds: $x + K = K + x$. A normal subhypergroup K of the weak canonical hypergroup $(R, +)$ is called (1) a *left H_v-ideal* of R if $x \cdot a \in K$ for all $x \in R$ and $a \in K$; (2) a *right H_v-ideal* of R if $(x + K) \cdot y - x \cdot y \subseteq K$ for all $x, y \in R$; (3) a *bilaterally H_v-ideal* of R if $(x+K) \cdot y - x \cdot y \cup z \cdot K \subseteq K$ for all $x, y, z \in R$. Let R_1 and R_2 be two H_v-near rings. The mapping $f : R_1 \to R_2$ is an *H_v-homomorphism* of H_v-near rings, if the following conditions hold: $f(x+y)\cap f(x)+f(y)\neq\emptyset$, $f(x\cdot y)=f(x)\cdot f(y)$ for all $x,y\in R$, and $f(0)=0$. Moreover, f is called a *strong homomorphism*, if $f(x+y)=f(x)+f(y)$, $f(x\cdot y)=f(x)\cdot f(y)$ for all $x,y\in R$, and $f(0)=0$. If f is one to one, onto and a strong homomorphism, then it is called an *isomorphism*.

If I is a bilaterally H_v-ideal of R, then we define the relation $x\equiv y\ mod(I)$ if and only if there exists a set $\{z_0,z_1,\dots,z_{k+1}\}\subseteq R$, where $z_0=x$, $z_{k+1}=y$ such that $(x-z_1)\cap I\neq\emptyset$, $(z_1-z_2)\cap I\neq\emptyset$, ..., $(z_k-y)\cap I\neq\emptyset$. This relation is called the *chain relation* and it is denoted by $x\sigma^*y$ if and only if $x\equiv y\ mod(I)$. The chain relation σ^* is an equivalence relation[3]. We denote $\sigma^*(x)$ the equivalence class with representative x. Let R be an H_v-nearring. If I is a bilaterally H_v-ideal of R, then on the set $R/I=\{\sigma^*(x)\mid x\in R\}$ we define the hyperoperation $\oplus$ and the multiplication $\odot$ as follows: $\sigma^*(x)\oplus\sigma^*(y)=\{\sigma^*(z)\mid z\in\sigma^*(x)+\sigma^*(y)\}$ and $\sigma^*(x)\odot\sigma^*(y)=\sigma^*(x\cdot y)$, what gives the factor H_v-near ring $(R/I,\oplus,\odot)$[3]. If f is a strong homomorphism from R_1 into R_2, the kernel of f is defined by $kerf=\{x\in R_1\mid f(x)=0\}$. It is easy to see that $kerf$ is a subhypergroup of R_1 but in general is not normal. If f is a strong homomorphism from R_1 into R_2 such that $f(x-x)=0$ for all $x\in R_1$, then $R_1/kerf\cong R_2$[3].

Similar to hypernearrings, we can define the relation γ^* as the smallest equivalence relation on an H_v-nearring R such that the quotient R/γ^*, the set of all equivalence classes, is a near-ring. In this case, γ^* called the *fundamental equivalence relation* on R and R/γ^* is called the *fundamental near-ring*. Let us denote $\widehat{\gamma}$ the transitive closure of γ. Then, we can rewrite the definition of $\widehat{\gamma}$ on R as follows: $a\widehat{\gamma}b$ if and only if there exist $z_1,z_2,\dots,z_{n+1}\in R$ with $z_1=a, z_{n+1}=b$ and $u_1,\dots,u_n\in\mathcal{U}$ such that $\{z_i,z_{i+1}\}\subseteq u_i$ $(i=1,\dots,n)$. The fundamental relation γ^* is the transitive closure of the relation γ[3]. The kernel of the canonical map $\varphi: R\to R/\gamma^*$ is called the *core* of R and is denoted by ω_R. Here we also denote by ω_R the zero element of R/γ^*. It is easy to prove that the following statements: (1)] $\omega_R=\gamma^*(0)$, (2) $\gamma^*(-x)=-\gamma^*(x)$ for all $x\in R$. Let R_1,R_2 be H_v-near rings and let γ_1^*,γ_2^* and γ^* be the fundamental equivalence relations on R_1,R_2 and $R_1\times R_2$ respectively. Then, $(R_1\times R_2)/\gamma^*\cong R_1/\gamma_1^*\times R_2/\gamma_2^*$[3].

Bibliography

1. V. Dasic, *Hypernearrings*, Proc. Fourth Int. Congress on Algebraic Hyperstructures and Applications (AHA 1990), World Scientific, (1991), 75–85.
2. B. Davvaz, *Hypernearrings and weak hypernearrings*, Proc. 11th Algebra Seminar of Iranian Math. Soc., Isfahan University of Technology, Isfahan, October 27–29, (1999), 68–78.
3. B. Davvaz, H_v-near rings, *Math. Japonica* **52**(3) (2000), 387–392.
4. B. Davvaz, A study on the structure of H_v-near ring modules, *Indian J. Pure Appl. Math.* **34**(5) (2003), 693–700.
5. B. Davvaz and K. H. Kim, Generalized hypernearring modules, *Southeast Asian Bull. Math.* **30** (2006), 621–630.
6. D. Freni, Une note sur le coeur d'un hypergroupe et sur la cloture β^* de β, *Riv. Mat. Pura Appl.* **8** (1991), 153–156.
7. V. M. Gontineac, *On hypernearrings and H-hypergroups*, Proc. Fifth Int. Congress on Algebraic Hyperstructures and Applications (AHA 1993), Hadronic Press, Inc., USA, (1994), 171–179.
8. K. H. Kim, B. Davvaz and E. H. Roh, On hyper R-subgroups of hypernearrings, *Sci. Math. Jpn.* **67** (2008), 413–420.

9. M. Koskas, Groupoides, demi-hypergroupes et hypergroupes, *J. Math. Pure Appl.* **49** (1970), 155–192.
10. T. Vougiouklis, Hypergroups, hyperrings. Fundamental relations and representations, *Quaderni del Seminario di Geometria Combinatoria* (1989), 1–20.
11. T. Vougiouklis, *The fundamental relation in hyperrings. The general hyperfield*, Algebraic hyperstructures and applications (Xanthi, 1990), 203–211, World Sci. Publishing, Teaneck, NJ, 1991.

Frontiers of fuzzy hypernearrings

B. Davvaz

Department of Mathematics, Yazd University,
Yazd, Iran
E-mail: davvaz@yazd.ac.ir

The theory of fuzzy set which was introduced by Zadeh is applied to many area of mathematics. In this paper, we review some results of fuzzy ideals in a nearring and then we generalize these results to hypernearrings. We study:

(1) Fuzzy subnearrings and ideals of a nearring;

(2) Hypernearrings;

(3) Fuzzy sets and hyperideals;

(4) Fuzzy hyper R-subgroups;

(5) Characterizations of sub-hypernearrings by using triangular norms.

Keywords: Nearring; Fuzzy set; Algebraic hyperstructure; Hypernearring; Hyperideal; Fuzzy hyperideal; Fuzzy Hyper R-subgroup; Triangular norm.

1. Fuzzy Subnearrings and Ideals of a Nearring

Zadeh[28] introduced the concept of fuzzy sets. Let X be a set. A *fuzzy subset* A in X is characterized by a membership function $\mu_A : X \longrightarrow [0,1]$ which associates with each point $x \in X$ its *grade* or *degree of membership* $\mu_A(x) \in [0,1]$. Let A and B be fuzzy sets in X. Then,

$$\begin{aligned}
&A = B \text{ if and only if } \mu_A(x) = \mu_B(x) \text{ for all } x \in X,\\
&A \subseteq B \text{ if and only if } \mu_A(x) \leq \mu_B(x) \text{ for all } x \in X,\\
&C = A \cup B \text{ if and only if } \mu_C(x) = \max\{\mu_A(x), \mu_B(x)\} \text{ for all } x \in X,\\
&D = A \cap B \text{ if and only if } \mu_D(x) = \min\{\mu_A(x), \mu_B(x)\} \text{ for all } x \in X.
\end{aligned}$$

The *complement* of A, denoted by A^c, is defined by $\mu_{A^c}(x) = 1 - \mu_A(x)$ for all $x \in X$. For the sake of simplicity, we shall show every fuzzy set by its membership function. Let f be a mapping from a set X to a set Y. Let μ be a fuzzy subset of X and λ be a fuzzy subset of Y. Then, the *inverse image* $f^{-1}(\lambda)$ of λ is the fuzzy subset of X defined by $f^{-1}(\lambda)(x) = \lambda(f(x))$ for all $x \in X$. The *image* $f(\mu)$ of μ is the fuzzy subset of Y defined by

$$f(\mu)(y) = \begin{cases} \sup\{\mu(t) \mid t \in f^{-1}(y)\} & \text{if } f^{-1}(y) \neq \emptyset \\ 0 & \text{otherwise} \end{cases}$$

for all $y \in Y$. It is not difficult to see that the following assertions hold:

(1) If $\{\lambda_i\}_{i \in I}$ be a family of fuzzy subsets of Y, then

$$f^{-1}(\bigcup_{i \in I} \lambda_i) = \bigcup_{i \in I} f^{-1}(\lambda_i) \text{ and } f^{-1}(\bigcap_{i \in I} \lambda_i) = \bigcap_{i \in I} f^{-1}(\lambda_i).$$

(2) If μ is a fuzzy subset of X, then $\mu \subseteq f^{-1}(f(\mu))$. Moreover, if f is one to one, then $f^{-1}(f(\mu)) = \mu$.
(3) If λ is a fuzzy subset of Y, then $f(f^{-1}(\lambda)) \subseteq \lambda$. Moreover, if f is onto, then $f(f^{-1}(\lambda)) = \lambda$.

After the introduction of fuzzy sets by Zadeh[28], there have been a number of generalizations of this fundamental concept. Abou-Zaid[1] introduced the notion of a fuzzy subnearring, and studied fuzzy ideals of a nearring. The concept was discussed further by many researchers, for example see the papers[9,18,19,26,29,30].

Let R be a nearring and μ be a fuzzy subset of R. We say μ a *fuzzy subnearring* of R if for all $x, y \in R$,

(1) $\mu(x-y) \geq \min\{\mu(x), \mu(y)\}$,
(2) $\mu(xy) \geq \min\{\mu(x), \mu(y)\}$.

μ is called a *fuzzy ideal* of R if μ is a fuzzy subnearring of R and

(3) $\mu(y+x-y) \geq \mu(x)$,
(4) $\mu(xy) \geq \mu(y)$,
(5) $\mu((x+a)y - xy) \geq \mu(a)$, for any $x, y, a \in R$.

It is not difficult to see that the condition (3) in the above definition is equivalent to the following condition:
(3′) $\mu(y+x-y) = \mu(x)$.

Note that μ is a fuzzy left ideal of R if it satisfies (1), (3) and (4), and μ is a fuzzy right ideal of R if it satisfies (1), (2) and (5). For any fuzzy subset μ of R and any $t \in (0,1]$, the set $\mu_t = \{x \in R \mid \mu(x) \geq t\}$ is called a *level subset* of R. Let R be a nearring and μ be a fuzzy subnearring (ideal) of R. Then, the level subset μ_t ($\neq \emptyset$) is a subnearring or (ideal) of R for all $t \in (0,1]$ if and only if μ is a fuzzy subnearring or (ideal), respectively[1]. Let R be a nearring and $a \in R$. The ideal generated by a (denoted by $\langle a \rangle$) is defined as the intersection of all ideals in R containing a. Let μ be a fuzzy ideal of a nearring R and $a \in R$. Then, $\mu(x) \geq \mu(a)$ for all $x \in \langle a \rangle$. Let μ and λ be two fuzzy subsets of a nearring. Then, the product $\mu \circ \lambda$ is defined by

$$\mu \circ \lambda(x) = \begin{cases} \sup\limits_{x=yz}\{\min\{\mu(y), \lambda(z)\}\} \\ 0 \quad \text{if } x \text{ is not expressible as } x = yz. \end{cases}$$

A fuzzy ideal ρ of a nearring R is said to be *prime* if ρ is not a constant function and for every fuzzy ideals μ, λ in R, $\mu \circ \lambda \subseteq \rho$ implies $\mu \subseteq \rho$ or $\lambda \subseteq \rho$. Let $P(\neq R)$ be an ideal in a nearring R. Then, P is a prime ideal if and only if χ_P is a fuzzy prime ideal. For any fuzzy subset μ of R, the set $\{x \in R \mid \mu(x) > 0\}$ is called the *support* of μ and is denoted by suppμ. A fuzzy set μ on R which takes the value $t \in (0,1]$ at some $x \in R$ and takes the value 0 for all $y \in R$ expect x is called a *fuzzy point* and is denoted by x_t, where the point x is called its *support point* and t is called its *value*. A fuzzy point x_t is said to belong to (resp. be quasi-coincident with) a fuzzy set A, written as $x_t \in A$ (resp. $x_t qA$) if $A(x) \geq t$ (resp. $A(x) + t > 1$). If $x_t \in A$ or $x_t qA$, then we write $x_t \in \vee qA$. The symbol $\overline{\in \vee q}$ means $\in \vee q$

does not hold. Based on Bhakat and Das[2], Davvaz[11] extended the concept of $(\in,\in\vee q)$-fuzzy subgroup to the concept of $(\in,\in\vee q)$-fuzzy sub-nearrings (ideals) in the following way. A fuzzy subset μ of a nearring R is said to be a $(\in,\in\vee q)$-*fuzzy subnearring* of R if for all $t,r\in(0,1]$ and $x,y\in R$,

(ia) $x_t,\ y_r\in\mu$ implies $(x+y)_{t\wedge r}\in\vee q\mu$,
(ib) $x_t\in\mu$ implies $(-x)_t\in\vee q\mu$,
(ii) $x_t,\ y_r\in\mu$ implies $(xy)_{t\wedge r}\in\vee q\mu$.

μ is called a $(\in,\in\vee q)$-*fuzzy ideal* of R if μ is a $(\in,\in\vee q)$-fuzzy subnearring of R and

(iii) $x_t\in\mu$ implies $(y+x-y)_t\in\vee q\mu$,
(iv) $y_r\in\mu$ and $x\in R$ implies $(xy)_r\in\vee q\mu$,
(v) $a_t\in\mu$ implies $((x+a)y-xy)_t\in\vee q\mu$ for any $x,y,a\in R$.

Conditions (i)–(v) are equivalent to the following conditions respectively[11].

(1a) $\mu(x+y)\geq\mu(x)\wedge\mu(y)\wedge 0.5$,
(1b) $\mu(-x)\geq\mu(x)\wedge 0.5$,
(2) $\mu(xy)\geq\mu(x)\wedge\mu(y)\wedge 0.5$,
(3) $\mu(y+x-y)\geq\mu(x)\wedge 0.5$,
(4) $\mu(xy)\geq\mu(y)\wedge 0.5$,
(5) $\mu((x+a)y-xy)\geq\mu(a)\wedge 0.5$,

for all $x,y,a\in R$. Note that μ is an $(\in,\in\vee q)$-fuzzy left ideal of R if it satisfies $(1a)$, $(1b)$, (3) and (4); and μ is an $(\in,\in\vee q)$-fuzzy right ideal of R if it satisfies $(1a)$, $(1b)$, (2) and (5). A fuzzy ideal is an $(\in,\in\vee q)$-fuzzy ideal. But the converse is not necessarily true. A non-empty subset I of R is a subnearring (ideal) of R if and only if χ_I is an $(\in,\in\vee q)$-fuzzy subnearring (ideal) of R[11]. An $(\in,\in\vee q)$-fuzzy ideal ρ of a nearring R is said to be *prime* if ρ is not a constant function and for every $(\in,\in\vee q)$-fuzzy ideals μ, λ in R, $\mu\circ\lambda\subseteq\rho$ implies $\mu\subseteq\rho$ or $\lambda\subseteq\rho$. Let P be a prime ideal of a nearring R. Then, χ_P is an $(\in,\in\vee q)$-fuzzy prime ideal of R[11]. Let μ be an $(\in,\in\vee q)$-fuzzy ideal of a nearring R. Then, for all $0<t\leq 0.5$, μ_t is a non-empty set or an ideal of R. Conversely, if μ is a fuzzy subset of R such that μ_t $(\neq\emptyset)$ is an ideal of R for all $0<t\leq 0.5$, then μ is an $(\in,\in\vee q)$-fuzzy ideal of R[11]. Naturally, a corresponding result should be considered when μ_t is an ideal (or subnearring) of R for all $(0.5,1]$. Let μ be a fuzzy subset of a nearring R. Then, μ_t $(\neq\emptyset)$ is a subnearring of R for all $t\in(0.5,1]$ if and only if

(1a) $\mu(x+y)\vee 0.5\geq\mu(x)\wedge\mu(y)$,
(1b) $\mu(-x)\vee 0.5\geq\mu(x)$,
(2) $\mu(xy)\vee 0.5\geq\mu(x)\wedge\mu(y)$.

Moreover, μ_t $(\neq\emptyset)$ is an ideal of R for all $t\in(0.5,1]$ if an only if μ satisfies the above conditions and satisfies the following conditions:

(3) $\mu(y+x-y)\vee 0.5\geq\mu(x)$,
(4) $\mu(xy)\vee 0.5\geq\mu(y)$,

(5) $\mu((x+a)y-xy) \vee 0.5 \geq \mu(a)$

for all $x,y,a \in R$. Yuan *et al.*[27] gave the definition of a fuzzy subgroup with thresholds which is a generalization of Rosenfeld's fuzzy subgroup, and Bhakat and Das's fuzzy subgroup. Then, Davvaz[11] extended the concept of a fuzzy subgroup with thresholds to the concept of fuzzy subnearring (ideal) with thresholds in the following way. Let $\alpha, \beta \in [0,1]$ and $\alpha < \beta$. Let A be a fuzzy subset of a nearring R. Then, μ is called a *fuzzy subnearring with thresholds of* R if for all $x,y \in R$,

(1a) $\mu(x+y) \vee \alpha \geq \mu(x) \wedge \mu(y) \wedge \beta$,
(1b) $\mu(-x) \vee \alpha \geq \mu(x) \wedge \beta$,
(2) $\mu(xy) \vee \alpha \geq \mu(x) \wedge \mu(y) \wedge \beta$.

Moreover, μ is a *fuzzy ideal with thresholds of* if μ satisfies the above conditions and satisfies the following conditions:

(3) $\mu(y+x-y) \vee \alpha \geq \mu(x) \wedge \beta$,
(4) $\mu(xy) \vee \alpha \geq \mu(y) \wedge \beta$,
(5) $\mu((x+a)y-xy) \vee \alpha \geq \mu(a) \wedge \beta$

for all $x,y,a \in R$. If μ is a fuzzy subnearring (ideal) with thresholds of R, then we can conclude that μ is an ordinary fuzzy subnearring (ideal) when $\alpha = 0$, $\beta = 1$; and μ is an $(\in, \in \vee q)$-fuzzy subnearring (ideal) when $\alpha = 0$, $\beta = 0.5$. A fuzzy subset μ of a nearring R is a fuzzy subnearring (ideal) with thresholds of R if and only if μ_t ($\neq \emptyset$) is a subnearring (ideal) of R for all $t \in (\alpha, \beta]$[11].

2. Hypernearrings

The theory of hypergroups has been introduced by Marty[21] in 1934 during the 8th Congress of the Scandinavian Mathematicians. Since the hypergroup is a very general hyperstructure, several researchers endowed it with more stronger or less strong axiom. As a result we are really dealing now with a big number of hypergroups. As an example, we can mention the canonical hypergroup introduced by Mittas[22]. A short review of this theory appears in several books[3,4,12,15,16]. Dasic[5] has introduced the notion of hypernearring in a particular case. Gontineac[17] called this zero symmetric hypernearring and studied the concept of hypernearrings in a general case, also see Refs. 7, 10, 13, 20, 25. Davvaz[8] introduced the notion of an H_v-nearring generalizing the notion of a hypernearring. The hypernearrings generalize the concept of nearrings, in the sense that instead of the operation $+$ the hyperoperation $+$ is defined on the set R. Let R be a non-empty set and let $\mathcal{P}^*(R)$ be the family of all non-empty subsets of R. A *hyperoperation* on R is a map $+ : R \times R \to \mathcal{P}^*(R)$ and the couple $(R,+)$ is called a *hypergroupoid*. If A and B are non-empty subsets of R, then we denote

$$A+B = \bigcup_{\substack{a\in A \\ b\in B}} a+b.$$

A hypergroupoid $(R,+)$ is called a *semihypergroup* if for all x,y,z of R, we have $(x+y)+z = x+(y+z)$. A semihypergroup $(R,+)$ is said to be a *canonical* if (1) for every $x,y \in R$, $x+y = y+x$; (2) there exists $0 \in R$ such that $0+x = x$ for all $x \in R$; (3) for every $x \in R$ there exists a unique element $x' \in R$ such that $0 \in x+x'$ (we shall write $-x$ for x' and we call it the opposite of x); (4) for every $x,y,z \in R$, $z \in x+y$ implies $y \in -x+z$ and $x \in z-y$. A *hypernearring* is an algebraic structure $(R,+,\cdot)$ which satisfies the following axioms.

(1) $(R,+)$ is a quasi canonical hypergroup (not necessarily commutative),
(2) with respect to the multiplication, $(R,\cdot)$ is a semigroup having absorbing element 0, i.e., $x \cdot 0 = 0$ for all $x \in R$,
(3) the multiplication $\cdot$ is distributive with respect to the hyperoperation $+$ on the left side, i.e., $x \cdot (y+z) = x \cdot y + x \cdot z$ for all $x,y,z \in R$.

Note that for all $x,y \in R$, we have $-(-x) = x, 0 = -0, -(x+y) = -y-x$ and $x(-y) = -xy$. Let $(R,+,\cdot)$ be a hypernear-ring. A non-empty subset H of R is called a *sub-hypernearring* if

(1) $(H,+)$ is a subhypergroup of $(R,+)$, i.e., $a,b \in H$ implies $a+b \subseteq H$, and $a \in H$ implies $-a \in H$,
(2) $a \cdot b \in H$ for all $a,b \in H$.

A subhypergroup N of R is called *normal* if for all $x \in R$ we have $x+N-x \subseteq N$. A normal sub-hypergroup N of the hypergroup $(R,+)$ is

(1) a *left hyperideal* of R if $x \cdot a \in N$ for all $x \in R$, $a \in N$,
(2) a *right hyperideal* of R if $(x+N) \cdot y - x \cdot y \subseteq N$ for all $x,y \in R$,
(3) a *hyperideal* of R if $(x+N) \cdot y - x \cdot y \cup z \cdot N \subseteq N$ for all $x,y,z \in R$.

Let R and S be two hypernearrings, the map $f : R \to S$ is called a *homomorphism of hypernearrings* if for all $x,y \in R$, the following relations hold: $f(x+y) = f(x)+f(y)$, $f(xy) = f(x)f(y)$ and $f(0) = 0$. From the above definition we get $f(-x) = -f(x)$, for all $x \in R$.

3. Fuzzy Sets and Hyperideals

Davvaz[8] introduced the concept of a fuzzy sub-hypernearring (hyperideal) of a hypernear-ring which are generalizations of the concept of a fuzzy subnearring (ideal) in a nearring as follows.

Definition 3.1. Let $(R,+,\cdot)$ be a hypernearring. A fuzzy subset μ in R is called a *fuzzy sub-hypernearring* of R if it satisfies

(1a) $\min\{\mu(x),\mu(y)\} \leq \inf_{\alpha \in x+y} \mu(\alpha)$,
(1b) $\mu(x) \leq \mu(-x)$,
(2) $\min\{\mu(x),\mu(y)\} \leq \mu(xy)$

for all $x,y \in R$. Furthermore, μ is called a *fuzzy hyperideal* of R if μ is a fuzzy sub-hypernearring of R and

(3) $\mu(y) \leq \inf_{\alpha \in x+y-x} \{\mu(\alpha)\}$ for all $x,y \in R$,
(4) $\mu(y) \leq \mu(x.y)$ for all $x,y \in R$,
(5) $\mu(i) \leq \inf_{\alpha \in (x+i).y-x.y} \{\mu(\alpha)\}$ for all $x,y,i \in R$.

Let μ be a fuzzy hyperideal of a hypernearring R. Then, $\mu(x) \leq \mu(0)$ for all $x \in R$. We define $R_\mu = \{x \in R \mid \mu(x) = \mu(0)\}$ and $R^*_\mu = \{x \in R \mid \mu(x) > 0\}$. Then, R_μ and R^*_μ are hyperideals of R. Let H be a non-empty subset of a hypernearring R. Then, H is a hyperideal of R if and only if χ_H is a fuzzy hyperideal of R, where χ_H is the characteristic function of H. Let μ be a fuzzy subset in a set X. For $t \in [0,1]$, the set $\mu_t = \{x \in X \mid \mu(x) \geq t\}$ is called a *level subset* of μ. Let R be a hypernearring and μ be a fuzzy subset of R. Then, the level subset μ_t is a sub-hypernearring (hyperideal) of R, for all $t \in \mathrm{Im}\mu$ if and only if μ is a fuzzy sub-hypernearring (hyperideal), respectively[8]. Let μ_1 and μ_2 be two fuzzy hyperideal of R. If $\mu_1(0) = \mu_2(0)$, then $R_{\mu_1 \cap \mu_2} = R_{\mu_1} \cap R_{\mu_2}$. Moreover, $R^*_{\mu_1 \cap \mu_2} = R^*_{\mu_1} \cap R^*_{\mu_2}$.

Davvaz and Corsini[14], by using the notion of "belongingness ($\in$)" and "quasi-coincidence (q)" of fuzzy points with fuzzy sets, introduced the concept of $(\in, \in \vee q)$-fuzzy sub-hypernear-ring (hyperideal of a hypernearring. A fuzzy subset μ of a hypernearring R is said to be an $(\in, \in \vee q)$*-fuzzy sub-hypernearring* of R if for all $t,r \in (0,1]$ and $x,y \in R$,

(ia) $x_t, y_r \in \mu$ implies $z_{t \wedge r} \in \vee q\mu$ for all $z \in x+y$,
(ib) $x_t \in \mu$ implies $(-x)_t \in \vee q\mu$,
(ii) x_t, $y_r \in \mu$ implies $(xy)_{t \wedge r} \in \vee \mu$.

μ is called an $(\in, \in \vee q)$*-fuzzy hyperideal of* R if μ is a $(\in, \in \vee q)$-fuzzy sub-hypernearring of R and

(iii) $x_t \in \mu$ implies $z_t \in \vee q\mu$ for all $z \in y+x-y$
(iv) $y_r \in \mu$ and $x \in R$ implies $(xy)_r \in \vee q\mu$,
(v) $a_t \in \mu$ implies $z_t \in \vee q\mu$ for any $x,y,a \in R$ and any $z \in (x+a)y-xy$.

Conditions (i)-(v) in this definition are equivalent to the following conditions respectively.

(1a) $\mu(x) \wedge \mu(y) \wedge 0.5 \leq \bigwedge_{z \in x+y} \mu(z)$,
(1b) $\mu(x) \wedge 0.5 \leq \mu(-x)$,
(2) $\mu(x) \wedge \mu(y) \wedge 0.5 \leq \mu(xy)$,
(3) $\mu(x) \wedge 0.5 \leq \bigwedge_{z \in y+x-y} \mu(z)$,
(4) $\mu(y) \wedge 0.5 \leq \mu(xy)$,
(5) $\mu(a) \wedge 0.5 \leq \bigwedge_{z \in (x+a)y-xy)} \mu(z)$,

for all $x,y,a \in R$.

Let μ be a fuzzy subset of a hypernearring R. Then,

(a) μ is an $(\in, \in \vee q)$-fuzzy sub-hypernearring of R if and only if the conditions $(1a)$, $(1b)$ and (2) in Theorem 3.3 hold.
(b) μ is an $(\in, \in \vee q)$-fuzzy hyperideal of R if and only if μ is an $(\in, \in \vee q)$-fuzzy sub-hypernearring, and the conditions (3), (4) and (5) in Theorem 3.3 hold.

Note that μ is an $(\in, \in \vee q)$-*fuzzy left hyperideal of* R if it satisfies $(1a)$, $(1b)$, (3) and (4); and μ is an $(\in, \in \vee q)$-*fuzzy right hyperideal of* R if it satisfies $(1a)$, $(1b)$, (2) and (5).

Let μ be an $(\in, \in \vee q)$-fuzzy hyperideal of a hypernearring R. Then, for all $0 < t \leq 0.5$, μ_t is a non-empty set or a hyperideal of R. Conversely, if μ is a fuzzy subset of R such that μ_t $(\neq \emptyset)$ is a hyperideal of R for all $0 < t \leq 0.5$, then μ is an $(\in, \in \vee q)$-fuzzy hyperideal of R. Naturally, a corresponding result should be considered when μ_t is a hyperideal of R for all $(0.5, 1]$. Let μ be a fuzzy subset of a hypernearring R. Then, μ_t $(\neq \emptyset)$ is a sub-hypernearring of R for all $t \in (0.5, 1]$ if and only if

(1a) $\mu(x) \wedge \mu(y) \leq \bigwedge_{z \in x+y} (\mu(z) \vee 0.5)$,
(1b) $\mu(x) \leq \mu(-x) \vee 0.5$,
(2) $\mu(x) \wedge \mu(y) \leq \mu(xy) \vee 0.5$.

Moreover, μ_t $(\neq \emptyset)$ is a hyperideal of R for all $t \in (0.5, 1]$ if and only if μ satisfies the above conditions and satisfies the following conditions:

(3) $\mu(x) \leq \bigwedge_{z \in y+x-y} (\mu(z) \vee 0.5)$,
(4) $\mu(y) \leq \mu(xy) \vee 0.5$,
(5) $\mu(a) \quad \bigwedge_{z \in (x+a)y-xy} (\mu(z) \vee 0.5)$

for all $x, y, a \in R$. Let μ be a fuzzy subset of a hypernearring R and

$$J = \{t \mid t \in (0, 1] \text{ and } \mu_t \text{ is an empty} - \text{set or a subhypernearring of } R\}.$$

Let $\alpha, \beta \in [0, 1]$ and $\alpha < \beta$. Let μ be a fuzzy subset of a hypernearring R. Then, μ is called a *fuzzy sub-hypernearring with thresholds of* R if for all $x, y \in R$,

(1a) $\mu(x) \wedge \mu(y) \wedge \beta \leq \bigwedge_{z \in x+y} (\mu(z) \vee \alpha)$,
(1b) $\mu(x) \wedge \beta \leq \mu(-x) \vee \alpha$,
(2) $\mu(x) \wedge \mu(y) \wedge \beta \leq \mu(xy) \vee \alpha$.

Moreover, μ is a *fuzzy hyperideal with thresholds of* R if and only if μ satisfies the above conditions and satisfies the following conditions:

(3) $\mu(x) \wedge \beta \leq \bigwedge_{z \in y+x-y} (\mu(z) \vee \alpha)$,
(4) $\mu(y) \wedge \beta \leq \mu(xy) \vee \alpha$,
(5) $\mu(a) \wedge \beta \leq \bigwedge_{z \in (x+a)y-xy} (\mu(z) \vee \alpha)$

for all $x, y, a \in R$. If μ is a fuzzy sub-hypernearring with thresholds of R, then we can conclude that μ is an ordinary fuzzy sub-hypernearring when $\alpha = 0$, $\beta = 1$; and μ is an $(\in, \in$

$\vee q$)-fuzzy sub-hypernearring when $\alpha = 0$, $\beta = 0.5$. A fuzzy subset μ of a hypernearring R is a fuzzy sub-hypernearring with thresholds of R if and only if μ_t ($\neq \emptyset$) is a sub-hypernearring of R for all $t \in (\alpha, \beta]$ [14]. A fuzzy subset μ of a hypernearring R is a fuzzy hyperideal with thresholds of R if and only if μ_t ($\neq \emptyset$) is a hyperideal of R for all $t \in (\alpha, \beta]$ [14]. Let μ be a fuzzy subset of a hypernearring R. Then, μ is said to be *normal* if and only if for all $x, y \in R$,

$$\mu(z) = \mu(z'),\ \forall z \in x+y,\ \forall z' \in y+x.$$

It is obvious that if μ is a fuzzy normal sub-hypernearring of R, then $\mu(z) = \mu(z')$, for all $z, z' \in x+y$ and for all $x, y \in R$. Let μ be a fuzzy sub-hypernearring of a hypernearring R. Then, the following conditions are equivalent.

(1) μ is a fuzzy normal sub-hypernearring of R,
(2) For all $x, y \in R$, $\mu(z) = \mu(x)$, $\forall z \in y+x-y$,
(3) For all $x, y \in R$, $\mu(z) \geq \mu(x)$, $\forall z \in y+x-y$,
(4) For all $x, y \in R$, $\mu(z) \geq \mu(x)$, $\forall z \in -y-x+y+x$.

An $(\in, \in \vee q)$-fuzzy sub-hypernearring of a hypernearring R is said to be $(\in, \in \vee q)$*-fuzzy normal* if for every $x, y \in R$ and $t \in (0, 1]$,

$$x_t \in \mu \text{ implies } z_t \in \vee q\mu \text{ for all } z \in y+x-y.$$

Let μ be an $(\in, \in \vee q)$-fuzzy sub-hypernearring of R. For any $x \in R$, $\overline{\mu_x}$ (resp. $\underline{\mu_x}$) is defined by

$$\overline{\mu_x}(a) = \Big(\bigwedge_{z \in a-x} \mu(z)\Big) \wedge 0.5,$$
$$\text{(respectively, } \underline{\mu_x}(a) = \Big(\bigwedge_{z \in -x+a} \mu(z) \wedge 0.5\Big)\text{, for all } a \in R$$

and is called an $(\in, \in \vee q)$*-fuzzy left* (respectively, *right*) *coset of R determined by x and μ*. Let μ be an $(\in, \in \vee q)$-fuzzy normal sub-hypernearring of R. Then, $\overline{\mu_x} = \underline{\mu_x}$ for all $x \in R$. Let $x, y \in R$ and for any $a \in R$, $\overline{\mu_x}(a) = \overline{\mu_y}(a)$. Then, for any non-empty subset S of R, we have

$$\Big(\bigwedge_{a \in S-x} \mu(a)\Big) \wedge 0.5 = \Big(\bigwedge_{a \in S-y} \mu(a)\Big) \wedge 0.5.$$

Let μ be an $(\in, \in \vee q)$-fuzzy normal sub-hypernearring of R. Then, $\overline{\mu_a} = \overline{\mu_b}$ if and only if $\mu_{0.5} + a = \mu_{0.5} + b$ for all $a, b \in R$. Let μ be an $(\in, \in \vee q)$-fuzzy hyperideal of a hypernearring R. Let R/μ be the set of all $(\in, \in \vee q)$-fuzzy left coset of μ in R. Then, R/μ is a hypernearring if

$$\overline{\mu_x} \oplus \overline{\mu_y} = \{\overline{\mu_z} \mid z \in \mu_{0.5} + x + y\},$$
$$\ominus \overline{\mu_x} = \overline{\mu_{-x}},$$
$$\overline{\mu_x} \odot \overline{\mu_y} = \overline{\mu_{xy}}.$$

Fuzzy logic is an extension of set theoretic multivalued logic in which the truth values are linguistic variables (or terms of the linguistic variable truth). Some operators, like $\wedge$, $\vee$, $\neg$, $\longrightarrow$ in fuzzy logic are also defined by using truth tables, the extension principle can be applied to derive definitions of the operators.

In fuzzy logic, truth value of fuzzy proposition P is denoted by $[P]$. In the following, we display the fuzzy logical and corresponding set-theoretical notions.

$$
\begin{aligned}
&[x \in \mu] = \mu(x),\\
&[x \notin \mu] = 1 - \mu(x),\\
&[P \wedge Q] = \min\{[P],[Q]\},\\
&[P \vee Q] = \max\{[P],[Q]\},\\
&[P \longrightarrow Q] = \min\{1, 1-[P]+[Q]\},\\
&[\forall x P(x)] = \inf[P(x)],\\
&\models P \text{ if and only if } [P] = 1 \text{ for all valuations.}
\end{aligned}
$$

Of course, various implication operators have been defined. We only show a selection of them in the next table. α denotes the degree of truth (or degree of membership) of the premise, β the respective values for the consequence, and I the resulting degree of truth for the implication. A fuzzy subset μ of a hypernearring R satisfies:

(1a) for any $x,y \in R$, $\models [\,[x\in\mu]\wedge[y\in\mu] \longrightarrow [\forall z\in x+y,\ z\in\mu]\,]$,
(1b) for any $x \in R$, $\models [[\,x\in\mu] \longrightarrow [-x\in\mu]]$,
(2) for any $x,y \in R$, $\models [\,[x\in\mu]\wedge[y\in\mu] \longrightarrow [xy\in\mu]\,]$,

then μ is called a *fuzzifying sub-hypernearring of R*. Moreover, μ is a *fuzzifying hyperideal of R* if and only if μ satisfies the above conditions and satisfies the following conditions:

(3) for any $x,y \in R$, $\models [\,[y\in\mu] \longrightarrow [\forall z\in x+y-x,\ z\in\mu]\,]$,
(4) for any $x,y \in R$, $\models [\,[y\in\mu] \longrightarrow [xy\in\mu]\,]$,
(5) for any $x,y,i \in R$, $\models [\,[i\in\mu] \longrightarrow [\forall z\in (x+i)y-xy,\ z\in\mu]\,]$.

Yuan *et al.*[27] introduced the concept of t-tautology as follows:

$$\models_t P \text{ if and only if } [P] \geq t \text{ for all valuations.}$$

Let μ be a fuzzy subset of a hypernearring R and $t \in (0,1]$ is a fixed number. If

(1a) for any $x,y \in R$, $\models_t [\,[x\in\mu]\wedge[y\in\mu] \longrightarrow [\forall z\in x+y,\ z\in\mu]\,]$,
(1b) for any $x \in R$, $\models_t [[\,x\in\mu] \longrightarrow [-x\in\mu]]$,
(2) for any $x,y \in R$, $\models_t [\,[x\in\mu]\wedge[y\in\mu] \longrightarrow [xy\in\mu]\,]$,

then μ is called a *t-implication-based fuzzy sub-hypernearring of R*. Moreover, μ is a *t-implication-based fuzzy hyperideal of R* if and only if μ satisfies the above conditions and satisfies the following conditions:

(3) for any $x,y \in R$, $\models_t [\,[y\in\mu] \longrightarrow [\forall z\in x+y-x,\ z\in\mu]\,]$,
(4) for any $x,y \in R$, $\models_t [\,[y\in\mu] \longrightarrow [xy\in\mu]\,]$,
(5) for any $x,y,i \in R$, $\models_t [\,[i\in\mu] \longrightarrow [\forall z\in (x+i)y-xy,\ z\in\mu]\,]$.

Now, let I be an implication operator. Then, μ is a t-implication-based fuzzy hyperideal of a hypernearring R if and only if

(ia) $I(\mu(x)\wedge\mu(y), \bigwedge_{z\in x+y} \mu(z)) \geq t$ for all $x,y \in R$,

(ib) for any $x \in R$, $I(\mu(x) \wedge \mu(-x) \geq t$,
(2) $I(\mu(x) \wedge \mu(y),\ \mu(xy)) \geq t$ for all $x, y \in R$,
(3) $I(\mu(y),\ \bigwedge_{z \in y+x-y} \mu(z)) \geq t$ for all $x, y \in R$,
(4) $I(\mu(y),\ \mu(xy)) \geq t$ for all $x, y \in R$,
(5) $I(\mu(y),\ \bigwedge_{z \in (x+a)y-xy} \mu(z)) \geq t$ for all $x, y \in R$.

4. Fuzzy Hyper R-subgroups

Kim, Davvaz and Roh[19] defined the notions of fuzzy hyper R-subgroup of a hypernearring and investigated some properties of fuzzy hyper R-subgroups with respect to the level subsets and the strong level subsets. Also, using the notion of fuzzy hyper R-subgroups, they constructed a uniform structure on a hypernearring R, which gives a topology on R. First, we define the definition of a two-sided fuzzy hyper R-subgroup and a two-sided anti fuzzy hyper R-subgroup of hypernear-ring R.

Let $(R, +, \cdot)$ be a hypernearring and μ a fuzzy subset of R. We say that μ is a *two-sided fuzzy hyper R-subgroup* of R if

(1) $\min\{\mu(x), \mu(y)\} \leq \inf_{\alpha \in x+y} \{\mu(\alpha)\}$ for all $x, y \in R$,
(2) $\mu(x) \leq \mu(-x)$ for all $x \in R$;
(3) $\mu(y) \leq \mu(x \cdot y)$ for all $x, y \in R$;
(4) $\mu(x) \leq \mu(x \cdot y)$ for all $x, y \in R$.

If μ satisfies (1), (2) and (3), then it is called a *left fuzzy hyper R-subgroup* of R. If μ satisfies (1), (2) and (4), then it is called a *right fuzzy hyper R-subgroup* of R. Let μ be a fuzzy hyper R-subgroup of a hypernearring R. Then, $\mu(x) \leq \mu(0)$ for all $x \in R$. If $\{\mu_i \mid i \in \Lambda\}$ is a family of fuzzy hyper R-subgroups of a hypernearring R, then so is $\bigwedge_{i \in \Lambda} \mu_i$. Let H be a non-empty subset of a hypernearring R and let μ be a fuzzy set in R defined by

$$\mu(x) = \begin{cases} t_1 & \text{if } x \in H \\ t_2 & \text{otherwise,} \end{cases}$$

where $t_1 > t_2$ in $[0, 1]$. Then, μ is a fuzzy right (resp. *left*) hyper R-subgroup of R if and only if H is a right hyper (resp. *left*) R-subgroup of R. If I is a normal hyper R-subgroup of a hypernearring R, then we define the relation $x \equiv y(mod\ I)$ if and only if $(x - y) \cap I \neq \emptyset$. This is a congruence relation on R. The class $x + I$ is represented by x and we denote it with $C(x)$. Then, $C(x) = C(y)$ if and only if $x \equiv y(mod\ I)$. We can define a hyperoperation $\oplus$ on R/I by $C(x) \oplus C(y) = \{C(z) \mid z \in C(x) + C(y)\}$, and an operation $\otimes$ on R/I by $C(x) \otimes C(y) = C(x \cdot y)$, for all $C(x), C(y) \in R/I$. Then, we obtain a hypernearring $(R/I, \oplus, \otimes)$, which is called the *factor hypernearring*. Let μ be a fuzzy hyper R-subgroup of a hypernearring R. We define $x \sim_\mu y$ if and only if there exists $r \in x - y$ such that $\mu(r) = \mu(0)$. Let μ be a fuzzy hyper R-subgroup of a hypernearring R with $\mu(y) \leq \inf_{\alpha \in x+y-x} \mu(\alpha)$ for all $x, y \in R$. Then, we have

(1) the relation $\sim_\mu$ is an equivalence relation,
(2) if $x \sim_\mu y$, then $\mu(x) = \mu(y)$.

Let X be any non-empty set and let U and V be any subsets of $X \times X$. Define
$U \circ V = \{(x,y) \in X \times X |$ for some $z \in X, (x,z) \in U$ and $(z,y) \in V\}$,
$U^{-1} = \{(x,y) \in X \times X \mid (y,x) \in U\}$,
$\triangle = \{(x,x) \in X \times X \mid x \in X\}$.

By a *uniformity* on a set X we mean a non-empty collection $\mathcal{K}$ of subsets of $X \times X$ which satisfies the following conditions:

(U1) $\triangle \subseteq U$ for any $U \in \mathcal{K}$,
(U2) if $U \in \mathcal{K}$, then $U^{-1} \in \mathcal{K}$,
(U3) if $U \in \mathcal{K}$, then there exists a $V \in \mathcal{K}$ such that $V \circ V \subseteq U$,
(U4) if $U, V \in \mathcal{K}$, then $U \cap V \in \mathcal{K}$,
(U5) if $U \in \mathcal{K}$ and $U \subseteq V \subseteq \mathcal{K} \times \mathcal{K}$, then $V \in \mathcal{K}$.

The pair $(M, \mathcal{K})$ is called a *uniform space*. Note that if μ is a fuzzy hyper R-subgroup of a hypernearring R then $\mu(x) \leq \mu(0)$ for all $x \in R$. If for any fuzzy hyper R-subgroups μ and ν in R, $\mu(0) = \nu(0)$ then we say that R satisfies the *strong condition*. From now on R is a hypernearring which satisfies the strong condition. Let μ be a fuzzy hyper R-subgroup of a hypernearring R and $U_\mu = \{(x,y) \in R \times R \mid x \sim_\mu y\}$. If

$$\mathcal{K}^* = \{U_\mu \mid \mu \text{ is a fuzzy hyper } R\text{-subgroup of } R\}$$

then $\mathcal{K}^*$ satisfies the conditions (U1)-(U4). Let $\mathcal{K} = \{U \subseteq X \times X \mid U_\mu \subseteq U$ for some $U_\mu \in \mathcal{K}^*\}$. Then, $\mathcal{K}$ satisfies a uniformity on R and the pair $(R, \mathcal{K})$ is a uniform structure. Given a $x \in R$ and $U \in \mathcal{K}$, we define $U[x] = \{y \in R \mid (x,y) \in U\}$. For any x in R, the collection $\mathcal{U}_x = \{U[x] \mid (x,y) \in U\}$, forms a neighborhood base at x, making R a topological space.

5. Characterizations of Sub-hypernearrings by Using Triangular Norms

By a *t-norm* T, we mean a function $T : [0,1] \times [0,1] \to [0,1]$ satisfying the following conditions:

(T1) $T(x,1) = x$,
(T2) $T(x,y) \leq T(x,z)$ if $y \leq z$,
(T3) $T(x,y) = T(y,x)$,
(T4) $T(x, T(y,z)) = T(T(x,y),z)$,

for all $x, y \in R$. Every t-norm T has a useful property:

$$T(\alpha, \beta) \leq \min\{\alpha, \beta\}, \text{ for all } \alpha, \beta \in [0,1].$$

This notion was introduced by Schweizer and Sklar[24] in order to generalize the ordinary triangle inequality in a metric space to the more general probabilistic metric space. Let T be a t-norm. A fuzzy set μ in R is called a *T-fuzzy sub-hypernearring*(for short, *TFS-ring*) of R if it satisfies

(TF1) $T(\mu(x),\mu(y)) \leq \inf_{\alpha \in x+y} \mu(\alpha)$,
(TF2) $\mu(x) \leq \mu(-x)$,
(TF3) $T(\mu(x),\mu(y)) \leq \mu(xy)$,

for all $x, y \in R$. If $\{\mu_i \mid i \in \Lambda\}$ is a family of TFS-rings of R, then so is $\bigcap_{i\in\Lambda} \mu_i$. Let T be a t-norm and μ be a fuzzy set of R. If R^t_μ is a sub-hypernearring of R for all $t \in Im(\mu)$, then μ is a TFS-ring of R[23]. Let T be a t-norm and H be a subhypernearring of R. Then, there exists a TFS-ring μ of R such that $R^t_\mu = H$ for some $t \in (0,1]$[23].

Let T be a t-norm and μ be a fuzzy set of R with $Im(\mu) = \{t_1, t_2, \cdots, t_n\}$, where $t_i < t_j$ whenever $i > j$. Suppose that there exists a chain of subhypernearrings of R:

$$H_0 \subseteq H_1 \subseteq \cdots \subseteq H_n = R$$

such that $\mu(H^*_k) = t_k$, where $H^*_k = H_k \backslash H_{k-1}, H_{-1} = \emptyset$ for $k = 0, 1, \cdots, n$. Then, μ is a TFS-ring of R[23]. For a t-norm T on $[0,1]$, denote by Δ_T the set of element $\alpha \in [0,1]$ such that $T(\alpha,\alpha) = \alpha$, i.e., $\Delta_T = \{\alpha \in [0,1] | T(\alpha,\alpha) = \alpha\}$. A fuzzy set μ in a set X is said to satisfy *imaginable property* if $Im(\mu) \subseteq \Delta_T$. A TFS-ring is said to be *imaginable* if it satisfies the imaginable property. For a sub-hypernearrings H of R, let μ be a fuzzy set in R given by

$$\mu(x) = \begin{cases} s & \text{if } x \in H, \\ t & \text{otherwise} \end{cases}$$

for all $s, t \in [0,1]$ with $s > t$. Then, μ is a $T_L FS$-ring of R, where T_L is the t-norm defined by $T_L(\alpha,\beta) = \max\{\alpha + \beta - 1, 0\}$, for all $\alpha, \beta \in [0,1]$. In particular, if $s = 1$ and $t = 0$ then μ is imaginable. Let T be a t-norm. Then, every imaginable TFS-ring of R is a fuzzy sub-hypernearring of R[23]. Let μ be a TFS-ring of R and let $t \in [0,1]$. Then,

(1) if $t = 1$ then R^t_μ is either empty or a sub-hypernearring of R
(2) if $T = min$, then R^t_μ is either empry or a sub-hypernearring of R.

Let T be a t-norm and let μ be an imaginable fuzzy set in R. If each non-empty level subset R^t_μ of μ is a sub-hypernearring of R, then μ is an imaginable TFS-ring of R[23]. Let T be a t-norm and let $f : R \to S$ be a homomorphism of hypernearrings. If μ is a TFS-ring of S, then $f^{-1}(\mu)$ is a TFS-ring of R. Let T be a t-norm and let μ and ν be fuzzy sets in R. Then, the *T-product* of μ and ν, written $[\mu \cdot \nu]_T$, is defined by $[\mu \cdot \nu]_T(x) = T(\mu(x), \nu(x))$ for all $x \in R$. Let T be a t-norm and let μ and ν be TFS-rings in R. If T^* is a t-norm which dominates, i.e., $T^*(T(\alpha,\beta), T(\gamma,\delta)) \geq T(T^*(\alpha,\gamma), T^*(\beta,\delta))$ for all $\alpha, \beta, \gamma, \delta \in [0,1]$, then T^*-product of μ and ν, $[\mu \cdot \nu]^*_T$ is a TFS-ring of R. Let $f : R \to S$ be a homomorphism of hypernearrings, and let T and T^* be t-norms such that T^* dominates T. If μ and ν is TFS-rings in S, then $[\mu \cdot \nu]^*_T$ is a TFS-ring of S. The inverse images $f^{-1}(\mu), f^{-1}(\nu)$ and $f^{-1}([\mu \cdot \nu]^*_T)$ are TFS-ring of R. If $[\mu \cdot \nu]^*_T$ is T^*-product of μ and ν, and $[f^{-1}(\mu) \cdot f^{-1}(\nu)]^*_T$ is the T^*-product of $f^{-1}(\mu)$ and $f^{-1}(\nu)$ then $f^{-1}([\mu \cdot \nu]^*_T) = [f^{-1}(\mu) \cdot f^{-1}(\nu)]^*_T$[23].

Bibliography

1. S. Abou-Zaid, On fuzzy subnearrings and ideals, *Fuzzy Sets and Systems* **44** (1991), 139–146.

2. S. K. Bhakat and P. Das, Fuzzy subrings and ideals redifind, *Fuzzy Sets and Systems* **81** (1996), 383–393.
3. P. Corsini, *Prolegomena of Hypergroup Theory*, Second Edition, Aviani Editor, 1993.
4. P. Corsini and V. Leoreanu, *Applications of Hyperstructures Theory*, Advances in Mathematics, Kluwer Academic Publishers, 2003.
5. V. Dasic, *Hypernearrings*, Proc. Fourth Int. Congress on Algebraic Hyperstructures and Applications (AHA 1990), World Scientific, (1991), 75–85.
6. B. Davvaz, On hypernearrings and fuzzy hyperideals, *J. Fuzzy Math.* **7**(3) (1999), 745–753.
7. B. Davvaz, *Hypernearrings and weak hypernearrings*, Proc. 11th Algebra Seminar of Iranian Math. Soc., Isfahan University of Technology, Isfahan, October 27–29, (1999), 68–78.
8. B. Davvaz, H_v-near rings, *Math. Japonica* **52**(3) (2000), 387–392.
9. B. Davvaz, Fuzzy ideals of near-rings with interval valued membership functions, *J. Sci. Islam. Repub. Iran* **12**(2) (2001), 171–175.
10. B. Davvaz, A study on the structure of H_v-near ring modules, *Indian J. Pure Appl. Math.* **34**(5) (2003), 693–700.
11. B. Davvaz, $(\in, \in \vee q)$-fuzzy subnearrings and ideals, *Soft Computing* **10** (2006), 206–211.
12. B. Davvaz, *Polygroup Theory and Related Systems*, World Scientific Publishing Co. Pte. Ltd., Hackensack, NJ, 2013.
13. B. Davvaz and K. H. Kim, Generalized hypernearring modules, *Southeast Asian Bull. Math.* **30** (2006), 621–630.
14. B. Davvaz and P. Corsini, Generalized fuzzy hyperideals of hypernearrings and many valued implications, *J. Intell. Fuzzy Syst.* **17**(3) (2006), 241–251.
15. B. Davvaz and V. Leoreanu-Fotea, *Hyperring Theory and Applications*, International Academic Press, USA, 2007.
16. B. Davvaz and I. Cristea, *Fuzzy Algebraic Hyperstructures–An Introduction*, Springer, 2015.
17. V. M. Gontineac, *On hypernearrings and H-hypergroups*, Proc. Fifth Int. Congress on Algebraic Hyperstructures and Applications (AHA 1993), Hadronic Press, Inc., USA, (1994), 171–179.
18. B. S. Kedukodi, S. P. Kuncham and S. Bhavanari, Equiprime, 3-prime and c-prime fuzzy ideals of nearrings, *Soft Computing* **13**(10) (2009), 933–944.
19. K. H. Kim, B. Davvaz and E. H. Roh, On fuzzy hyper R-subgroups of hypernearrings, *Ital. J. Pure Appl. Math.* **20** (2006), 177–192.
20. K. H. Kim, B. Davvaz and E. H. Roh, On hyper R-subgroups of hypernearrings, *Sci. Math. Jpn.* **67** (2008), 413–420.
21. F. Marty, *Sur une generalization de la notion de group*, 8th Congress Math. Scandenaves, Stockholm (1934), 45–49.
22. J. Mittas, Hypergroupes canoniques, *Math. Balkanica*, Beograd **2** (1972), 165–179.
23. E. H. Roh, B. Davvaz and K. H. Kim, T-fuzzy subhypernearrings of hypernearrings, *Sci. Math. Jpn.* **61**(3) (2005), 535–545.
24. B. Schweizer and A. Sklar, Statistical metric spaces, *Pacifc J. Math.* **10** (1960), 313–334.
25. M. Stefanescu, On hypernearrings, *An. St. Univ. Al. I. Cuza*, Iasi, 1992.
26. S. Yamak, O. Kazancı and B. Davvaz, Normal fuzzy hyperideals in hypernearrings, *Neural Computing and Applications* **20** (2011), 25–30.
27. X. Yuan, C. Zhang and Y. Ren, Generalized fuzzy groups and many-valued implications, *Fuzzy Sets and Systems* **138** (2003), 205–211.
28. L. A. Zadeh, Fuzzy sets, *Inform. Control* **8** (1965), 338–353.
29. J. Zhan and B. Davvaz, Generalized fuzzy ideals of near-rings, *Appl. Math. J. Chinese Univ. Ser. B* **24**(3) (2009), 343–349.
30. J. Zhan and Y. B. Jun, Fuzzy ideals of near-rings based on the theory of falling shadows, *Politehn. Univ. Bucharest Sci. Bull. Ser. A Appl. Math. Phys.* **74**(3) (2012), 67–74.

Weakly automorphism invariant modules

C. Selvaraj, S. Santhakumar

Department of Mathematics, Periyar University, Salem,
Tamilnadu, India - 636 011
E-mail: selvavlr@yahoo.com

In this paper, we introduce weakly automorphism N-invariant module, which is a generalization of both the concepts weakly N-injective and automorphism N-invariant. We obtain some basic properties and results of such modules. Also we characterize semisimple ring, CEP-ring and injective module by properties of weakly automorphism N-invariant module.

Keywords: Automorphism invariant module, semisimple ring, weakly injective module; H_v-nearring.

1. Introduction

Dickson and Fuller studied modules that are invariant under automorphisms of their injective envelopes in Ref. 2 for the particular case of finite-dimensional algebras over fields F with more than two elements. But recently this notion has been studied for modules over any ring by Lee and Zhou in Ref. 7. The automorphism invariant modules are studied in several papers; e.g., see Refs. 1, 3, 4, 7 and 10. Recently, Quynh and Kosan introduced automorphism relative invariant modules in Ref. 12. Selvaraj and Santhakumar investigated some more properties of dual automorphism invariant modules in Ref. 8. The concept of weakly relative injective modules was introduced by Jain and Lopez-Permouth in Ref. 6. More properties of weakly injective modules were studied by Jain and Lopez-Permouth in Refs. 5 and 6. In this paper we generalize weakly N-injective modules by using automorphism relative invariant modules.

All rings are assumed to be associative with identity and all modules are assumed to be right unital. Let B be a submodule of a module A. We say that B is an essential submodule of A, and that A is an essential extension of B, if whenever C is a nonzero submodule of A, then $C \cap B$ is also nonzero. A monomorphism $f : N \to M$ is said to be an essential monomorphism if $f(N)$ is an essential submodule of M. For modules M and N, M is said to be *N-injective* if every homomorphism from each submodule of N to M extends to a homomorphism from N to M. A module M is called *injective* if M is N-injective for all R module N. An injective hull of a module M is an injective essential extension of M. If M is right R-module then $E(M)$ denotes its injective hull. A module M is called *quasi injective* or *self injective* if M is M-injective. Equivalently, a right R-module M is called a quasi injective module if M is invariant under any endomorphism of $E(M)$. Clearly, any injective module is quasi-injective. A right R-module M is called a *pseudo injective* module if M is invariant under any monomorphism of $E(M)$.

A right R-module M is called *automorphism invariant* if it is invariant under automophisms of its injective envelope, that is, if $\phi(M) \subseteq M$ for every $\phi \in Aut(E(M))$ (equivalently, if $\phi(M) = M$ for every $\phi \in Aut(E(M))$). Equivalently, M is called an automorphism invariant module if every isomorphism between two essential submodules of M extends to an automorphism of M. Clearly, any quasi injective or pseudo injective module is automorphism-invariant.

Let M and N be R-modules. Then M is called *weakly N-injective* if for each homomorphism $f: N \to E(M)$, $f(N) \subseteq X \cong M$, for some submodule X of $E(M)$. Equivalently, M is weakly N-injective if for each homomorphism $f: N \to E(M)$, there exist a monomorphism $\sigma: M \to E(M)$ and a homomorphism $f': N \to M$ such that the diagram

$$\begin{array}{ccc} N & \xrightarrow{f} & E(M) \\ {\scriptstyle f'} \downarrow & \nearrow {\scriptstyle \sigma} & \\ M & & \end{array}$$

is commutative. Every N-injective module is weakly N-injective.

Let M and N be R-modules. We say M is automorphism N-invariant if for any essential submodule A of N, any essential monomorphism $f: A \to M$ can be extended to some $g \in Hom(N,M)$. Equivalently, M is automorphism N-invariant if every isomorphism between an essential submodule of N and essential submodule of M extends to a homomorphism from N to M. Every N-injective module is automorphism N-invariant. We call a ring R a right (left) *CEP*-ring if each cyclic right (left) R-module is essentially embeddable in a projective module.

In this paper, we generalize both the concepts weakly N-injective and automorphism N-invariant as weakly automorphism N-invariant. Also we discuss some properties of a weakly automorphism N-invariant module and we characterize the automorphism invariant envelope. Also we characterize semisimple ring and *CEP*-ring by properties of weakly automorphism N-invariant modules.

2. Automorphism Invariant Envelope

In this section, we prove that the intersection of automorphism invariant modules is again automorphism invariant and also we give an equivalent form for automorphism invariant envelope. Throughout this section all the automorphism invariant modules are submodules of some large module.

In general, the intersection of two injective modules need not be injective, but over a semisimple artinian ring the intersection of two injective modules is injective. Here we discuss this property for automorphism invariant modules in a general ring.

Theorem 2.1. *The intersection of two automorphism invariant modules is again an automorphism invariant module.*

Proof. Let M_1 and M_2 be two automorphism invariant modules and $\sigma: E(M_1 \cap M_2) \to E(M_1 \cap M_2)$ be an automorphism. Since $M_1 \cap M_2 \subseteq M_1 \subseteq E(M_1)$, $E(M_1 \cap M_2) \subseteq^{\oplus} E(M_1)$. We can write $N_1 \oplus E(M_1 \cap M_2) = E(M_1)$, for a submodule N_1 of $E(M_1)$, σ can be extended to $\sigma_1: E(M_1) \to E(M_1)$ by $\sigma_1(n_1 + m) = \sigma(n_1) + m$, $m \in E(M_1 \cap M_2)$ and $n_1 \in N_1$. Similarly, $M_1 \cap M_2 \subseteq M_2 \subseteq E(M_2)$, $E(M_1 \cap M_2) \subseteq^{\oplus} E(M_2)$. We can write

$N_2 \oplus E(M_1 \cap M_2) = E(M_2)$, for a submodule N_2 of $E(M_2)$, then σ can be extended to $\sigma_2 : E(M_2) \to E(M_2)$ by $\sigma_2(n_2 + m) = \sigma(n_2) + m$, $m \in E(M_1 \cap M_2)$ and $n_2 \in N_2$. Since M_1 and M_2 are automorphism invariant modules, then $\sigma_1(M_1) \subseteq M_1$ and $\sigma_2(M_2) \subseteq M_2$. Thus $\sigma(M_1 \cap M_2) = \sigma_1(M_1 \cap M_2) \subseteq M_1$ and $\sigma(M_1 \cap M_2) = \sigma_2(M_1 \cap M_2) \subseteq M_2$. Hence $\sigma(M_1 \cap M_2) \subseteq M_1 \cap M_2$. □

As a consequence, we have the following.

Corollary 2.1. *An arbitrary intersection of automorphism invariant modules is again an automorphism invariant module.*

Proof. Let $M_\alpha, \alpha \in A$ be automorphism invariant modules and $\sigma : E(\cap M_\alpha) \to E(\cap M_\alpha)$ an automorphism. Since $\cap M_\alpha \subseteq M_\beta \subseteq E(M_\beta)$, $E(\cap M_\alpha) \subseteq^{\oplus} E(M_\beta)\ \ \forall\, \beta$. We can write $N_\beta \oplus E(\cap M_\alpha) = E(M_\beta)$ for some submodules N_β of $E(M_\beta)$, Then σ can be extended to $\sigma_\beta : E(M_\beta) \to E(M_\beta)$ by $\sigma_\beta(n_\beta + m) = \sigma(n_\beta) + m$, $m \in E(\cap M_\alpha)$ and $n_\beta \in N_\beta\ \ \forall\, \beta$. Since $M_\alpha, \alpha \in A$ are automorphism invariant modules, $\sigma_\beta(M_\beta) \subseteq M_\beta$ this implies $\sigma(\cap M_\alpha) = \sigma_\beta(\cap M_\alpha) \subseteq M_\beta\ \ \forall\, \beta$. Hence $\sigma(\cap M_\alpha) \subseteq \cap M_\alpha$. □

The above result is also true for quasi injective modules when they are considered as submodules of some large module.

Theorem 2.2. *An arbitrary intersection of quasi injective modules is again a quasi injective module.*

Proof. Let $M_\alpha, \alpha \in A$ be quasi injective modules and $\sigma : E(\cap M_\alpha) \to E(\cap M_\alpha)$ an endomorphism. Since $\cap M_\alpha \subseteq M_\beta \subseteq E(M_\beta)$, $E(\cap M_\alpha) \subseteq^{\oplus} E(M_\beta)\ \forall\, \beta$. We can write $N_\beta \oplus E(\cap M_\alpha) = E(M_\beta)$ for some submodules N_β of $E(M_\beta)$, then σ can be extended to $\sigma_\beta : E(M_\beta) \to E(M_\beta)$ by $\sigma_\beta(n_\beta + m) = \sigma(n_\beta) + m$, $m \in E(\cap M_\alpha)$ and $n_\beta \in N_\beta\ \ \forall\, \beta$. Since M_α's are quasi injective modules, then $\sigma_\beta(M_\beta) \subseteq M_\beta$ and so that $\sigma(\cap M_\alpha) = \sigma_\beta(\cap M_\alpha) \subseteq M_\beta\ \ \forall\, \beta$. Hence $\sigma(\cap M_\alpha) \subseteq \cap M_\alpha$. □

Alahmadi *et al.* introduced the automorphism invariant envelope in Ref. 1, for any module M, the module $AI(M) = \sum_{\phi \in Aut(E(M))} \phi(M)$ is an *automorphism invariant envelope* of M. Clearly, every module has an automorphism invariant envelope.

Theorem 2.3. *Let M be an R-module. Then $AI(M)$ is the intersection of all automorphism invariant module containing M.*

Proof. Let A be the intersection of all automorphism invariant modules containing M. By Corollary 2.1, we have A is an automorphism invariant module and $M \subseteq A$. Then for any $\phi \in Aut(E(M))$ $\phi(M) \subseteq A$ and so $AI(M) \subseteq A$. Also we have $A \subseteq AI(M)$. Hence $AI(M)$ is the intersection of all automorphism invariant modules containing M. □

Theorem 2.4. *Let M be an R-module which is invariant under automorphism of its automorphism invariant envelope. Then M itself an automorphism invariant module.*

Proof. Let $\phi : M_1 \to M_2$ be an isomorphism and M_1 and M_2 be essential submodule of M. Since M is an essential submodule of $AI(M)$, M_1 and M_2 are essential submodules of $AI(M)$. Then ϕ can be extended to $\sigma : AI(M) \to AI(M)$ and σ is an automorphism, $\sigma(M) \subseteq M$. Hence $\sigma|_M$ is the required extension of ϕ. □

3. Submodules of an Automorphism Invariant Module

In this section we discuss the submodules of automorphism invariant modules. From the definition of an automorphism invariant module we obtain the following theorem obviously.

Theorem 3.1. *Fully invariant submodules of an automorphism invariant module are automorphism invariant.*

Lee and Zhou discussed that a submodule of an automorphism invariant module is again automorphism invariant which is a direct summand.

Theorem 3.2.[7] *A Direct summand of an automorphism invariant module is automorphism invariant.*

Recall that, a module M is called a CS-module if every submodule of M is essential in a direct summand of M.

Theorem 3.3. *Let M be a CS-module. Then every submodule of M is automorphism invariant if and only if every essential submodule of M is automorphism invariant.*

Proof. Let N be a submodule of M. Since M is a CS-module, N is an essential submodule of a direct summand M_1 of M. Let M_2 be the complement of M_1. Then $N \oplus M_2$ is an essential submodule of M. This gives $N \oplus M_2$ is automorphism invariant and by Theorem 3.2, N is an automorphism invariant module. □

From the above proof, if M is a CS-module and p is a property which is closed under direct summands, then every submodule of M has property p if and only if every essential submodule of M has property p. We have the following corollary.

Corollary 3.1. *Let M be a quasi injective module. Then every submodule of M is quasi injective if and only if every essential submodule of M is quasi injective.*

Proof. We know that every CS-automorphism invariant module is a quasi injective module. Then the proof of the corollary is immediate. □

Theorem 3.4. *If every essential submodule of an automorphism invariant module is automorphism invariant then R is semi-simple.*

Proof. Since every module can be written as an essential submodule of an injective module, then every module is an essential submodule of an automorphism invariant module. So every module is automorphism invariant. This gives the direct sum of an automorphism invariant module is automorphism invariant. Then, by Ref. 11, R is semi-simple. □

4. Weakly Automorphism Invariant

In this section we introduce weakly automorphism relative invariant modules as a generalization of automorphism relative invariant and weakly injective relative modules.

Lemma 4.1. *Let M and N be two R-modules, then M is automorphism N-invariant if and only if for every essential monomorphism $f : N \to E(M)$, $f(N) \subseteq M$.*

Proof. (1) $\Rightarrow$ (2): Let $f : N \to E(M)$ be an essential monomorphism. Then by injectivity of $E(M)$, there exists a homomorphism $f' : E(N) \to E(M)$ such that $f'|_N = f$ and the diagram

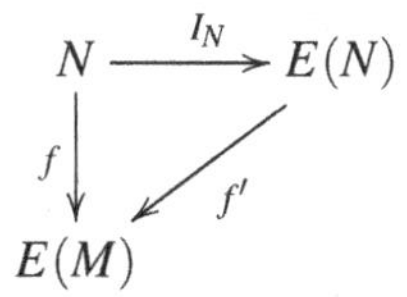

$$\begin{array}{ccc} N & \xrightarrow{I_N} & E(N) \\ {\scriptstyle f}\downarrow & \swarrow{\scriptstyle f'} & \\ E(M) & & \end{array}$$

commutes, where I_N is an inclusion map. Also we have $Im(I_N) \cap kerf' = \{0\}$. Since $Im(I_N)$ is essential, $kerf' = \{0\}$. Hence f' is an essential monomorphism. By Ref. 12 Theorem 2.2, $f'(N) \subseteq M \Rightarrow f(N) \subseteq M$.
(2) $\Rightarrow$ (1): Let A be an essential submodule of N and $f : A \to M$ be an essential monomorphism. Then by injectivity of $E(M)$, there exists a homomorphism $f' : N \to E(M)$. Hence by hypothesis $f'(N) \subseteq M$, then there exists a homomorphism $g : N \to M$, which is the required extension of f. $\square$

The above lemma motivates the following definition as a generalization of weakly N-injective.

Definition 4.1. A module M is called weakly automorphism N-invariant if for each essential monomorphism $f : N \to E(M)$, $f(N) \subseteq X \cong M$ for some essential submodule X of $E(M)$.

Example 4.1.[10] Consider the example $R = \begin{pmatrix} \mathbb{F} & \mathbb{F} & \mathbb{F} \\ 0 & \mathbb{F} & 0 \\ 0 & 0 & \mathbb{F} \end{pmatrix}$, where $\mathbb{F}$ is a field of order 2. Let $A = e_{12}\mathbb{F} \oplus e_{13}\mathbb{F}$ and $B = e_{11}R$, both of them have equivalent injective hull, $E(A) \cong^f E(B)$. Let $\phi_1 : A \to E(A)$ and $\phi_2 : B \to E(B)$ be essential monomorphisms. Here the only essential monomorphism from A to $E(B)$ is $f \circ \phi_1$. Then we have $f \circ \phi_1(A) \subseteq \phi_2(B) \cong B$. Hence B is weakly automorphism A-invariant.

Theorem 4.1. *A module M is weakly automorphism N-invariant iff for each essential monomorphism $f : N \to E(M)$, there exists an essential monomorphism $\sigma : M \to E(M)$ and a homomorphism $f' : N \to M$ such that the diagram*

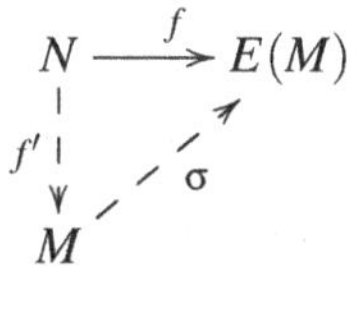

$$\begin{array}{ccc} N & \xrightarrow{f} & E(M) \\ {\scriptstyle f'}\downarrow & \nearrow{\scriptstyle \sigma} & \\ M & & \end{array}$$

commutes.

Proof. Assume that M is weakly automorphism N-invariant. Let $f : N \to E(M)$ be an essential monomorphism, then by definition we have $f(N) \subseteq X \cong M$ for some essential submodule X of $E(M)$. Let $\sigma_1 : M \to X$ be an isomorphism and $\sigma : M \to E(M)$ a monomorphism induced from σ_1. Let $f' = \sigma_1^{-1} \circ f$, then we have the following commutative diagram;

$$\begin{array}{ccc} N & \xrightarrow{f} & E(M) \\ {\scriptstyle f'}\downarrow & \nearrow{\scriptstyle \sigma} & \\ M & & \end{array}$$

Conversely, let $f : N \to E(M)$ be an essential monomorphism. Then there is an essential monomorphism $\sigma : M \to E(M)$ and homomorphism $f' : N \to M$ such that the diagram

$$\begin{array}{ccc} N & \xrightarrow{f} & E(M) \\ {\scriptstyle f'}\downarrow & \nearrow{\scriptstyle \sigma} & \\ M & & \end{array}$$

is commutative. Let $X = \sigma(M)$. Clearly, $f(N) \subseteq X$ and $X \cong M$. Hence M is weakly automorphism N-invariant. □

Proposition 4.1. *The module M is weakly automorphism N-invariant iff for each isomorphism $f : E(N) \to E(M)$, there exist an essential monomorphism $\sigma : M \to E(M)$ and a homomorphism $f' : N \to M$ such that the diagram*

$$\begin{array}{ccc} E(N) & \xrightarrow{f} & E(M) \\ {\scriptstyle i_N}\uparrow & & \uparrow{\scriptstyle \sigma} \\ N & \xrightarrow[f']{} & M \end{array}$$

commutes.

Proof. Let $f : N \to E(M)$ be an essential monomorphism, then by injectivity of $E(M)$, f can be extended to $g : E(N) \to E(M)$. By hypothesis, there exist an essential monomorphism $\sigma : M \to E(M)$ and a homomorphism $f' : N \to M$ such that the diagram

$$\begin{array}{ccc} E(N) & \xrightarrow{g} & E(M) \\ {\scriptstyle i_N}\uparrow & & \uparrow{\scriptstyle \sigma} \\ N & \xrightarrow[f']{} & M \end{array}$$

commutes. Therefore,

$$\begin{array}{ccc} N & \xrightarrow{f} & E(M) \\ {\scriptstyle f'}\downarrow & \nearrow{\scriptstyle \sigma} & \\ M & & \end{array}$$

commutes. Hence, by above theorem, M is a weakly automorphism N-invariant module.

Conversely, Let $f : E(N) \to E(M)$ be an isomorphism, then $f|_N : N \to E(M)$ is an essential monomorphism. By above theorem there exists an essential monomorphism $\sigma : M \to E(M)$ and a homomorphism $f' : N \to M$ such that the diagram

$$\begin{array}{ccc} N & \xrightarrow{f|_N} & E(M) \\ {\scriptstyle f'}\downarrow & \nearrow{\scriptstyle \sigma} & \\ M & & \end{array}$$

commutes. Then,

$$\begin{array}{ccc} E(N) & \xrightarrow{f} & E(M) \\ \uparrow{\scriptstyle i_N} & & \uparrow{\scriptstyle \sigma} \\ N & \xrightarrow[f']{} & M \end{array}$$

commutes. □

Proposition 4.2. *If M is weakly automorphism N-invariant and L is an essential extension of M, then L is weakly automorphism N-invariant.*

Proof. Let $f : N \to E(M)$ be an essential monomorphism. Since M is weakly automorphism N-invariant, then there is an essential monomorphism $\sigma : M \to E(M)$ and homomorphism $f' : N \to M$ such that the diagram

$$\begin{array}{ccc} N & \xrightarrow{f} & E(M) \\ {\scriptstyle f'}\downarrow & \nearrow{\scriptstyle \sigma} & \\ M & & \end{array}$$

commutes. By the injectivity of E(M), σ can be extend to $\sigma_1 : E(M) \to E(M)$. Hence we have the following commutative diagram

$$\begin{array}{ccc} N & \xrightarrow{f} & E(M) \\ {\scriptstyle I_M \circ f'}\downarrow & \nearrow{\scriptstyle \sigma_1|_L} & \\ L & & \end{array}$$

where $I_M : M \to L$ is an inclusion map. Hence L is weakly automorphism N-invariant. □

Proposition 4.3. *Let M be an R-module. If for all $x_1, x_2, ..., x_n \in E(M)$ with $Ann(x_i) = \{0\}, i = 1, 2, .., n$, there exists a submodule X of $E(M)$ such that $x_i \in X \cong M, i = 1, 2, .., n$, then M is weakly automorphism R^n-invariant.*

Proof. Since every homomorphism $\phi : R^n \to E(M)$ is determined by choosing arbitrary elements $x_1, x_2, ..., x_n \in E(M)$, let $f : R^n \to E(M)$ be an essential monomorphism. Then f can be determined by $x_1, x_2, ..., x_n$ for some $x_i \in E(M), i = 1, 2, .., n$. Since f is monomorphism,

$Ann(x_i) = \{0\}$. Hence by hypothesis there exists an essential submodule X of $E(M)$ such that $x_i \in X \cong M, i = 1,2,..,n$. Therefore, M is weakly automorphism R^n-invariant. □

Proposition 4.4. *Let M be an R-module and $E(R^n) \cong E(M)$. Then M is weakly automorphism R^n-invariant iff for a linearly independent subset $\{x_1,x_2,...,x_n\}$ of $E(M)$, there exists a essential submodule X of $E(M)$ such that $x_i \in X \cong M, i = 1,2,..,n$.*

Proof. Assume that M is weakly automorphism R^n-invariant. Let $\{x_1,x_2,...,x_n\}$ be linearly independent subset of $E(M)$. Define a map $f : R^n \to E(M)$ by $f(e_i) = x_i, i = 1,2,...,n$, where $\{e_1,e_2,...,e_n\}$ is basis of R^n. Clearly, f is a monomorphism and essential. Hence by definition there exists an essential submodule X of $E(M)$ such that $x_i \in X \cong M, i = 1,2,..,n$.

Conversely, let $f : R^n \to E(M)$ be an essential monomorphism. Then f can be determined by $x_1,x_2,...,x_n$ for some $x_i \in E(M), i = 1,2,..,n$. Since f is a monomorphism then $\{x_1,x_2,...,x_n\}$ is linearly independent. Hence, by hypothesis there exists an essential submodule X of $E(M)$ such that $x_i \in X \cong M, i = 1,2,..,n$. Therefore, by definition, M is weakly automorphism R^n-invariant. □

Remark 4.1. Let M and N be R-modules and $E(N) \ncong E(M)$. Then, by uniqueness of injective hull, we have there is no essential monomorphism from N to $E(M)$. Hence, M is weakly automorphism N-invariant.

Proposition 4.5. *Let M and N be R-modules and $E(N) \cong E(M)$. Then M is weakly automorphism N-invariant iff N can be essentially embedded into M.*

Proof. Assume that M is weakly automorphism N-invariant. Since $E(N) \cong E(M)$, then there exists a monomorphism $f : N \to E(M)$. Hence, by definition there exists an essential submodule X of $E(M)$ such that $f(N) \subseteq X \cong M$. Therefore, N can be essentially embedded into M.

Conversely, assume that N can be essentially embedded into M. Let $f : N \to E(M)$ be an essential monomorphism and $g : N \to M$ be an embedding of N into M. Then by injectivity of $E(M)$ there exists an isomorphism $h : E(M) \to E(M)$ such that the diagram

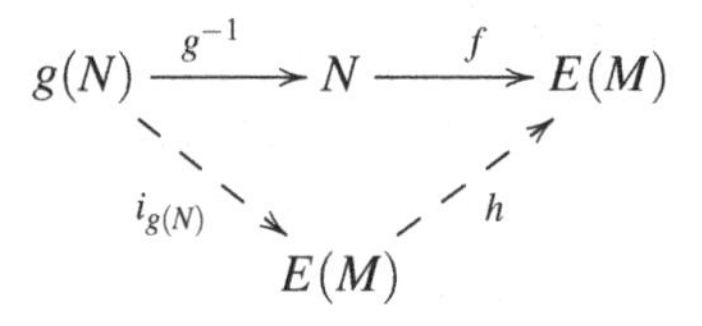

is commutative. Consider the restriction $h' : M \to E(M)$ of g we get the following commutative diagram

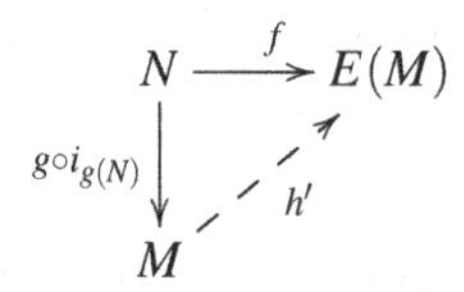

Clearly, h' is a monomorphism. Hence by Theorem 4.1, M is weakly automorphism N-invariant. □

Example 4.2. In example 4.1, consider the essential monomorphism $f^{-1} \circ \phi_2 : B \to E(A)$. Clearly, there is no submodule X of $E(A)$ such that $f^{-1} \circ \phi_2(B) \subseteq X \cong A$. Hence, A is weakly automorphism B-invariant. Also B can't be essentially embedded into A.
In the same example take $C = e_{22}\mathbb{F} \oplus e_{33}\mathbb{F}$. $A \cong C$, A is weakly automorphism C-invariant and C is weakly automorphism A-invariant.

Theorem 4.2. *Let M be an R-module. Then M is injective if and only if M is weakly automorphism N-invariant for all R-module N.*

Proof. The first part of the proof is straight froward.

Conversely, suppose M is weakly automorphism N-invariant for all R-module N, then M is weakly automorphism $E(M)$-invariant. Also, the injective hull of M and $E(M)$ are same. Hence, by Proposition 4.5, $E(M)$ can be essentially embedded into M which is possible only when $M = E(M)$. Therefore, M is injective. □

Corollary 4.1. *Let M be an R-module. Then M is injective if and only if M is automorphism N-invariant for all R-module N.*

Proof. Since every automorphism N-invariant module is also weakly automorphism N-invariant, by Theorem 4.2, the proof is immediately follows. □

Corollary 4.2. *A ring R is semisimple if and only if for any pair of R-modules M and N, M is weakly automorphism N-invariant.*

Corollary 4.3. *A ring R is semisimple if and only if for any pair of R-modules M and N, M is weakly N-injective.*

Corollary 4.4. *A ring R is semisimple if and only if for any pair of R-modules M and N, M is automorphism N-invariant.*

Proposition 4.6. *Let M be an R-module. Take $\mathscr{A}_M = \{N \mid M$ is weakly automorphism N-invariant$\}$, then the following hold:*

(1) $\mathscr{A}_M$ is closed under essential submodules.
(2) If $\mathscr{A}_M$ is closed under essential extension, then M is injective.
(3) If $\mathscr{A}_M$ is closed under essential extension for all M, then R is semisimple.

Proof. (1) Let M be weakly automorphism N-invariant and N_1 an essential submodule of M. Suppose $E(M) \not\cong E(N)$, then there is nothing to prove. Otherwise, N can be essentially embedded into M, then N_1 also can be essentially embedded into M. Hence, M is weakly automorphism N_1-invariant.
(2) We already know that M is weakly automorphism M-invariant. By hypothesis M is weakly automorphism $E(M)$-invariant. Then $E(M)$ can be essentially embedded into M. Thus $M \cong E(M)$. Hence, M is injective.
(3) From (2), every R-module M is injective. Hence, R is semisimple. □

Lemma 4.2. *Let R be a right artinian ring, and let M and N be finitely generated R-modules. If M is weakly automorphism N-invariant, N is weakly automorphism M-invariant, and $Soc(M) \cong Soc(N)$, then $M \cong N$.*

Proof. M and N are artinian implies $Soc(M) \ll M$ and $Soc(N) \ll N$. Let $\sigma : Soc(M) \to Soc(N)$ be an isomorphism. Since $Soc(M) \ll M$, there is an essential monomorphism $\sigma_1 : Soc(M) \to E(M)$ induced from σ. Because of injectivity of $E(N)$ there exists an essential monomorphism $\phi_1 : M \to E(M)$. Since N is weakly automorphism M-invariant, M can be essentially embedded into N. Similarly, N can be essentially embedded into M. Because M and N are finitely generated modules over a right artinian ring, it follows that $M \cong N$. □

Lemma 4.3. *Let R be a right artinian ring such that all indecomposable projective R-modules are uniform and weakly automorphism N-invariant, for every indecomposable projective N. Then*

(1) every simple R-module is isomorphic to the socle of an indecomposable projective module
(2) every simple R-module is embeddable in $Soc(R)$
(3) if P and Q are projective modules with $Soc(P) \cong Soc(Q)$, then $P \cong Q$.

Proof. Because R is artinian, we can write $R = \oplus_{i=1}^{n} e_iR$ as a direct sum of indecmopos-able right ideals, where $\mathscr{P} = \{e_1R, e_2R, ..., e_kR\}$, $(k < n)$ is an irredudant complete set of representatives for the projective decomposable modules. Take $S_i = Soc(e_iR)$. Clearly, $e_iR \cong e_jR$ implies $S_i \cong S_j$, by Proposition 4.2, $e_iR \cong e_jR$. Since $\mathscr{L} = \{S_1, S_2, ..., S_K\}$ is an irredundant set of simple R-modules containing $k = |\mathscr{P}|$ members, $\mathscr{L}$ must be a complete set of representatives.

This proves (i) and (ii), and also (iii) when P and Q are indecoposable. In general, $P = \oplus_{i=1}^{n}(e_iR)^{(A_i)}$ and $Q = \oplus_{i=1}^{n}(e_iR)^{(B_i)}$. So $Soc(P) \cong Soc(Q)$ yields $|A_i| = |B_i|$, this proving $P \cong Q$. □

Theorem 4.3. *A semiperfect ring R is CEP if and only if the following all hold*

(1) R is a right artinian ring.
(2) Every indecomposable projective module is uniform and weakly automorphism N-invariant, for every indecomposable projective N.
(3) Every projective module is weakly automorphism C-invariant for all cyclic modules C.

Proof. Assume that R is *CEP* ring. By Ref. 6 Theorem 5.2, (1) and (2) are immediate . Let P be a projective module, then P can be written as $P = \oplus(P_i)^{\alpha_i}$, P_i are indecomposable projective module and $\alpha_i \geq 0$. By Ref. 6 Proposition 4.1 and Lemma 4.3, P is weakly C-injective for all cyclic modules C. This proves that every projective module is weakly automorphism C-invariant for all cyclic modules C.

Conversely, assume the hypothesis. Let $R = \oplus_{i=1}^{n} e_iR$ as a direct sum of indecmoposable right ideals, where $\mathscr{P} = \{e_1R, e_2R, ..., e_kR\}$ $(k < n)$ is an irredudant complete set of representatives for the projective indecomposable modules. Let I be an ideal of R, by 4.3(i),

$Soc(R/I) \cong Soc(\oplus_{i=1}^{n}(e_iR)^{\beta_i}), \beta_i \geq 0$. Let $P = \oplus_{i=1}^{n}(e_iR)^{\beta_i}$, then $Soc(R/I) \cong Soc(P) \cong (E(P))$ and $E(R/I) \cong E(P)$. Because P is weakly automorphism R/I-invariant we get R/I essentially embedded into P. Hence R is CEP ring. □

Acknowledgments

The second author was supported by CSIR - Human Resource Development Group, CSIR Complex, Library Avenue, Pusa, New Delhi - 110 012, India, Grant No.: 09/810(0020)/2012-EMR-I.

Bibliography

1. A. Alahmadi, A. Facchini and N. K. Tung, *Automorphism-invariants modules*, Rend. Sem. Mat. Univ. Padova **133** (2015), 241–259.
2. S. E. Dickson and K. R. Fuller, *Algebras for which every indecomposable right module is invariant in its injective envelope*, Pacific J. Math. **31**, 3 (1969), 655–658.
3. N. Er, S. Singh and A. K. Srivastava, *Rings and modules which are stable under automorphisms of their injective hulls*, J. Algebra **379** (2013), 223–229.
4. P. A. Guil Asensio and A. K. Srivastava, *Automorphism-invariant modules satisfy the exchange property*, J. Algebra **388** (2013), 101–106.
5. S. K. Jain and S. R. Lopez-Permouth, *A survey on the theory of weakly-injective modules*, Comm. Alg. (1994), 205–232.
6. S. K. Jain and S. R. Lopez-Permouth, *Rings whose cyclics are essentially embeddable in projective modules* , J. Algebra **128** (1990), 257–269.
7. T. K. Lee and Y. Zhou, *Modules which are invariant under automorphisms of their injective hulls*, J. Algebra Appl. **12**(2) (2013), 1250159.
8. C. Selvaraj and S. Santhakumar, *A note on dual automorphism-invariant module*, J. Algebra Appl. **16**(1) (2017), 1750024.
9. C. Selvaraj and S. Santhakumar, *On Dual automorphism relative invariant module*, Communicated to Mathematical Proceedings of the Cambridge Philosophical Society.
10. S. Singh and A. K. Srivastava, *Rings of invariant module type and automorphism invariant modules*, Ring Theory and Its Applications, Contemp. Math., Amer. Math. Soc. **609** (2014), 299–311.
11. A. K. Tiwary and B. M. Pandeya, *Pseudo Projective and Pseudo Injective Modules*, Indian J. of Pure Appl. Math. Vol. 9, No. 9 (1978), 941–949.
12. T. C. Quynh and M. T. Kosan, *On automorphism-invariant modules*, J. Algebra Appl. **14**(5) (2015), 1550074.

Trivial generalized matrix rings

G. F. Birkenmeier*, B. J. Heider

Department of Mathematics, University of Louisiana at Lafayette, Lafayette, LA 70504-1010, USA
E-mail: gfb1127@louisiana.edu, bheider@louisiana.edu

Recall that an n-by-n generalized matrix ring R is defined in terms of sets of rings $\{R_i\}_{i=1}^n$, (R_i,R_j)-bimodules $\{M_{ij}\}$, and bimodule homomorphisms $\theta_{ijk}: M_{ij}\otimes_{R_j} M_{jk} \to M_{ik}$, where the set of diagonal matrix units $\{E_{ii}\}$ form a complete set of orthogonal idempotents. Moreover, an arbitrary ring R with a complete set of orthogonal idempotents $\{e_i\}_{i=1}^n$ has a Peirce decomposition which can be arranged into an n-by-n generalized matrix ring R^π which is isomorphic to R. In this paper, we focus on the subclass $\mathcal{T}_n$ of n-by-n generalized matrix rings with $\theta_{iji}=0$ for $i \neq j$. $\mathcal{T}_n$ contains all upper and all lower generalized triangular matrix rings and is called the class of n-by-n trivial generalized matrix rings. This paper is primarily an expository paper based on a plenary talk presented at the 24th International Conference on Nearrings, Nearfields, and Related Topics. However some new results are presented at the end of the paper.

Keywords: idempotent, generalized matrix ring, formal matrix ring, Morita Context, Peirce trivial, annihilator, bimodule, essential, dense.

Introduction

Throughout this paper all rings are associative with a unity and modules are unital unless explicitly indicated otherwise.

Given a complete set of orthogonal idempotents, $\{e_i\}_{i=1}^n$, of a ring R, we can form a group direct sum,

$$R = e_1Re_1 \oplus \cdots \oplus e_1Re_n \oplus e_2Re_1 \oplus \cdots \oplus e_2Re_n \oplus \cdots \oplus e_nRe_1 \oplus \cdots \oplus e_nRe_n,$$

called the Peirce decomposition of R. This decomposition can be arranged into an n-by-n square array, called R^π, with

$$R^\pi = \begin{bmatrix} e_1Re_1 & e_1Re_2 & \cdots & e_1Re_n \\ e_2Re_1 & e_2Re_2 & \ddots & \vdots \\ \vdots & \ddots & \ddots & e_{n-1}Re_n \\ e_nRe_1 & \cdots & e_nRe_{n-1} & e_nRe_n \end{bmatrix}.$$

The array R^π forms a ring, where addition is componentwise and multiplication is the usual row-column matrix multiplication. Moreover, there is a ring isomorphism $h: R \to R^\pi$ defined by $h(x) = [e_ixe_j]$ for all $x \in R$. Observe that the e_iRe_i are rings with unity and the

*Corresponding author

e_iRe_j are (e_iRe_i, e_jRe_j)-bimodules. Note that the bimodule product $e_iRe_j \cdot e_jRe_k$, arising in the row-column multiplication, may be thought of as a bimodule homomorphism θ_{ijk} : $e_iRe_j \otimes_{e_jRe_j} e_jRe_k \to e_iRe_k$ determined by the multiplication of R.

The above discussion motivates the following well known definition:

an *n-by-n generalized (or formal) matrix ring R* is a square array

$$R = \begin{bmatrix} R_1 & M_{12} & \cdots & M_{1n} \\ M_{21} & R_2 & \ddots & \vdots \\ \vdots & \ddots & \ddots & M_{n-1,n} \\ M_{n1} & \cdots & M_{n,n-1} & R_n \end{bmatrix}$$

where each R_i is a ring, each M_{ij} is an (R_i, R_j)-bimodule and there exist (R_i, R_k)-bimodule homomorphisms $\theta_{ijk} : M_{ij} \otimes_{R_j} M_{jk} \to M_{ik}$ for all $i, j, k = 1, \ldots, n$ (with $M_{ii} = R_i$). For $m_{ij} \in M_{ij}$ and $m_{jk} \in M_{jk}$, $m_{ij}m_{jk}$ denotes $\theta_{ijk}(m_{ij} \otimes m_{jk})$. The homomorphisms θ_{ijk} must satisfy the associativity relation: $(m_{ij}m_{jk})m_{k\ell} = m_{ij}(m_{jk}m_{k\ell})$ for all $m_{ij} \in M_{ij}$, $m_{jk} \in M_{jk}$, $m_{k\ell} \in M_{k\ell}$ and all $i, j, k, \ell = 1, \ldots, n$. Observe that θ_{iii} is determined by the ring multiplication in R_i, while θ_{ijj} and θ_{jjk} are determined by the bimodule scalar multiplications.

With these conditions, addition on R is componentwise and multiplication on R is row-column matrix multiplication. A Morita context is a 2-by-2 generalized matrix ring. Other examples of generalized matrix rings are: incidence algebras of directed graphs with a finite number of vertices, endomorphism rings of finite direct sums of modules, and structural matrix rings. *An n-by-n generalized upper (lower) triangular matrix ring* is a generalized matrix ring with $M_{ij} = 0$ for $j < i$ ($M_{ij} = 0$ for $i < j$). Note that $\{E_{ii} \in R \mid E_{ii}$ is the matrix with $1 \in R_i$ in the (i,i)-position and 0 elsewhere, $i = 1, \ldots, n\}$ is a complete set $\{E_{ii}\}_{i=1}^n$ of orthogonal idempotents in the above constructed generalized matrix ring R.

As indicated in [ABvW] the foregoing observations allow one to consider a generalized matrix ring in two ways:

(1) given a ring R and a complete set of orthogonal idempotents, $\{e_i\}_{i=1}^n$, then R^π is an "internal" representation of R as a generalized matrix ring in terms of substructures of R; whereas

(2) given collections $\{R_i\}$, $\{M_{ij}\}$, and $\{\theta_{ijk}\}$, we construct a new ring from these "external" components via the generalized matrix ring notion.

An important problem in the study of generalized matrix rings is: given a collection of rings $\{R_i \mid i = 1, \ldots, n\}$ and bimodules $\{M_{ij} \mid i, j = 1, \ldots, n,\ i \neq j$, and each M_{ij} is an (R_i, R_j)-bimodule$\}$ determine the θ_{iji} $(i \neq j)$ and the θ_{ijk} $(i, j, k$ distinct) to produce an n-by-n generalized matrix ring. We can simplify this problem by trivializing the θ_{ijk} in the following three ways (note that for $n = 2$, all three conditions coincide):

(I) Define $\theta_{iji} = 0$, for all $i \neq j$.

(II) Define $\theta_{ijk} = 0$, for all i, j, k pairwise distinct.

(III) Define $\theta_{ijk} = 0$, for all $i \neq j$ and $j \neq k$ (I and II combined).

Two questions immediately arise:

(A) Are there significant examples of generalized matrix rings with trivialized θ_{ijk}?

(B) How can the theory of generalized matrix rings with trivialized θ_{ijk} be used to gain insight into the theory of arbitrary generalized matrix rings?

In [ABvW], the class of n-by-n ($n > 1$) generalized matrix rings satisfying condition (I) (i.e., $\theta_{iji} = 0$, for all $i \neq j$) is considered. This class is called the class of trivial generalized matrix rings, denoted by $\mathcal{T}_n$.

For each generalized matrix ring R,

one uses $\overline{R}$ to denote the ring in $\mathcal{T}_n$ which has the same corresponding $R_i, M_{ij}, \theta_{ijk}$ as R, except that for all $i \neq j$ the homomorphisms θ_{iji} are taken to be 0 in $\overline{R}$.

Thus R and $\overline{R}$ are the same ring if and only if $R \in \mathcal{T}_n$. Note that the classes of n-by-n generalized upper and lower triangular matrix rings form significant proper subclasses of $\mathcal{T}_n$ (see Question A). Further examples are provided throughout this paper.

Next, let R be a generalized n-by-n matrix ring and take

$$\mathcal{D}(R) = \{[r_{ij}] \in R \mid r_{ij} = 0 \text{ for all } i \neq j\}$$

and

$$\mathcal{D}(R)^- = \{[r_{ij}] \in R \mid r_{ii} = 0 \text{ for all } i = 1, \ldots, n\}.$$

It appears to the authors that the systematic study of the external generalized matrix rings begins in 1958 with the introduction of the Morita context in [M]. The first appearance of the concept of a trivial generalized matrix ring that we have found is in 1976 in [S], wherein he considered the injective hulls of modules over a trivial generalized matrix ring. In [H], the hopfian, cohopfian, and strongly π-regular conditions are investigated for 2-by-2 trivial generalized matrix rings. The strongly π-regular condition was characterized for 2-by-2 trivial generalized matrix rings in [C]. In considering, left self distributively generated algebras in [L], examples of trivial generalized matrix rings without unity are provided. The module theory over trivial generalized matrix rings is considered in [KT]. In 2014, Section 3 of [TLZ] is devoted to 2-by-2 trivial generalized matrix rings. An immediate consequence of their results is the following theorem.

Theorem A 1. Let $R = \begin{bmatrix} A & M \\ N & B \end{bmatrix}$ be a trivial generalized matrix ring. Then $\mathcal{D}(R)$ has each of the following properties if and only if R does so:

(1) semilocal,
(2) semiperfect,
(3) left (or right) perfect,

(4) strongly π-regular,
(5) potent,
(6) clean,
(7) directly finite, or
(8) 2-good.

Note that in Theorem A, (3) and (4) have already appeared in [KT] and [C], respectively. It seems that in the literature on trivial generalized matrix rings many results only focus on the 2-by-2 case (i.e., trivial Morita Contexts) and on information transfer between $\mathcal{D}(R)$ and R where R is a 2-by-2 generalized matrix ring.

Recently (2016), in contrast to much of the previous work on trivial generalized matrix rings, the authors of [ABvW] consider n-by-n ($n \geq 2$) trivial generalized matrix rings and show that every n-by-n generalized matrix ring has subrings S maximal with respect to being in $\mathcal{T}_n$ such that S is essential in R as an (S,S)-bimodule. This fact allows for a two-step transfer of information from $\mathcal{D}(R)$ to S (Theorem 1.16) and from S to R (Theorem 2.12 and Corollary 2.14). Moreover, they show that the triviality of the θ_{iji} motivates three new types of idempotents which appear in the internal (Peirce decomposition) generalized matrix ring representation of a ring in $\mathcal{T}_2$. For $e = e^2 \in R$,

(1) e is *inner Peirce trivial* if $eR(1-e)Re = 0$;
(2) e is *outer Peirce trivial* if $(1-e)ReR(1-e) = 0$;
(3) e is *Peirce trivial* if e is both inner and outer trivial.

These idempotents are important tools in characterizing and developing properties of trivial generalized matrix rings.

In the remainder of this paper, we survey some of the results in [ABvW] and provide some new applications of their results. Note that all of the definitions, examples and results in Section 1 and 2 are from [ABvW] except where indicated otherwise.

Notation and Terminology

(1) R is Abelian - means every idempotent is central.
(2) $\mathcal{B}(R), \mathcal{P}(R)$ and $\mathcal{J}(R)$ denote the central idempotents of R, the prime radical of R and the Jacobson radical of R respectively.
(3) $\mathcal{S}_\ell(R) = \{e = e^2 \in R \mid Re = eRe\}$, $\mathcal{S}_r(R) = \{e = e^2 \in R \mid eR = eRe\}$.
(4) $\mathrm{Cen}(R)$ is the center of R.
(5) $\mathrm{U}(R)$ is the group of units of R.
(6) $< - >_R$ is the subring of R generated by $-$, and $(-)_R$ is the ideal of R generated by $-$.
(7) $X \trianglelefteq R$ means X is an ideal of R.
(8) $\underline{r}_A(B)$ and $\underline{\ell}_A(B)$ denote the right and left annihilator of B in A, respectively.
(9) $\mathbb{Z}$ and $\mathbb{Z}_n$ denote the ring of integers and the ring of integers modulo n, respectively.
(10) $\mathbb{Z}^+$ means the positive integers.

(11) Let X and Y be right R-modules. Then $X \leq^{ess} Y$ and $X \leq^{den} Y$ mean X is essential in Y and X is dense in Y, respectively.
(12) Let R be a ring. Then $Q(R)$ (respectively, $Q_\ell(R)$) denote the maximal right (respectively, left) ring of quotients of R.

1. Basic Properties of Peirce trivial Idempotents

Definition 1.1. Let R be a ring, not necessarily with a unity, and let $e = e^2 \in R$. We say e is *inner Peirce trivial* (respectively, *outer Peirce trivial*) if $exye = exeye$ (respectively, $xey + exeye = xeye + exey$) for all $x, y \in R$. If e is both inner and outer Peirce trivial, we say e is Peirce trivial.

For a ring R with a unity, e is inner (respectively, outer) Peirce trivial if and only if $eR(1-e)Re = \{0\}$ (respectively, $(1-e)ReR(1-e) = \{0\}$); moreover, e is inner Peirce trivial if and only if $f = 1 - e$ is outer Peirce trivial. Let

$$\mathfrak{P}_{\rm it}(R), \mathfrak{P}_{\rm ot}(R) \text{ and } \mathfrak{P}_{\rm t}(R)$$

denote the set of all inner Peirce trivial idempotents, all outer Peirce trivial idempotents and all Peirce trivial idempotents of R, respectively. Note that $\mathcal{B}(R) \subseteq \mathfrak{P}_t(R)$.

Example 1.1. Inner and outer Peirce trivialities are independent properties of idempotents. Let $R_1 = \mathbb{Z}, R_2 = \mathbb{Z}/8\mathbb{Z} = \mathbb{Z}_8, M_{12} = \mathbb{Z}_4, M_{21} = \mathbb{Z}_2$, together with tensor products $M_{12} \otimes_{R_2} M_{21} \cong \mathbb{Z}_2 \mapsto 0 \in R_1$ and $M_{21} \otimes_{R_1} M_{12} \cong \mathbb{Z}_2 \cong 4R_2$, respectively. Then in $R = \begin{bmatrix} R_1 & M_{12} \\ M_{21} & R_2 \end{bmatrix}$ the elements $e = \begin{bmatrix} 1 & 0 \\ 0 & 0 \end{bmatrix}$ and $f = \begin{bmatrix} 0 & 0 \\ 0 & 1 \end{bmatrix}$ are idempotents. Moreover, e is inner Peirce trivial, but not outer Peirce trivial, and f is outer Peirce trivial, but not inner Peirce trivial.

Proposition 1.1. *Let* $R = \begin{bmatrix} R_1 & M_{12} \\ M_{21} & R_2 \end{bmatrix} \in \mathcal{T}_2$ *and assume* $\alpha = \begin{bmatrix} e & m \\ n & f \end{bmatrix} \in R$.

(1) Then $\alpha = \alpha^2$ *if and only if* $e = e^2, f = f^2, em + mf = m$ *and* $ne + fn = n$.
(2) If $\alpha = \alpha^2$ *and* e *and* f *are central idempotents, then* $\alpha \in \mathfrak{P}_{\rm t}(R)$. *In particular, if* R_1 *and* R_2 *are commutative, then* $\mathfrak{P}_{\rm t}(R) = \{\alpha \in R \mid \alpha = \alpha^2\}$.

Note that Proposition 1.3(2) is, in general, not true when $R \in \mathcal{T}_n$ for $n > 2$ (see Example 1.9).

As a consequence of Definition 1.1, one has the following descriptions:

Lemma 1.1. *For* $e^2 = e \in R$ *the following conditions are equivalent:*

(1) e *is inner Peirce trivial.*

(2) $e\underline{\ell}_R(e) = eR(1-e)$ *is a right ideal of* R.
(3) $\underline{r}_R(e)e = (1-e)Re$ *is a left ideal of* R.
(4) $efge = efege$ *for all idempotents* $f, g \in R$.
(5) $h : R \to eRe$, *defined by* $h(x) = exe$, *is a surjective ring homomorphism.*
(6) $eRtRe = 0$ *for all* $t \in R$ *such that* $ete = 0$.
(7) $ReR \subseteq \underline{\ell}_R((1-e)Re)$.

Lemma 1.2. *For* $e^2 = e \in R$ *the following conditions are equivalent:*

(1) *e is outer Peirce trivial.*
(2) $e\underline{\ell}_R(e) = eR(1-e)$ *is a left ideal of* R.
(3) $\underline{r}_R(e)e = (1-e)Re$ *is a right ideal of* R.
(4) $feg + efege = fege + efeg$ *for all idempotents* $f, g \in R$.

Corollary 1.1. *For* $e^2 = e \in R$ *the following conditions are equivalent:*

(1) *e is Peirce trivial.*
(2) $e\underline{\ell}_R(e) = eR(1-e)$ *is an ideal of* R.
(3) $\underline{r}_R(e)e = (1-e)Re$ *is an ideal of* R.
(4) $e, 1-e \in \mathfrak{P}_{\mathrm{it}}(R)$.

From the above results, one can see that if R is semiprime, then $\mathfrak{P}_{\mathrm{it}}(R) = \mathfrak{P}_{\mathrm{ot}}(R) = \mathfrak{P}_{\mathrm{t}}(R) = \mathcal{B}(R)$.

Lemma 1.3. *Let* $e, f \in R$ *such that* $e = e^2$ *and* $f = f^2$.

(1) $e \in \mathfrak{P}_{\mathrm{it}}(R)$ *implies* $efe = (efe)^2$, $(ef)^2 = (ef)^3$ *and* $(fe)^2 = (fe)^3$.
(2) $e \in \mathfrak{P}_{\mathrm{t}}(R)$ *implies* $fef = (fef)^2$.
(3) $e, f \in \mathfrak{P}_{\mathrm{it}}(R)$ *implies* $efe, fef \in \mathfrak{P}_{\mathrm{it}}(R)$.
(4) *If R is a generalized matrix ring and* $[e_{ij}] \in \mathfrak{P}_{\mathrm{it}}(R)$ *(resp.* $\mathfrak{P}_{\mathrm{ot}}(R), \mathfrak{P}_{\mathrm{t}}(R)$*), then* $e_{ii} \in \mathfrak{P}_{\mathrm{it}}(R_i)$ *(resp.* $\mathfrak{P}_{\mathrm{ot}}(R_i), \mathfrak{P}_{\mathrm{t}}(R_i)$*).*

Lemma 1.4. *Let* $c, e \in R$ *such that* $c = c^2$ *and* $e = e^2$.

(1) $e \in \mathfrak{P}_{\mathrm{it}}(R)$ *if and only if* $\mathfrak{P}_{\mathrm{it}}(eRe) = eRe \cap \mathfrak{P}_{\mathrm{it}}(R)$.
(2) $\mathfrak{P}_{\mathrm{t}}(R) \cap cRc \subseteq \mathfrak{P}_{\mathrm{t}}(cRc)$.
(3) *Assume* $I \trianglelefteq R$. *Then* $\mathfrak{P}_{\mathrm{it}}(I) = I \cap \mathfrak{P}_{\mathrm{it}}(R)$.

Example 1.2. In general, for $c \in \mathfrak{P}_{\mathrm{t}}(R)$, $\mathfrak{P}_{\mathrm{t}}(R) \cap cRc \subsetneq \mathfrak{P}_{\mathrm{t}}(cRc)$. Let R be the 3-by-3 upper triangular matrix ring over a ring A with $e = E_{22} \in R$ and $c = E_{22} + E_{33}$. Then $c \in \mathfrak{P}_{\mathrm{t}}(R)$ and $e \in \mathfrak{P}_{\mathrm{t}}(cRc)$, but $e \notin \mathfrak{P}_{\mathrm{ot}}(R)$. Thus, $\mathfrak{P}_{\mathrm{t}}(R) \cap cRc \subsetneq \mathfrak{P}_{\mathrm{t}}(cRc)$.

In [BHKP] (also see [AvW1] and [AvW2]), it is shown that a ring R has a generalized triangular matrix form if and only if it has a set of left (or right) triangulating idempotents. Such a set is an ordered complete set of orthogonal idempotents which are constructed from $\mathcal{S}_\ell(R)$ and $\mathcal{S}_r(R)$.

The next result and results from Section 2 show that $\mathfrak{P}_{\mathrm{it}}(R)$ and $\mathfrak{P}_{\mathrm{t}}(R)$ can be used to naturally extend the notion of a generalized triangular matrix ring to that of a trivial generalized matrix ring. Moreover, the inherent symmetry in the definitions of $\mathfrak{P}_{\mathrm{it}}(R)$ and $\mathfrak{P}_{\mathrm{t}}(R)$ frees us from the "ordered" condition on sets of idempotents when characterizing these natural extensions.

Proposition 1.2. *(1) $\mathcal{S}_l(R) \cup \mathcal{S}_r(R) \subseteq \mathfrak{P}_{\mathrm{t}}(R)$.*

(2) Let $\{e_1, \ldots, e_n\}$ be a set of left or right triangulating idempotents of R. Then $\{e_1, \ldots, e_n\} \subseteq \mathfrak{P}_{\mathrm{it}}(R)$.

From Proposition 1.10, $\mathcal{B}(R) \subseteq \mathcal{S}_l(R) \cup \mathcal{S}_r(R) \subseteq \mathfrak{P}_{\mathrm{t}}(R) \subseteq \mathfrak{P}_{\mathrm{it}}(R)$ ($\mathfrak{P}_{\mathrm{ot}}(R)$). If R is semiprime these containment relations become equalities by Lemmas 1.4 and 1.5.

Example 1.3. Let A be a ring whose only idempotents are 0 and 1. Assume $0 \neq X, Y \trianglelefteq A$. Using Proposition 1.3 we obtain:

(1) Let $R = \begin{bmatrix} A & X \\ 0 & A \end{bmatrix}$. Then $\mathcal{S}_l(R) \cup \mathcal{S}_r(R) = \mathfrak{P}_{\mathrm{t}}(R) = \{e \mid e = e^2 \in R\}$.

(2) Let $R = \begin{bmatrix} A & X \\ Y & A \end{bmatrix}$ and $XY = 0 = YX$. Then $\mathcal{S}_r(R) = \mathcal{S}_l(R) = \{0, 1\} \subsetneq \mathfrak{P}_{\mathrm{t}}(R) = \{e \mid e = e^2 \in R\}$.

(3) Let B be a subring of A with $X \trianglelefteq B$ and $R = \begin{bmatrix} A & X \\ A/X & B \end{bmatrix}$. Then $\mathcal{S}_r(R) = \mathcal{S}_l(R) = \{0, 1\} \subsetneq \mathfrak{P}_{\mathrm{t}}(R) = \{e \mid e = e^2 \in R\}$.

Section 1 concludes with showing in the next results (1.12–1.16) that given a base ring R, an overring T, and a set $\mathcal{E}$ contained in $\mathfrak{P}_{\mathrm{it}}(T) \cup \mathfrak{P}_{\mathrm{ot}}(T)$, there is a significant transfer of information between R and S, where S is the subring of T generated by R and $\mathcal{E}$. These results indicate the importance of the inner and outer Peirce trivial idempotents.

Let S be an overring of R. We consider the following properties between prime ideals of R and S (see [BPR2, pp. 295–296] or [R1, p. 292]).

(1) *Lying over (LO)*. For any prime ideal P of R, there exists a prime ideal Q of S such that $P = Q \cap R$.

(2) *Going up (GO)*. Given prime ideals $P_1 \subseteq P_2$ of R and Q_1 of S with $P_1 = Q_1 \cap R$, there exists a prime ideal Q_2 of S with $Q_1 \subseteq Q_2$ and $P_2 = Q_2 \cap R$.

(3) *Incomparable (INC)*. Two different prime ideals of S with the same contraction in R are not comparable.

Lemma 1.5. *Let T be a ring, R a subring of T,*

$$\mathcal{E}_{\mathcal{P}} = \{e = e^2 \in T \mid e + \mathcal{P}(T) \text{ is central in } T/\mathcal{P}(T)\},$$

and $S = \langle R \cup \mathcal{E} \rangle_T$, where $\emptyset \neq \mathcal{E} \subseteq \mathcal{E}_{\mathcal{P}}$. Then:

(1) $\mathfrak{P}_{\mathrm{it}}(T) \cup \mathfrak{P}_{\mathrm{ot}}(T) \subseteq \mathcal{E}_{\mathcal{P}}$.
(2) If K is a prime ideal of S, then $R/(K \cap R) \cong S/K$.
(3) LO, GU and INC hold between R and S.

Recall that a ring R is *strongly π-regular* if for each x there is a positive integer n (depending on x) such that $x^n \in x^{n+1}R$.

Theorem 1.1. *Let C be a property of rings such that a ring A has property C if and only if every prime factor of A has property C. Assume T is a ring, R is a subring of T and $S := \langle R \cup \mathcal{E} \rangle_T$, where $\emptyset \neq \mathcal{E} \subseteq \mathcal{E}_{\mathcal{P}}$, with $\mathcal{E}_{\mathcal{P}}$ as in Lemma 1.12. Then R has property C if and only if S has property C. In particular, R is strongly π-regular if and only if S is strongly π-regular.*

See [GW] for the definition of a special radical. Observe that the prime, Jacobson, and nil radicals are included in the collection of special radicals.

Theorem 1.2. *Let R be a subring of a ring T, $\emptyset \neq \mathcal{E} \subseteq \mathcal{E}_{\mathcal{P}}$, and $S = \langle R \cup \mathcal{E} \rangle_T$. Then:*

(1) $\rho(R) = \rho(S) \cap R$, where ρ is any special radical.
(2) The classical Krull dimensions of both S and R are equal.
(3) If S is a von Neumann regular ring, then so is R.

Lemma 1.6. *Let $T \in \mathcal{T}_n$ $(n > 1)$ and*

$$\mathcal{E}_k = \{[t_{ij}] \in T \,|\, t_{kj} \in M_{kj} \text{ for } j \neq k, t_{kk} = 1 \in T_k \text{ and all other entries are zero}\}.$$

Then $\cup_{k=1}^{n} \mathcal{E}_k \subseteq \mathfrak{P}_{\mathrm{it}}(T)$.

Theorem 1.3. *Let $T \in \mathcal{T}_n$ $(n > 1)$, $R = \mathcal{D}(T)$, and $\mathcal{E} = \cup_{k=1}^{n} \mathcal{E}_k$ (as in Lemma 1.15). Then:*

(1) $\langle R \cup \mathcal{E} \rangle_T = S = T$.
(2) R has property C (as in Theorem 1.13) if and only if T has property C.
(3) $\rho(R) = \rho(T) \cap R$.
(4) The classical Krull dimension of both R and T are equal.

This result extends [TLZ, Corollary 3.6] and results in [C].

2. Characterization of $\mathcal{T}_n$

Lemma 2.1. *Let $\{e_1, \ldots, e_n\}$ be a complete set of orthogonal idempotents of R.*

(1) $e_i \in \mathfrak{P}_{\mathrm{it}}(R)$ if and only if $e_iRe_jRe_i = 0$ for all $j \neq i$.
(2) $e_j \in \mathfrak{P}_{\mathrm{ot}}(R)$ if and only if $e_iRe_jRe_k = 0$ for all $i \neq j$ and $k \neq j$.
(3) $\{e_1, \ldots, e_n\} \subseteq \mathfrak{P}_{\mathrm{ot}}(R)$ if and only if $\{e_1, \ldots, e_n\} \subseteq \mathfrak{P}_{\mathrm{t}}(R)$.

Lemma 2.1 shows remarkably that inner and outer Peirce trivial idempotents behave quite differently when they are considered together as a complete set of idempotents although their definition seems very symmetric! Lemma 2.1 shows clearly the equivalence of the first three statements in the next result.

Theorem 2.1. *Let $\{e_1, \ldots, e_n\}$ be a complete set of orthogonal idempotent elements of R. The following conditions are equivalent:*

(1) $R^\pi \in \mathcal{T}_n$.
(2) $e_iRe_jRe_i = 0$, *for all* $i \neq j$.
(3) $\{e_1, \ldots, e_n\} \subseteq \mathfrak{P}_{it}(R)$.
(4) $\mathcal{D}(R^\pi)^-$ *is a right ideal of* R^π.

Observe that from Lemma 1.4 and Theorem 2.2, any property that is preserved by a surjective ring homomorphism passes from a ring in $\mathcal{T}_n$ to its diagonal rings.

Corollary 2.1. *Let* R *be an n-by-n generalized matrix ring. Then the following conditions are equivalent:*

(1) $R \in \mathcal{T}_n$.
(2) *Let* $[a_{ij}], [b_{ij}] \in R$ *with* $[c_{ij}] = [a_{ij}][b_{ij}]$. *Then* $c_{ii} = a_{ii}b_{ii}$ *for all* i *and* j.
(3) $\{E_{ii} \in R \mid i = 1, \ldots, n\} \subseteq \mathfrak{P}_{it}(R)$.

Thus $\mathcal{T}_n$ is exactly the class of n-by-n generalized matrix rings in which the diagonal entries of the product of two matrices is completely determined by the corresponding entries of the diagonals of the factor matrices.

Note that the idempotents in a generalized matrix ring are not characterized. However, for $R \in \mathcal{T}_n$, $e = e^2 \in R$ if and only if $e = [e_{ij}]$, where $e_{ii} = e_{ii}^2$ and $e_{ij} = \sum_{k=1}^n e_{ik}e_{kj}$ for $i \neq j$.

Proposition 2.1. *Assume* $\{e_1, \ldots, e_n\}$ *is a complete set of orthogonal idempotents of* R, *and* $X = \mathcal{D}(R^\pi)^-$.

(1) *If* $\{e_1, \ldots, e_n\} \subseteq \mathfrak{P}_{it}(R)$, *then* $X^n = 0$ *and* $\oplus_{i=1}^n R_i$ *is a homomorphic image of* R^π *with kernel* X, *where* $R_i = e_iRe_i$.
(2) *If* $\{e_1, \ldots, e_n\} \subseteq \mathfrak{P}_t(R)$, *then* $X^2 = 0$.

Example 2.1. Let A be a ring and $X, Y \trianglelefteq A$ such that $X^2 \subseteq Y$. Take

$$R = \begin{bmatrix} A & X & Y \\ X & A & X \\ Y & X & A \end{bmatrix}.$$

Then routine calculation yields:

(1) $E_{22} \in \mathfrak{P}_t(R)$ if and only if $X^2 = 0$.
(2) $\{E_{11}, E_{22}, E_{33}\} \subseteq \mathfrak{P}_{it}(R)$ if and only if $X^2 = 0 = Y^2$.
(3) $\{E_{11}, E_{22}, E_{33}\} \subseteq \mathfrak{P}_t(R)$ if and only if $X^2 = 0 = Y^2$ and $XY = 0 = YX$.

For an illustration of (2) and (3), let B be a ring and $A = B[x,y]/(x^2,y^2)$ and $A = B[x,y]/(x^2,y^2,xy)$, respectively.

The next three results (2.6–2.8) indicate a transfer of important ring properties between $\mathcal{D}(R)$ and $R \in \mathcal{T}_n$.

Corollary 2.2. *Let* $R \in \mathcal{T}_n$ $(n > 1)$. *Then* $\mathcal{D}(R)$ *satisfies each of the following conditions if and only if* R *does so:*

(1) semilocal,
(2) semiperfect,
(3) left (or right) perfect,
(4) semiprimary,
(5) bounded index (of nilpotence).

Corollary 2.6(3) extends [TLZ, Corollary 3.8]. In [ABP], the authors determine several generalizations of the condition that a ring satisfies a polynomial identity. With these generalizations they were able to extend classical theorems by Armendariz and Steenberg, Fisher, Kaplansky, Martindale, Posner and Rowen. Two of these generalizations are: (1) a ring R is an *almost PI-ring* if every prime factor ring of R is a PI-ring; (2) R is an *instrinsically PI-ring* if every nonzero ideal contains a nonzero PI-ideal of R.

Corollary 2.3. *Let $R \in \mathcal{T}_n$ $(n > 1)$. Then:*

(1) $\mathcal{D}(R)$ satisfies a PI if and only if R does so.
(2) If $\mathcal{D}(R)$ is commutative, then R satisfies $(xy - yx)^n = 0$ for all $x, y \in R$.
(3) $\mathcal{D}(R)$ is almost PI if and only if R is almost PI.
(4) If $\mathcal{D}(R)$ is intrinsically PI, then R is instrinsically PI.

Let R be an n-by-n generalized matrix ring, and let

$$\mathrm{UT}(R) \text{ and } \mathrm{LT}(R)$$

be the n-by-n upper and lower generalized triangular matrix rings, respectively, formed from R. The next result shows that elements of $\mathcal{T}_n$ are subdirect products of generalized triangular matrix rings.

Proposition 2.2. *Let $R \in \mathcal{T}_n$ $(n > 1)$. Then there is a ring monomorphism $\psi : R \to \mathrm{UT}(R) \times \mathrm{LT}(R)$ such that R is a subdirect product of $UT(R)$ and $LT(R)$.*

Definition 2.1. Let R be an n-by-n generalized matrix ring. Let R^{la} denote the *lower annihilating* subring

$$\begin{bmatrix} R_1 & M_{12} & \cdots & M_{1n} \\ \underline{r}_{M_{21}}(M_{12}) \cap \underline{\ell}_{M_{21}}(M_{12}) & R_2 & \ddots & \vdots \\ \vdots & \ddots & \ddots & M_{n-1,n} \\ \underline{r}_{M_{n1}}(M_{1n}) \cap \underline{\ell}_{M_{n1}}(M_{1n}) & \cdots & \underline{r}_{M_{n,n-1}}(M_{n-1,n}) \cap \underline{\ell}_{M_{n,n-1}}(M_{n-1,n}) & R_n \end{bmatrix}$$

of R, and let R^{ua} denote the *upper annihilating* subring

$$\begin{bmatrix} R_1 & \underline{r}_{M_{12}}(M_{21}) \cap \underline{\ell}_{M_{12}}(M_{21}) & \cdots & \underline{r}_{M_{1n}}(M_{n1}) \cap \underline{\ell}_{M_{1n}}(M_{n1}) \\ M_{21} & R_2 & \ddots & \vdots \\ \vdots & \ddots & \ddots & \underline{r}_{M_{n-1,n}}(M_{n,n-1}) \cap \underline{\ell}_{M_{n-1,n}}(M_{n,n-1}) \\ M_{n1} & \cdots & M_{n,n-1} & R_n \end{bmatrix}$$

of R.

Note that R^{la} and R^{ua} are subrings of both R and $\overline{R}$. Moreover, if R is the n-by-n matrix ring over a ring A, then R^{la} and R^{ua} are the n-by-n upper and lower triangular matrix rings over A, respectively.

Example 2.2. Let A and B be rings. Let $R = \begin{bmatrix} A\times B & A\times\{0\} \\ A\times B & A\times B \end{bmatrix}$. Then $R^{\mathrm{la}} = \begin{bmatrix} A\times B & A\times\{0\} \\ \{0\}\times B & A\times B \end{bmatrix}$, and $R^{\mathrm{ua}} = \begin{bmatrix} A\times B & \{0\} \\ A\times B & A\times B \end{bmatrix}$.

Lemma 2.2. *Let R be an n-by-n generalized matrix ring. Then* $\mathrm{Cen}(R) = \mathrm{Cen}(\overline{R}) = \{[c_{ij}] \in R \mid c_{ij} = 0$ *if* $i \neq j$, $c_{ii} \in \mathrm{Cen}(R_i)$ *for all* i, *and* $c_{ii}m_{ij} = m_{ij}c_{jj}$ *for all* $m_{ij} \in M_{ij}$ *if* $i \neq j\}$.

Let S be a subring of a ring R. It is well known (see [W, p. 26]) that the (S,S)-bimodule structure of S and R is equivalent to the right T-module structure of S and R, respectively, where $T = S^{\mathrm{op}} \otimes_{\mathbb{Z}} S$, with S^{op} denoting the opposite ring of S.

The next three results (2.12–2.14) indicate the transfer of significant information between an n-by-n generalized matrix ring and certain subrings which are maximal with respect to being in $\mathcal{T}_n$.

In particular, Theorem 2.12 shows that for any ring R with a complete set of orthogonal idempotents $\{e_i\}_{i=1}^n$ $(n > 1)$ there are subrings S containing $\{e_i\}_{i=1}^n$ which are maximal with respect to S^{π} being in $\mathcal{T}_n$ and S_T is right essential in R_T, where $T = S^{\mathrm{op}} \otimes_{\mathbb{Z}} S$. Moreover, this result and its consequences provide a connection between the structure of an arbitrary generalized matrix ring and the structure of rings in $\mathcal{T}_n$ (see Question B in the introduction).

Theorem 2.2. *Let R be an n-by-n generalized matrix ring, and S denotes either R^{la} or R^{ua}.*

(1) *S is a subring of R maximal with respect to being in $\mathcal{T}_n$.*
(2) *Let $0 \neq y \in R$. Then either $0 \neq syE_{jj} \in S$ or $0 \neq E_{ii}yt \in S$ for some $s, t, E_{ii}, E_{jj} \in S$.*
(3) *Every nonzero (S,S)-bisubmodule of R has nonzero intersection with S. Thus every nonzero ideal of R has nonzero intersection with S, and S_T is right essential in R_T where $T = S^{\mathrm{op}} \otimes_{\mathbb{Z}} S$.*
(4) $\mathrm{Cen}(R) = \mathrm{Cen}(R^{\mathrm{la}}) \cap \mathrm{Cen}(R^{\mathrm{ua}}) \subseteq \mathrm{Cen}(\mathcal{D}(R))$.

(5) $\mathrm{U}(\overline{R}) = \{u+x \mid u \in \mathrm{U}(\mathcal{D}(R)) \text{ and } x \in \mathcal{D}(\overline{R})^-\}$, *and*
$\mathrm{U}(S) = \{u+y \mid u \in \mathrm{U}(\mathcal{D}(R)) \text{ and } y \in \mathcal{D}(S)^-\} \subseteq \mathrm{U}(R)$.

(6) *If S is a subdirectly irreducible ring (i.e., the intersection of all nonzero ideals of S is nonzero), then so is R.*

Note that if $n = 2$, then in Theorem 2.12(2), there is no need for the E_{ii} and E_{jj}. Also, Theorem 2.12(6) is an addition by the authors. It is an immediate consequence of Theorem 2.12(3).

Corollary 2.4. *Let R be an n-by-n generalized matrix ring, S denotes R^{la} or R^{ua} and $T = S^{\mathrm{op}} \otimes_{\mathbb{Z}} S$. Then:*

(1) *S is maximal among subrings Y of R for which $\{E_{ii}\}_{i=1}^n \subseteq \mathfrak{P}_{\mathrm{it}}(R)$.*
(2) *The sum of the minimal ideals of S equals $\mathrm{Soc}(S_T) = \mathrm{Soc}(R_T)$.*
(3) *The uniform dimension of ${}_SS_S$ equals the uniform dimension of ${}_SR_S$ equals the uniform dimension of S_T equals the uniform dimension of R_T.*

The next result demonstrates that useful information can be transferred from the diagonal rings R_i of a generalized matrix ring R to R itself via Theorems 1.16 and 2.12. Recall from [R2] and [CR] that an n-by-n ($n > 1$) matrix ring over a strongly π-regular ring is not, in general, a strongly π-regular ring.

Corollary 2.5. *Let R be an n-by-n generalized matrix ring, and $S = R^{\mathrm{la}}$ or R^{ua}. If $\mathcal{D}(R)$ is strongly π-regular, then for each $0 \neq y \in R$ either:*

(1) *$y \in S$, in which case $y^n \in y^{n+1}S \subseteq y^{n+1}R$ for some positive integer n; or*
(2) *$y \notin S$, in which case either $0 \neq syE_{jj} \in S$ and $(syE_{jj})^m \in (syE_{jj})^{m+1}S \subseteq (syE_{jj})^{m+1}R$, or $0 \neq E_{ii}yv \in S$ and $(E_{ii}yv)^k \in (E_{ii}yv)^{k+1}S \subseteq (E_{ii}yv)^{k+1}R$ for some $s, v, E_{ii}, E_{jj} \in S$ and positive integers k, m, n.*

Thus if $\mathcal{D}(R)$ is strongly π-regular, then R is "almost" strongly π-regular.

The final two results are by the authors [BH], but were motivated by Theorem 2.12.

Proposition 2.3. *Let $R = \begin{bmatrix} A & M \\ N & B \end{bmatrix}$ be a generalized matrix ring . Take $T = \begin{bmatrix} A & 0 \\ N & B \end{bmatrix}$ and $T' = \begin{bmatrix} A & M \\ 0 & B \end{bmatrix}$.*

(1) *If $\ell_M(N) = 0_M$, then $T = R^{ua}$ and $T_T \leq^{ess} R_T$. Additionally, if $\ell_B(N) = 0_B$ (i.e., ${}_BN$ is faithful), then $T_T \leq^{den} R_T$, so $Q(T) = Q(R)$.*
(2) *If $r_M(N) = 0_M$, then $T = R^{ua}$ and ${}_TT \leq^{ess} {}_TR$. Additionally, if $r_A(N) = 0_A$ (i.e., N_A is faithful), then ${}_TT \leq^{den} {}_TR$, so $Q_\ell(T) = Q_\ell(R)$.*
(3) *If $\ell_N(M) = 0_N$, then $T' = R^{la}$ and $T'_{T'} \leq^{ess} R_{T'}$. Additionally, if $\ell_A(M) = 0_A$ (i.e., ${}_AM$ is faithful), then $T'_{T'} \leq^{den} R_{T'}$, so $Q(T') = Q(R)$.*

(4) *If* $\underline{r}_N(M) = 0_N$, *then* $T' = R^{la}$ *and* ${}_{T'}T' \leq^{ess} {}_{T'}R$. *Additionally, if* $\underline{r}_B(M) = 0_B$ *(i.e.,* M_B *is faithful), then* ${}_{T'}T' \leq^{den} {}_{T'}R$, *so* $Q_\ell(T') = Q_\ell(R)$.

Corollary 2.6. *Let* R, T, *and* T' *be as in Proposition 2.14. Assume* $M = Hom_A(N,A)$ *and* $B = End_A(N)$. *Then:*

(1) $T = R^{ua}$ *and* $T_T \leq^{den} R_T$, *so* $Q(T) = Q(R)$.
(2) *If* $\underline{r}_A(N) = 0_A$ *(i.e.,* N_A *is faithful), then* $T = R^{ua}$ *and* ${}_TT \leq^{den} {}_TR$, *so* $Q_\ell(T) = Q_\ell(R)$.
(3) *If* N_A *is a generator, then* $T' = R^{la}$ *and* $T'_{T'} \leq^{den} R_{T'}$, *so* $Q(T') = Q(R)$.
(4) *If* N_A *is torsionless, then* $T' = R^{la}$ *and* ${}_{T'}T' \leq^{ess} {}_{T'}R$.

Acknowledgements

The first author wishes to thank the organizers of the 24th International Conference on Nearrings, Nearfields, and Related Topics for inviting him to give a plenary talk and to submit a paper for the proceedings of that conference.

References

[ABP] E. P. Armendariz, G. F. Birkenmeier and J. K. Park, Ideal intrinsic extensions with connections to PI-rings, J. Pure Appl. Algebra **213** (2009), 1756–1776. Corrigendum **215** (2011), 99–100.

[ABvW] P. N. Anh, G. F. Birkenmeier and L. van Wyk, Idempotents and structures of rings, Linear and Multilinear Algebra, DOI: 10.1080/03081087.2015.1134429.

[AvW1] P. N. Ánh and L. van Wyk, Automorphism groups of generalized triangular matrix rings, Linear Algebra Appl. **434** (2011), 1018–1025.

[AvW2] P. N. Ánh and L. van Wyk, Isomorphisms between strongly triangular matrix rings, Linear Algebra Appl. **438** (2013), 4374–4381.

[BH] G. F. Birkenmeier and B. J. Heider, Essential subrings of Morita contexts, preprint.

[BHKP] G. F. Birkenmeier, H. E. Heatherly, J. Y. Kim and J. K. Park, Triangular matrix representations, J. Algebra **230** (2000), 558–595.

[BPR2] G. F. Birkenmeier, J. K. Park, S. T. Rizvi, Extensions of Rings and Modules, Birkhäuser/Springer, New York, 2013.

[C] H. Chen, Strongly π-regular Morita contexts, Bull. Korean Math. Soc. **40** (2003), 91–99.

[CR] F. Cedo and L. H. Rowen, Addendum to "Examples of semiperfect rings", Israel J. Math. **107** (1998), 343–348.

[GW] B. J. Gardner and R. Wiegandt, Radical Theory of Rings, Marcel Dekker, New York, 2004.

[H] A. Haghany, Hopficity and co-hopficity for Morita contexts, Comm. Algebra **27** (1999), 477–492.

[KT] P. A. Krylov and A. A. Tuganbaev, Modules over formal matrix rings (Russian), Fundam. Prikl. Mat. **15**(8) (2009), 145–211; translation in J. Math. (N.Y.) **171**(2) (2010), 248–295.

[L] J. Lewallen, Left self distributively generated structural matrix rings, Comm. Algebra **33** (2005), 2865–2877.

[M] K. Morita, Duality for modules and its application to the theory of rings with minimum conditions, Science Reports of the Tokyo Kyoiku Daigoku Sect. A.6 (1958), 83–142

[R1] L. H. Rowen, Ring Theory I, Academic Press, Boston, 1988.

[R2] L. H. Rowen, Examples of semiperfect rings, Israel J. Math. **65** (1989), 273–283.

[S] B. Stenström, The maximal ring of quotients of a generalized matrix ring, in "Universale Algebren und Theorie der Radikale, Studien zur Algebra und ihre Anwendungen", Akademie-Verlag, Berlin, 1976, 65–67.

[TLZ] G. Tang, C. Li and Y. Zhou, Study of Morita contexts, Comm. Algebra **42** (2014), 1668–1681.

[W] R. Wisbauer, Modules and Algebras: Bimodule Structure and Group Actions on Algebras, Addison Wesley Longman, Harlow, 1996.

Introduction to matrix nearrings

J. H. Meyer

Department of Mathematics & Applied Mathematics,
University of the Free State,
PO Box 339, Bloemfontein, 9300, South Africa,
E-mail: meyerjh@ufs.ac.za

1. Preliminary remarks

The rationale behind the style in which this chapter is written is simply to provide the reader with an overview of ideas and results regarding matrix nearrings. All the material in this chapter have already been published and an extensive list of references is given where these results and their proofs can be found. It was attempted to put together some of the most important and useful results, and in such a way that the interested reader might easily pursue new avenues of research — and there are certainly many opportunities in this regard.

2. Introduction

If one simply generalises the concept of a matrix ring to that of a matrix nearring (the set $\mathcal{M}_n(R)$ of square arrays with entries from a nearring R, endowed with the usual operations of matrix addition and multiplication), it turns out that if R has an identity, then $\mathcal{M}_n(R)$, $n > 1$, is a nearring if and only if R is a ring[3]. The only interesting case left was to consider nearrings without identity. But even in this case, the nearring R has to be very special for $\mathcal{M}_n(R)$, $n > 1$, to be a nearring in this sense. See Refs. 5 and 9.

In 1986, Meldrum and van der Walt[14] used a different strategy, namely to rather consider matrices (over nearrings) as mappings in order to define the notion of a matrix nearring over an arbitrary nearring with identity. Certain elementary maps were used to generate these matrix nearrings. These elementary maps imitate the well-known elementary matrices

$$rE_{ij} = \begin{pmatrix} 0 \cdots 0 \cdots 0 \\ \cdot \quad \cdot \quad \cdot \\ \cdot \quad \cdot \quad \cdot \\ 0 \cdots 0 \cdots 0 \\ \cdot \quad \cdot \quad \cdot \\ \cdot \quad \cdot \quad \cdot \\ 0 \cdots 0 \cdots 0 \end{pmatrix}$$

where r (from a ring R) occupies the (i, j)-th entry of a square $n \times n$ array, and the other entries are zero. The idea in Ref. 14 was to consider the elementary matrices rE_{ij} as maps $f_{ij}^r : R^n \to R^n$; $f_{ij}^r \alpha = \iota_i(r\pi_j\alpha)$, where, in this case, R^n denotes the direct sum of n copies of the additive group of a right nearring R with identity, and ι_i and π_j denote the usual i-th co-ordinate injection function and the j-th co-ordinate projection function, respectively, and $r \in R$.

Definition 2.1. Let $n \geq 1$ be an integer and R a right nearring with identity. The $n \times n$ *matrix nearring over* R, denoted $\mathcal{M}_n(R)$, is defined to be the subnearring of the nearring $M(R^n)$, generated by $\{f_{ij}^r : r \in R, 1 \leq i, j \leq n\}$. (Here, $M(R^n)$ denotes the nearring of all selfmaps of the group $(R^n, +)$, with pointwise addition and composition.)

It should be noted that this approach can be modified for nearrings without identity. See Ref. 17 p. 12 and Ref. 7 Definition 1.1. We will, however, focus only on right near-rings with identity for the remainder of this chapter.

3. Some Basic Results

A summary of a few immediate and elementary consequences of the definition is given below. The proofs are all straight forward.

Theorem 3.1.[17] *Let R be a right near-ring with identity* 1. *Then*

(1) $\mathcal{M}_n(R)$ is a right nearring with identity $f_{11}^1 + f_{22}^1 + \cdots + f_{nn}^1$.
(2) If R happens to be a ring, then $\mathcal{M}_n(R)$ is (isomorphic to) the usual full $n \times n$ matrix ring over R.
(3) R is zero-symmetric if and only if $\mathcal{M}_n(R)$ is zero-symmetric.
(4) R is abelian if and only if $\mathcal{M}_n(R)$ is abelian.
(5) R is abstract affine (aa) if and only if $\mathcal{M}_n(R)$ is aa.
(6) If R is distributively generated (dg), then $\mathcal{M}_n(R)$ is dg.

A first approach to a new theory is to see which results in the existing theory carry over to the new one. The following result (which is known to be true in ring theory) was an open problem for quite a number of years, and only solved (positively) in 1997:

Theorem 3.2.[18] *Let R be a nearring. Then $\mathcal{M}_{mn}(R) \cong \mathcal{M}_n(\mathcal{M}_m(R))$ for all positive integers m and n.*

4. Modules over matrix nearrings and primitivity

In this section, all nearrings will be zero-symmetric. Given a (left) nearring module $_RG$ (sometimes called an R-group), it is natural to ask how we could define an action of $\mathcal{M}_n(R)$ on $G^n = \oplus_{i=1}^n (G, +)$ that will turn G^n into a left $\mathcal{M}_n(R)$-module. It turns out that, in order to have an action as natural as possible, it helps if the $_RG$ also satisfies the following property, which is a generalisation of a *monogenic* module:

Definition 4.1. (Ref. 12 Definition 1.2) For a nearring R, the module $_RG$ is called *locally monogenic* if, for each finite subset $H \subseteq G$, there exists $g \in G$ such that $H \subseteq Rg$.

Note that if G is finite, then the concepts 'locally monogenic' and 'monogenic' are identical.

For a locally monogenic module ${}_RG$ we can (naturally) define an action of $\mathcal{M}_n(R)$ on G^n as follows: Let $U \in \mathcal{M}_n(R)$ and $\langle g_1, g_2, \ldots, g_n \rangle \in G^n$. Then there are $g \in G$ and $r_1, r_2, \ldots, r_n \in R$ such that $r_i g = g_i$, $i = 1, 2, \ldots, n$. Then

$$\begin{aligned} &U\langle g_1, g_2, \ldots, g_n \rangle \\ &= U\langle r_1 g, r_2 g, \ldots, r_n g \rangle \\ &:= \langle s_1 g, s_2 g, \ldots, s_n g \rangle, \text{ where } U\langle r_1, r_2, \ldots, r_n \rangle = \langle s_1, s_2, \ldots, s_n \rangle \in R^n. \end{aligned}$$

This action ensures that G^n is a (left) $\mathcal{M}_n(R)$-module. We agree to call it the 'natural action'. The next two results concern type 2 modules:

Theorem 4.1. *(Ref. 28 Proposition 3.2) If Γ is an $\mathcal{M}_n(R)$-module of type 2, then $\Gamma \cong_{\mathcal{M}_n(R)} G^n$, where G is an R-module of type 2, and the action of $\mathcal{M}_n(R)$ on G^n is the natural one.*

Theorem 4.2. *(Ref. 27 Theorem 3.10) For any $n \geq 1$, $\mathcal{M}_n(R)$ is 2-primitive if and only if R is 2-primitive.*

It is well-known that 2-primitive nearrings are dense subnearrings of suitable centralizer nearrings (Ref. 10 Theorem 3.35). This phenomenon carries over to matrix nearrings:

Theorem 4.3. *(Ref. 17 Propositions 2.22 and 2.24) If R is not a ring, and 2-primitive on G, then $M_n(R)$ is dense in $M_A(G^n)$, where $A = Aut_{\mathcal{M}_n(R)}(G^n)$. Furthermore, if A has only finitely many orbits on G^n, then $\mathcal{M}_n(R) = M_A(G^n)$.*

Theorem 4.4. *(Ref. 17 Corollary 2.25) If F is a finite nearfield (not a division ring), then $\mathcal{M}_n(F) = M_F(F^n)$, where the action of F on F^n is considered to be multiplication on the right ($\langle \alpha_1, \alpha_2, \ldots, \alpha_n \rangle \alpha := \langle \alpha_1 \alpha, \alpha_2 \alpha, \ldots, \alpha_n \alpha \rangle$, for all $\alpha, \alpha_1, \alpha_2, \ldots, \alpha_n \in F$).*

Unfortunately, the same results do not hold in general for the type 0 (resp. 0-primitive) situations. There do exist type 0 modules ${}_{\mathcal{M}_n(R)}\Gamma$ which cannot be obtained via the natural action (Ref. 12 Example 2.3). Because of the "unpredictable" behaviour of type 0 modules, it is possible to construct a nearring R for which $\mathcal{J}_0(R) = \mathcal{J}_2(R)$, but $\mathcal{J}_0(\mathcal{M}_n(R)) \subsetneq \mathcal{J}_2(\mathcal{M}_n(R))$ for $n \geq 2$ (Ref. 12 Example 3.3).

It is still an open question whether the following theorem holds for general (zero-symmetric) nearrings:

Theorem 4.5. *(Ref. 6 Theorem 1) If $\mathcal{M}_n(R)$ has the DCC on left $\mathcal{M}_n(R)$ subgroups, then $\mathcal{M}_n(R)$ is 0-primitive if and only if R is 0-primitive.*

One of the useful applications of matrix nearrings was to resolve a long standing open problem of G. Betsch. In the early 1970's, Betsch proved the following theorem:

Theorem 4.6.[4] *Suppose R is 2-primitive on the R-module G. If R contains an idempotent e of rank 1, then R has a minimal left ideal, or else we have*

$$e \neq 1 \text{ and } Ann_R(G \setminus eG) = \{0\}. \qquad (*)$$

It was not known whether the exceptional case (*) can actually occur. In 1987, the construction of an example in matrix nearrings showed that (*) can indeed occur:

Example 4.1.[19] Consider the right near-field $(R,+,\circ)$, where $R = \mathbb{Q}$ (the rational functions over $\mathbb{Q}$), $+$ is defined in the standard way, and $\circ$ is defined by

$$\frac{f(x)}{g(x)} \circ \frac{h(x)}{k(x)} = \begin{cases} \frac{f(x+\partial h-\partial k)}{g(x+\partial h-\partial k)} \cdot \frac{h(x)}{k(x)} & \text{if } \frac{h(x)}{k(x)} \neq 0 \\ 0 & \text{otherwise.} \end{cases}$$

Here, $\cdot$ denotes the standard multiplication in the field $(Q(x),+,\cdot)$. See (Ref. 20 Example 8.29) for further details on this near-field.

It follows that the matrix nearring $\mathcal{M}_2(R)$ is 2-primitive on $G := R^2$, $e = f_{11}^1$ is an idempotent of rank 1, and it is not equal to the identity $f_{11}^1 + f_{22}^1$. But still, $\mathrm{Ann}_{\mathcal{M}_2(R)}(G \setminus eG) = \{0\}$.

5. Left and two-sided ideals

In order to study the 0-primitivity of a nearring R, it is useful to have as much information as possible about the minimal and maximal left ideals of R.

When we consider a matrix ring $\mathcal{M}_n(D)$ over a division ring D, it follows easily that those matrices with arbitrary elements in a specific column (say the k-th column), and zeroes elsewhere, form a minimal left ideal in $\mathcal{M}_n(D)$. This minimal left ideal can be described as the left ideal generated by the matrix E_{1k} with 1 in position $(1,k)$, and zeroes elsewhere. In our more general notation for nearrings, this would be the left ideal generated by f_{1k}^1, denoted by $\mathrm{li}\langle f_{1k}^1 \rangle$. We get the following surprising result:

Theorem 5.1. *(Ref. 16 Theorems 15 and 16) If the nearring R contains at least one non-distributive element, then $li\langle f_{1k}^1 \rangle$ is not a minimal left ideal of $\mathcal{M}_n(R)$ if $n \geq 2$.*

Related to this, we also have

Theorem 5.2. *(Ref. 17 Proposition 2.29) Let $n \geq 2$ and let F be an infinite nearfield that is not a division ring. Then $\mathcal{M}_n(F)$ does not satisfy the DCC for left ideals.*

In fact, it is conjectured that given the conditions of this theorem, then $\mathcal{M}_n(F)$ does not contain any minimal left ideals. See Example 3.8, for which this conjecture is known to be true (see Ref. 19).

We turn to maximal left ideals. The maximal left ideals of a matrix ring is characterized as follows:

Theorem 5.3.[24] *Let L be a maximal left ideal of a ring R and let $\alpha \in R^n \setminus L^n$. Then $(L^n : \alpha) := \{U \in \mathcal{M}_n(R) \ : \ U\alpha \in L^n\}$ is a maximal left ideal of $\mathcal{M}_n(R)$. Moreover, every maximal left ideal of $\mathcal{M}_n(R)$ can be obtained in this way.*

The best generalization to nearrings regarding maximal left ideals, is given by

Theorem 5.4. *(Ref. 12 Theorems 4.2 and 4.3) Let* $\alpha = \langle s_1, \ldots, s_n \rangle \in R^n$. *Let M be the R-subgroup of R generated by* $\{s_1, \ldots, s_n\}$ *and let K be a maximal R-ideal of M. (Note that K need not be an R-ideal of R.) Then* $(K^n : \alpha)$ *is a maximal left ideal of* $\mathcal{M}_n(R)$.

Conversely, suppose $\mathcal{L}$ *is a maximal left ideal of* $\mathcal{M}_n(R)$. *Then* $\mathcal{L}$ *is of the form described above, or* $\mathcal{L}$ *has the following density property: there exists an* $m \geq 1$ *such that for any m elements* $\alpha_1, \ldots, \alpha_m \in R^n$, *and any* $U \in \mathcal{M}_n(R)$, *there exists* $L \in \mathcal{L}$ *such that* $U\alpha_i = L\alpha_i$, $i = 1, \ldots, m$.

It is conjectured that the density part in the second paragraph is not necessary, i.e., all maximal left ideals can be obtained as described in the first paragraph.

We now turn to (two-sided) ideals. For an ideal I of R, there are two natural ways (see Ref. 26) to construct a corresponding ideal in $\mathcal{M}_n(R)$, namely

$$I^+ := id\langle f_{ij}^a : \ a \in I \ \ and \ \ 1 \leq i,j \leq n \rangle$$

and

$$I^* := (I^n : R^n) = \{U \in \mathcal{M}_n(R) \ : \ U\alpha \in I^n \ \ for \ \ all \ \ \alpha \in R^n\}.$$

It is easy to see that $I^+ \subseteq I^*$ and proper containment is possible (see, for example, Refs. 26 and 12). It is known that $\mathcal{J}_2(\mathcal{M}_n(R)) = (\mathcal{J}_2(R))^*$ and $\mathcal{J}_0(\mathcal{M}_n(R)) \subseteq (\mathcal{J}_0(R))^*$ (Ref. 27 Theorem 4.4 and Ref. 17 Theorem 2.34). In Ref. 12 an example is given to show that the containment for the $\mathcal{J}_0$-case can be proper. The number of ideals (if any) that are situated strictly between I^+ and I^* could be an indication of the "size" of the gap $I^+ \subsetneqq I^*$. Hence the notion of an intermediate ideal was introduced:

Theorem 5.5. *(Ref. 11 Definition 2.1) An ideal* $\mathcal{I}$ *of* $\mathcal{M}_n(R)$ *such that* $I^+ \subsetneqq \mathcal{I} \subsetneqq I^*$ *for some ideal I of R, is called an intermediate ideal of* $\mathcal{M}_n(R)$.

An immediate consequence of this definition is that $\mathcal{I}$ *is an intermediate ideal of* $\mathcal{M}_n(R)$ *if and only if* $\mathcal{I} \notin \{I^+ \ : I \ an \ ideal \ of R\} \cup \{I^* \ : I \ an \ ideal \ of R\}$.

Several examples of intermediate ideals are known (see Refs. 15 and 11). In particular, an example of a finite, abelian (zero-symmetric) nearring R is constructed in Ref. 13 for which

$$(\mathcal{J}_0(R))^+ \subsetneqq \mathcal{J}_0(\mathcal{M}_2(R)) \subsetneqq (\mathcal{J}_0(R))^*$$

holds, i.e., it is possible for the $\mathcal{J}_0$-radical of a matrix nearring to be intermediate.

6. Conclusive remarks

The roles that matrix nearrings play in some other algebraic structures have been investigated. For example, some relations between matrix nearrings and certain centraliser nearrings have been investigated in Refs. 23 and 21. In Ref. 25, van der Walt describes an important connection between matrix nearrings (and variants thereof) and near-vector spaces, as defined by André in Ref. 1.

Group nearrings and polynomial nearrings have also been defined using a functional approach such as for matrix nearrings. The interested reader should consult Refs. 8 and 2 where the groundwork has been laid for these respective areas of research.

Some substantial and interesting results have also been obtained in the area of *generalised matrix nearrings*, a concept that was coined by Smith, in Ref. 22.

Bibliography

1. J. André, Algebra über fastkörpern, *Math. Z.* **136** (1974), 295–313.
2. S. W. Bagley, Polynomial near-rings, distributor and $\mathcal{J}_2$ ideals of generalised centraliser near-rings. Dissertation (Texas A&M University, 1993).
3. J. C. Beidleman, On near-rings and near-ring modules, Dissertation (Pennsylvania State University, 1964).
4. G. Betsch, Some structure theorems on 2-primitive near-rings. In "Rings, Modules and Radicals, Hungary, 1971," *Colloq. Math. Soc. János Bolyai*, North-Holland, **6** (1973), 73–102.
5. H. E. Heatherly, Matrix near-rings, *J. London Math. Soc.* **7** (1973), 355–356.
6. W.-F. Ke and J. H. Meyer, Matrix near-rings and 0-primitivity, *Monatsh. Math.* **165** (2012), 353–363.
7. W.-F. Ke, J. H. Meyer and G. Wendt, Matrix maps over planar nearrings, *Proc. Royal Soc. Edinburgh* (Series A) **140** (2010), 83–99.
8. L. R. le Riche, J. D. P. Meldrum and A. P. J. van der Walt, On group near-rings, *Arch. Math.* **52** (1989), 132–139.
9. S. Ligh, A note on matrix near-rings, *J. London Math. Soc.* **11** (1975), 383–384.
10. J. D. P. Meldrum, Near-rings and their links with groups, Pitman Advanced Publishing Program, 1985.
11. J. D. P. Meldrum and J. H. Meyer, Intermediate ideals in matrix near-rings, *Comm. Algebra* **24**(5) (1996), 1601–1619.
12. J. D. P. Meldrum and J. H. Meyer, Modules over matrix near-rings and the $\mathcal{J}_0$-radical, *Monatsh. Math.* **112** (1991), 125–139.
13. J. D. P. Meldrum and J. H. Meyer, The $\mathcal{J}_0$-radical of a matrix nearring can be intermediate. *Canad. Math. Bull.* **40**(2) (1997), 198–203.
14. J. D. P. Meldrum and A. P. J. van der Walt, Matrix near-rings, *Arch. Math.* **47** (1986), 312–319.
15. J. H. Meyer, Chains of intermediate ideals in matrix near-rings, *Arch. Math.* **63** (1994), 311–315.
16. J. H. Meyer, Left ideals in matrix near-rings, *Comm. Algebra* **17** (1989), 1315–1335.
17. J. H. Meyer, Matrix near-rings, Dissertation, University of Stellenbosch, South Africa, 1986.
18. J. H. Meyer, On the near-ring counterpart of the matrix ring isomorphism $\mathcal{M}_{mn}(R) \cong \mathcal{M}_n(\mathcal{M}_m(R))$, *Rocky Mountain J. Math.* **27(1)** (1997), 231–240.
19. J. H. Meyer and A. P. J. van der Walt, Solution of an open problem concerning 2-primitive near-rings, Near-rings and Near-fields (G. Betsch, ed.), Elsevier Science Publishers B.V. (North-Holland), 1987, 185–191.
20. G. Pilz, Near-rings, Revised edition, North-Holland, 1983.
21. R. S. Rao, On near-rings with matrix units, *Quaestiones Math.* **17** (1994), 321–332.
22. K. C. Smith, Generalized matrix near-rings, *Comm. Algebra* **24**(6) (1996), 2065–2077.
23. K. C. Smith and L. van Wyk, When is a centralizer near-ring isomorphic to a matrix near-ring?, *Comm. Algebra* **24**(14) (1996), 4549–4562.
24. D. R. Stone, Maximal left ideals and idealizers in matrix rings, *Canad. J. Math.* **32** (1980), 1397–1410.
25. A. P. J. van der Walt, Matrix near-rings contained in 2-primitive near-rings with minimal subgroups, *J. Algebra* **148** (1992), 296–304.

26. A. P. J. van der Walt, On two-sided ideals in matrix near-rings. In *Near-rings and Near-fields*, G. Betsch, ed., North-Holland, 1987, 267–272.
27. A. P. J. van der Walt, Primitivity in matrix near-rings, *Quaestiones Math.* **9** (1986), 459–469.
28. A. P. J. van der Walt and L. van Wyk, The $\mathcal{J}_2$-radical in structural matrix nearrings, *J. Algebra* **123** (1989), 248–261.

Recent developments in Boolean nearrings

Y. V. Reddy

Retd. Professor, Department of Mathematics, Acharya Nagarjuna University, Guntur A.P., India
E-mail: yvreddy47@gmail.com

1. Preliminaries

A (right) near-ring is a nonempty set N together with two binary operations addition and multiplication such that $(N,+)$ is a group (not necessarily abelian), $(N,\cdot)$ is a semigroup, and $(k+l)m = km + lm$ (right distributive law) for all $k,l,m \in N$. Throughout we consider right near-rings. So, by a near-ring we mean a right near-ring. For notation, terminology and basic concepts we refer to Pilz[4].

2. Boolean near-rings

Boolean near-rings were studied and basic properties developed in Clay[2].

Definition 2.1. A near-ring N is called a Boolean near-ring if $n = n^2$ for all $n \in N$.

Example 2.1. 1) Every Boolean ring is a Boolean near-ring.
2) Let N be a near-ring such that for all $x,y \in N : xy = x$ if $y \notin N_c$ and $xy = x_c$ if $y \in N_c$ (x_c is the constant part of x), where N_c is the constant part of N. This is a Boolean near-ring. Such a near-ring is called an **almost trivial near-ring.**
2(i) Every constant near-ring N (That is, $xy = x$ for all $x,y \in N$) is an almost trivial near-ring. Hence N is a Boolean near-ring.
2(ii) Every (zero-symmetric) trivial near-ring N (ie. for all $x,y \in N : xy = x$ if $y \neq 0$ and $xy = 0$ if $y = 0$) is an almost trivial near-ring. So it is a Boolean near-ring.
2(iii) We will give an example of an almost trivial near-ring which is neither constant nor trivial.
3) [Clay-Lawer[2]] Let $(R,+,\wedge)$ be a Boolean ring. Let x be an arbitrary but a fixed element in R. Define the multiplication '$\circ$' on R by $a \circ b = a \wedge (b + x + b \wedge x)$. Then $(R,+,\circ)$ is a Boolean near-ring and it is a ring if and only if $x = 0$

Lemma 2.1. (Reddy and Murty[12]): If N is a Boolean near-ring, then $ab = aba$ for all $a,b \in N$.

Proof. For $a,b \in N$, consider $(ab - aba)a = aba - aba^2 = aba - aba = 0$.
From this it follows $a(ab - aba) = [a(ab - aba)]^2 = a(ab - aba)a(ab - aba) = a0$.
Similarly, we can prove that $ab(ab - aba) = ab0$.
Then $ab - aba = (ab - aba)^2 = (ab - aba)(ab - aba) = ab(ab - aba) - aba(ab - aba) = ab0 - aba0 = (ab - aba)0 = (ab - aba)(ab - aba)a = (ab - aba)a = 0$. Hence $ab = aba$.
□

Definition 2.2. (Subrahmanyam[9]): A near-ring N is said to be (right) weakly commutative if $abc = acb$ for all $a, b, c \in N$.

Theorem 2.1. Every Boolean near-ring is weakly commutative.

Proof. Let N be a Boolean near-ring and let $a, b, c \in N$. Then by repeated use of Lemma 2.1, we get $abc - acb = (a - ac)bc = (a - ac)b(a - ac)c = (a - ac)b0 = ab0 - acb0$.

Replacing b by bc and using Lemma 2.1 we get, $abc - acb = abc0 - acb0$. Therefore $ab0 = abc0$. Taking $b = a$, we get that $aa0 = aac0$. This implies that $a0 = ac0$ for all $c \in N$. Therefore $abc - acb = a0 - a0 = 0$. Hence $abc = acb$. □

Corollary 2.1. A Boolean near-ring with multiplicative identity 1 is a Boolean ring.

Remark 2.1. Theorem 1.1 was proved for zero-symmetric Boolean near-rings by Murty[3].

3. One sided ideals in Boolean near-rings

It is well-known that every one sided ideal in a Boolean ring is a two sided ideal. In this section we prove that every left ideal of a Boolean near-ring is a two sided ideal. We present an example to establish that the above result is not true, in general, for right ideals. We show that every $N-$subgroup of N, where N is a zero-symmetric Boolean near-ring, is an ideal of N.

Theorem 3.1. Every left ideal of a Boolean near-ring N is an ideal of N.

Proof. Let L be a left ideal of N. To show that L is an ideal it suffices to show that $LN \subseteq L$. Let $l \in L, n \in N$. Decompose l and n as $l = l_0 + l_c, n = n_0 + n_c$, where $l_0, n_0 \in N_0$, $l_c, n_c \in N_c(say(i))$. Since L is a left ideal, we have that $N_0L \subseteq L$, for $m_0l' = m_0(0 + l') - m_00 \in L$ for all $m_0 \in N_0, l' \in L(say(ii))$. Now $l = l^2 = (l_0 + l_c)l = l_0l + l_c$. By (ii) $l, l_0l \in L$ and it follows that $l_c \in L$ and hence $l_0 \in L$. But $ln = (l_0 + l_c)n = l_0n + l_cn = l_0n + l_c$. By weak commutativity, we have that $ln = l_0n + l_c = l_0(n_0 + n_c) + l_c = l_0(n_0 + n_c)l_0 + l_c = l_0(n_0l_0 + n_c) + l_c$. Again by weak commutativity, $l_0n_c = 0$. Since $n_0l_0 \in L$, we have that $l_0(n_0l_0 + n_c) = l_0(n_0l_0 + n_c) - l_0n_c \in L$. Thus, $ln = l_0(n_0l_0 + n_c) + l_c \in L$, since $l_0(n_0l_0 + n_c), l_c \in L$. From this it follows that L is an ideal. □

We provide an example to show that Theorem 3.1 is not true, in general, for right ideals. The following example is taken from Pilz[4].

Example 3.1. Let $N = D_8 = \{0, a, 2a, 3a, b, a+b, 2a+b, 3a+b\}$ be the dihedral group of order 8. Then N is a Boolean near-ring with the following addition and multiplication.

+	0	a	$2a$	$3a$	b	$a+b$	$2a+b$	$3a+b$
0	0	a	$2a$	$3a$	b	$a+b$	$2a+b$	$3a+b$
a	a	$2a$	$3a$	0	$a+b$	$2a+b$	$3a+b$	b
$2a$	$2a$	$3a$	0	a	$2a+b$	$3a+b$	b	$a+b$
$3a$	$3a$	0	a	$2a$	$3a+b$	b	$a+b$	$2a+b$
b	b	$3a+b$	$2a+b$	$a+b$	0	$3a$	$2a$	a
$a+b$	$a+b$	b	$3a+b$	$2a+b$	a	0	$3a$	$2a$
$2a+b$	$2a+b$	$a+b$	b	$3a+b$	$2a$	a	0	$3a$
$3a+b$	$3a+b$	$2a+b$	$a+b$	b	$3a$	$2a$	a	0

.	0	a	$2a$	$3a$	b	$a+b$	$2a+b$	$3a+b$
0	0	0	0	0	0	0	0	0
a	0	a	a	a	0	a	a	a
$2a$	0	$2a$	$2a$	$2a$	0	$2a$	$2a$	$2a$
$3a$	0	$3a$	$3a$	$3a$	0	$3a$	$3a$	$3a$
b	b	b	b	b	b	b	b	b
$a+b$	b	$a+b$	$a+b$	$a+b$	b	$a+b$	$a+b$	$a+b$
$2a+b$	b	$2a+b$	$2a+b$	$2a+b$	b	$2a+b$	$2a+b$	$2a+b$
$3a+b$	b	$3a+b$	$3a+b$	$3a+b$	b	$3a+b$	$3a+b$	$3a+b$

N is a Boolean near-ring. $(N,+)$ is non-abelian. $N_0 = \{0,a,2a,3a\}$ is an ideal. Write $R = \{0,2a\}$. R is a normal subgroup (since it is the center of $(N,+)$) and $RN \subseteq R$. Therefore R is a right ideal. But R is not an ideal, since $(3a+b)(0+2a)-(3a+b)0 = 3a+b-b = 3a \notin R$. Therefore N is subdirectly irreducible and almost trivial near-ring. Here N_0 is the only non trivial ideal of N.

We now show that every N-subgroup of N, where N is a zero-symmetric Boolean near-ring, is an ideal. We first prove a lemma.

Lemma 3.1. Every Boolean near-ring N has the IFP-property.

Proof. Let $a,b,n \in N$ with $ab = 0$. By weak commutativity, $anb = abn = 0n = 0$. Hence N has the IFP. □

Theorem 3.2. Let N be a zero-symmetric Boolean near-ring. Then for every $e \in N$, Ne is an ideal of N.

Proof. By Lemma 2.1, N has the IFP-property. Therefore $(0:e) = \{x \in N \mid xe = 0\}$is an ideal. We show that $(0:(0:e)) = \{a \in N \mid ax = 0 \text{ for all } x \in (0:e)\} = Ne$. By Proposition 9.3 of G. Pilz[4], $(0:(0:e))$ is an ideal of N. For any $x \in (0:e), ex = exe = e0 = 0$, by weak commutativity. Hence $e \in (0:(0:e))$and so $Ne \subseteq (0:(0:e))$. Now for any $n \in (0:(0:e))$, $n-ne = (n-ne)^2 = (n-ne)(n-ne) = n(n-ne)-ne(n-ne) = n(n-ne)-n(n-ne)e = 0$, since $n-ne \in (0:e)$. Therefore $n = ne \in Ne$. Then $(0:(0:e)) \in Ne$. Thus $(0:(0:e)) = Ne$. Therefore Ne is an ideal of N. □

Theorem 3.3. Let N be a zero-symmetric Boolean near-ring. If M is an N-subgroup of N, then M is an ideal of N.

Proof. By Theorem 2.2, for each $m \in M$, Nm is an ideal of N. But the sum of family $\{Nm\}_{m\in M}$ of ideals is M (since $NM \subseteq M$). By Corollary 2.3 of G. Pilz[4] M is an ideal. □

4. Subdirect product representation of Boolean near-rings

In this section we obtain a subdirect product representation of Boolean near-rings. Let us recall that (Example 2.1) a near-ring N is called an almost trivial near-ring if for all $x, y \in N$:
$xy = x$ if $y \notin N_c$
$\quad = x_c$ if $y \in N_c$.

Theorem 4.1. Every subdirectly irreducible Boolean near-ring is an almost trivial near-ring.

Proof. Let N be a subdirectly irreducible Boolean near-ring. If N is a constant near-ring, clearly N is almost trivial. So, assume that N is non-constant. Then $N_0 \neq (0)$.
(i) Let $J = \{n \in N/(0:n) \neq (0)\}.J \neq \phi$, since $N_0 = (0:0) \neq \{0\}$. Since N is subdirectly irreducible, $\bigcap_{n\in J}(0:n) \neq (0)$
Let $0 \neq e \in \cap_{n\in J}(0:n)$. Since $e \notin (0:e)$, it follows that $e \notin J$ and hence $(0:e) = (0)$. For any $a \in N, (a-ae)e = 0 \implies a - ae \in (0:e) = (0) \implies a = ae$.
Thus e is a right identity of N. On the other hand, if $j \in J$ then $e \in (0:j)$ and $ej = 0$. Therefore $j = je = (je)^2 = jeje = j0 \implies j \in N_c$. Thus $J \subseteq N_c$. Moreover, if $n \notin N_c$, then $n \notin J$ and $(0:n) = (0)$. By the above argument, we can prove that n is a right identity. Also if $n_c \in N_c$ then $xn_c = xn_c0 = x0n_c = x0 = x_c$, for any $x \in N$. Therefore we get that $xy = x$ if $y \notin N_c$ and $xy = x_c$ if $y \in N_c$. □

We now prove the main theorem of the section.

Theorem 4.2. Every Boolean near-ring N is a subdirect product of almost trivial near-rings.

Proof. Let N be the subdirect product of subdirectly irreducible near-rings $\{N_i\}_{i\in I}$. For each $i \in I$, N_i is a subdirectly irreducible Boolean near-ring. By Theorem 4.1, each N_i is an almost trivial near-ring. Therefore N is a subdirect product of almost trivial near-rings. □

Note. Let N be a zero-symmetric almost trivial near-ring. Since N is zero-symmetric, $N = N_0$ and $N_c = (0)$. Therefore for any $x, y \in N, xy = x$ if $y \neq 0$ (hence $y \notin N_c$) and $xy = x_c = 0$ if $y = 0$ (that is, $y \in N_c$). Thus N is a trivial near-ring.

Corollary 4.1. Every zero-symmetric Boolean near-ring is a subdirect product of trivial near-rings.

Proof. Let N be a zero-symmetric Boolean near-ring. By Theorem 4.2, N is a subdirect product of almost trivial near-rings $\{N_i\}$. For each i, N_i is a zero-symmetric almost trivial near-ring. By Note 4.1 each N_i is a trivial near-ring. □

Note. Let $R \neq (0)$ be a Boolean ring. Then $R \cong Z_2$.

By Corollary 4.1 and Note 4.2 we get the following.

Corollary 4.2. Every Boolean ring is a subdirect product of two element fields.

5. Radical of a zero-symmetric Boolean near-ring

In this section we prove that if N is a zero-symmetric Boolean near-ring then $J_0(N) = (0)$.

We first present below the required prerequisites from Pilz[4] those are necessary for proving the main theorem.

Notation 5.1. Let N be a zero-symmetric near-ring. For $z \in N$, denote the left ideal generated by the set $\{n - nz \mid n \in N\}$ by L_z.

Definition 5.1. Let N be a zero-symmetric near-ring. Then (i) $z \in N$ is called quasiregular if $z \in L_z$ (ii) $S \subseteq N$ is called quasiregular if each element of S is quasiregular.

Result 5.1. If N is a zero-symmetric near-ring then $z \in N$ is quasiregular iff $L_z = N$.

Result 5.2. Let N be a zero-symmetric near-ring. Then $J_0(N)$ is the greatest quasiregular ideal of N.

Theorem 5.1. Let N be a zero-symmetric Boolean near-ring. Then $J_0(N) = (0)$.

Proof. Let $0 \neq e \in J_0(N)$. By Result 5.1, e is quasiregular. Since N has the IFP, $(0:e) = \{x \in N / xe = 0\}$ is an ideal of N. Since N is a Boolean near-ring, we have that e is an idempotent. Now for all $n \in N, (n - ne)e = ne - ne^2 = ne - ne = 0$ and hence $n - ne \in (0 : e)$. Therefore $L_e \subseteq (0:e) \neq N$, since $e \notin (0:e)$. By Result 5.1, e cannot be quasiregular, a contradiction. Therefore $J_0(N) = (0)$. □

If R is a ring, then $J_0(R) = J(R)$, where $J(R)$ is the Jacobson radical of R.

Corollary 5.1. If R is a Boolean ring, then $J(R) = (0)$.

6. Special Boolean near-rings

In this section, we introduce special Boolean near-rings and several properties of special Boolean near-rings are proved. We obtain subdirect product representation of special Boolean near-rings in terms of two element fields and constant near-rings.

Definition 6.1. An abelian Boolean near-ring N is called a special Boolean near-ring if for all $m, n \in N, m_0 n = n_0 m$, where $n = n_0 + n_c m = m_0 + m_c, n_0, m_0 \in N_0, n_c, m_c \in N_c$. Such n_0 and m_0 are called the zero symmetric parts of n and m respectively.

Example 6.1. (i) Every Boolean ring is a special Boolean near-ring and every special Boolean near-ring is a Boolean near-ring.

(ii) Let $N = \{0, a, b, c\}$ be the Klein's four group. Then the addition and multiplication tables are as follows:

$+$	0	a	b	c
0	0	a	b	c
a	a	0	c	b
b	b	c	0	a
c	c	b	a	0

$.$	0	a	b	c
0	0	0	0	0
a	a	a	a	a
b	0	0	b	b
c	a	a	c	c

N is a special Boolean near-ring, which is neither a Boolean ring nor a constant near-ring.

Following example is due to Clay-Lawver[2].

(iii) Let $(R, +, \wedge)$ be a Boolean ring. Let x be an arbitrary but a fixed element in R. Define multiplication '$\circ$' on R by $a \circ b = a \wedge (b + x + b \wedge x)$. Then $(R, +, \circ)$ is a special Boolean near-ring and it is a ring if and only if $x = 0$.

(iv) Examples 2.1 (iii) are special Boolean near-rings.

Lemma 6.1. If N is a special Boolean near-ring, then
(i) N_0 is a Boolean ring.
(ii) N_0 is an ideal of N.

Proof. (i) Clearly N_0 is a subnear-ring of N. If $m, n \in N_0$, then $m = m_0, n = n_0$. Since N is a special boolean near-ring, $mn = m_0 n = n_0 m = nm$. Thus multiplication in N_0 is commutative. As $(N_0, +)$ is abelian, it follows that N_0 is a boolean ring.
(ii) Following a Result from Hansen and Luh[5], N satisfies the weak commutative law. N has the IFP property. So, for every $n \in N, (0 : n) = \{x \in N / xn = 0\}$ is an ideal of N. In particular, $(0 : 0) = N_0$ is an ideal of N. □

Theorem 6.1. Let N be a Special Boolean near-ring and let N_d be the set of all distributive elements in N. Then
(i) $(x + y)_0 = x_0 + y_0, (x + y)_c = x_c + y_c$ for all $x, y \in N$
(ii) $N_0 = N_d$
(iii) $(xy)_0 = x_0 y_0$ and $(xy)_c = x_c y_c$ for all $x, y \in N$
(iv) N_c is an ideal of N.

Proof. (i) If $x, y \in N$, then $x = x_0 + x_c, y = y_0 + y_c$, where $x_0, y_0 \in N_0, x_c, y_c \in N_c$. Now $x + y = x_0 + x_c + y_0 + y_c = x_0 + y_0 + x_c + y_c$ and $x_0 + y_0 \in N_0, x_c + y_c \in N_c$. Hence $(x + y)_0 = x_0 + y_0$ and $(x + y)_c = x_c + y_c$.

(ii) For $n, m \in N, nm = (n_0 + n_c)m = n_0 m + n_c m = n_0 m + n_c$ where $n = n_0 + n_c, n_0 \in N_0, n_c \in N_c$. Similarly, $mn = m_0 n + m_c$ where $m = m_0 + m_c m_0 \in N_0, m_c \in N_c$. Hence $nm - mn = (n_0 m + n_c) - (m_0 n + m_c) = n_0 m - m_0 n + n_c - m_c = n_c - m_c = (n - m)_c$ by

hypothesis and (i). To show that every element of N_0 is distributive, let $n = n_0 + n_c \in N$ and $m_0 \in N_0$. Then $nm_0 - m_0 n = (n - m0)_c = (n_0 + n_c - m_0)_c = (n_0 - m_0 + n_c)_c = n_c$. So, $n_c + m_0 n = nm_0 = (n_0 + n_c)m_0 = n_0 m_0 + n_c \implies m_0 n = n_0 m_0$.
Hence, for $n' = n'_0 + n'_c, m_0(n + n') = (n + n')_0 m_0 = (n_0 + n'_0)m_0 = n_0 m_0 + n'_0 m_0 = m_0 n + m_0 n'$. Thus $N_0 \subseteq N_d$. Trivially, $N_d \subseteq N_0$ .Hence $N_0 = N_d$.

(iii) For $x, y \in N, (xy)_c = xy0 = x0y = x0 = x_c = x_c y_c$, by weak commutativity. Since x_0 is distributive, $x_0 y_0 = x_0(y - y_c) = x_0 y - x_0 y_c = x_0 y - x_0 y_c 0 = x_0 y - x_0 0 y_c = x_0 y = (x - x_c)y = xy - x_c y = xy - x_c = xy - x_c y_c = xy - (xy)_c = (xy)_0$. Therefore $(xy)_0 = x_0 y_0$.

(iv) Clearly N_c is a subgroup of $(N, +)$. To show that, for $n, n' \in N, x_c \in N_c, n(n' + x_c) - nn' \in N_c$, it suffices to show that its zero symmetric part is 0 .
Now $[n(n' + x_c) - nn']_0 = [n(n' + x_c)]_0 - [nn']_0 = n_0 n'_0 - n_0 n'_0 = 0$. Thus $n(n' + x_c) - nn' \in N_c$. Further $(x_c n)_c = x_c n_c = x_c n_c 0 = x_c 0 n_c = x_c 0 = x_c = x_c n$. Therefore $x_c n \in N_c$. Hence N_c is an ideal of N. □

Corollary 6.1. If N is a special Boolean near-ring, then for all $a_0, b_0 \in N_0$ for $n_c \in N_c, a_0(b_0 + n_c) = a_0 b_0$.

Proof. By Theorem 3.1, a_0 is a distributive element.
Therefore $a_0(b_0 + n_c) = a_0 b_0 + a_0 n_c = a_0 b_0 + a_0 n_c 0 = a_0 b_0 + a_0 0 n_c = a_0 b_0 + 0 = a_0 b_0$. □

Pilz[4], a near-ring N is the called an abstract affine near-ring if (a) N is an abelian near-ring and (b) $N_0 = N_d$, where N_d is the set of all distributive elements in N.

By Theorem 6.1, we have the following

Corollary 6.2. Every special Boolean near-ring is an abstract affine near-ring.

Theorem 6.2. Let N be a Boolean near-ring. Then N is a special Boolean near-ring if and only if it is an abstract affine near-ring.

Proof. Suppose N is an abstract affine near-ring. Since N is a Boolean near-ring, N_0 is also Boolean near-ring. Since $N_0 = N_d, N_0$ is distributive near-ring. Since N_0 is a Boolean ring, we have N_0 is a commutative ring. Let $m, n \in N$ with $m = m_0 + m_c, n = n_0 + n_c, m_0, n_0 \in N_0$ and $m_c, n_c \in N_c$. Then $m_0 n = m_0(n_0 + n_c) = m_0 n_0 + m_0 n_c = m_0 n_0$, since $m_0 n_c = 0$. Similarly we get that $n_0 m = n_0 m_0$. Hence $m_0 n = n_0 m$. Since $(N, +)$ is abelian, N is a special Boolean near-ring. Converse follows from Corollary 6.2. □

We now prove another characterization theorem.

Theorem 6.3. Every one sided ideal of a special Boolean near-ring is two sided (that is, every special Boolean near-ring is a duo-near-ring).

Definition 6.2. Let N be a near-ring. An ideal I of N is called a strong ideal if $NI \subseteq I$.

Theorem 6.4. Let N be a special Boolean near-ring and let I be a strong ideal of N. Then I is a prime ideal of N iff I is a maximal ideal of N.

Theorem 6.5. Let N be a special Boolean near-ring. Then N is a subdirect product of near-rings $\{N_i\}$, where each N_i is either a two element field or an abelian constant near-ring.

Theorem 6.6. Let N be a special Boolean near-ring. Then $J_0(N) = J_1(N) = J_2(N) =$ the intersection of all maximal subgroups of $(N_c, +)$.

Proof. $nm = (n_0 + n_c)(m_0 + m_c) = n_c + n_0(m_0 + m_c) + n_c(m_0 + m_c) = n_0(m_0 + m_c) + n_c = n_0m_0 + n_0m_c + n_c = n_0m_0 + n_c = n_0m_0 + n_cm_c$. □

Lemma 6.2. Let N be a special Boolean near-ring. If a near-ring N' is a homomorphic image of N and if N' does not contain a nonzero distributive element, then N' is a constant near-ring.

Proof. Let $D(N)$ and $D(N')$ be the sets of distributive elements in N and N' respectively. Let $h : N \to N'$ be an epimorphism. Clearly $h(D(N)) \subseteq D(N')$. By Theorem 6.3, $N_0 = D(N)$. By hypothesis, $D(N') = (0)$ and hence $h(D(N)) = (0)$. Therefore $D(N) \subseteq kerh$ and hence $N_0 \subseteq kerh$. Suppose $n = n_0 + n_c, m = m_0 + m_c \in N$, where $n_0, m_0 \in N_0, n_c, m_c \in N_c$. By Theorem 6.6, $nm = n_0m_0 + n_cm_c$. Hence $h(nm) = h(n_0m_0) + h(n_cm_c) = h(n_cm_c)$, since $n_0m_0 \in N_0$ and this implies that $h(n)h(m) = h(nm) = h(n_cm_c) = h(n_c) = h(n)$. Therefore N' is a constant near-ring. □

Lemma 6.3. Every homomorphic image of a special Boolean near-ring is a Special Boolean near-ring.

Proof. Let $\phi : N \to N'$ be an epimorphism of a Special Boolean near-ring N onto a near-ring N'. If $x \in N'$, there exists $n \in N$ such that $\phi(n) = x$. Then $x^2 = \phi(n)^2 = \phi(n^2) = \phi(n) = x$, since $n^2 = n$ in N. Therefore N' is a Boolean near-ring. Since $(N, +)$ is abelian, $(N', +)$ is also abelian. Hence N' is an abelian Boolean near-ring. Let $m', n' \in N'$ with $m' = m'_0 + m'_c, n' = n'_0 + n'_c$, where $m'_0, n'_0 \in N'_0$ and $m'_c, n'_c \in N'_c$. Since ϕ is onto, there exists $m, n \in N$ such that $\phi(m) = m', \phi(n) = n'$. Suppose $m = m_0 + m_c, n = n_0 + n_c$, where $m_0, n_0 \in N_0, m_c, n_c \in N_c$ Now $m' = \phi(m) = \phi(m_0 + m_c) = \phi(m_0) + \phi(m_c)$. Let 0 and $0'$ be the zero elements of N and N' respectively. $\phi(m_0)0' = \phi(m_0)\phi(0) = \phi(m_00) = \phi(0) = 0'$ and hence $\phi(m_0) \in N'_0$. Also $\phi(m_c)0' = \phi(m_c)\phi(0) = \phi(m_c0) = \phi(m_c)$ and this implies that $\phi(m_c) \in N'_c$. Thus $m' = \phi(m_0) + \phi(m_c)$, where $\phi(m_0) \in N'_0, \phi(m_c) \in N'_c$. By the uniqueness of representation, $\phi(m_0) = m'_0, \phi(m_c) = m'_c$. Similarly, $\phi(n_0) = n'_0$ and $\phi(n_c) = n'_c$. Finally, $m'_0n' = \phi(m_0)\phi(n) = \phi(m_0n) = \phi(n_0m) = \phi(n_0)\phi(m) = n'_0m$. So N' is a special Boolean near-ring. □

Lemma 6.4. If N is a subdirectly irreducible special Boolean near-ring containing a nonzero distributive element, then N is a two element field.

Proof. By Lemma 6.1 and Theorem 6.3, N_0 and N_c are ideals of N. Since $N_0 \cap N_c = \{0\}$ and since N is subdirectly irreducible, either $N_0 = (0)$ or $N_c = (0)$. By hypothesis $N_d \neq \{0\}$

and by Theorem 6.3, $N_0 = N_d$. Therefore $N_0 \neq (0)$ and hence $N_c = (0)$. Thus $N = N_0$ and by Lemma 6.1, N is a Boolean ring. Let $0 \neq e \in N$. Then $(0:e) = \{n/ne = 0\}$ is an ideal of N. Clearly $(0:e) \cap Ne = (0)$. Since $Ne \neq (0)$ and it is an ideal , it follows that $(0:e) = (0)$. For any $n \in N, e(n-en) = 0$ and hence $n - en \in (0:e) = (0)$. Thus $n = en$. So e is the identity element of N. For any $0 \neq m \in N, m(m+e) = m+m = 0$ and hence $m+e = 0 \implies m = -e = e$. Therefore $N = \{0,e\}$ is a two element field. □

We now obtain the subdirect product representation of special Boolean near-rings.

Theorem 6.7. Let N be a special boolean near-ring. Then N is a subdirect product of near-rings $\{N_i\}$, where each N_i is a two element field or a constant near-ring.

Proof. N is a subdirect product of subdirectly irreducible near-rings $\{N_i\}$. Then each N_i is a homomorphic image of N. By Lemma 6.3, N_i is a special Boolean near-ring. If N_i does not contain a nonzero distributive element, then N_i is a constant near-ring. If N_i contains a nonzero distributive element, N_i is a two element field. Hence N_i is either a two element field or a constant near-ring. □

Theorem 6.8. Let N be a special Boolean near-ring. Then
(i) $(0:N_c) = N_0$
(ii) If I is an ideal of N, then $I = I \cap N_0 + I \cap N_c = I_0 + I_c$ with I_0 and I_c are ideals of I.

Proof. (i) For $n_0 \in N_0, n_c \in N_c, n_0 n_c = n_0 n_c 0 = n_0 0 n_c = n_0 0 = 0$. Therefore $N_0 N_c = \{0\}$ and hence $N_0 \subseteq (0:N_c)$. Let $n \in (0:N_c)$ and $n = n_0 + n_c$. Then $nN_c = (0)$ and hence $nn_c = 0$.
Therefore , $nn_c = (n_0 + n_c)n_c = n_0 n_c + n_c n_c = n_c = 0$ and so, $n \in N_0$. Hence $N_0 = (0:N_c)$.
(ii) If I is an ideal of N, then by Proposition 2.18 of Pilz[4], $I = I \cap N_0 + I \cap N_c = I_0 + I_c$. Since N_0 and N_c are ideals of N, it follows that I_0 and I_c are ideals of I. □

Pilz[4], a subnear-ring M of a near-ring N is called invariant if $MN \subseteq M$ and $NM \subseteq M$.

Theorem 6.9. Let N be a special Boolean near-ring. Then
(i) Every invariant subnear-ring is an ideal.
(ii) If M is an invariant subnear-ring, then $M = (M \cap N_0) + N_c$.

Proof. (i) Suppose M is an invariant subnear-ring of N. Trivially M is a right ideal of N. or $n, n' \in N, m \in M$, Consider $n(n'+m) - nn' = (n_0 + n_c)(n'+m) - (n_0 + n_c)n' = n_0(n' + m) + n_c - n_0 n' - n_c = n_0 n' + n_0 m - n_0 n' = n_0 m \in M$, where $n = n_0 + n_c n_0 \in N_0, n_c \in N$. Therefore M is an ideal of N.

(ii) By (i), M is an ideal of N. By Theorem 6.8, $M = M \cap N_0 + M \cap N_c$. Now $N_c = N_c M \subseteq NM \subseteq M$, since M is an invariant subnear-ring. Therefore, $M = (M \cap N_0) + N_c$.
□

Corollary 6.3. Let N be a special Boolean near-ring. If L is a subgroup of $(N,+)$ such that $N_0 L \subseteq L$ then L is a left ideal of N.

Corollary 6.4. Let N be a Special Boolean near-ring. Then all subgroups S of $(N_c, +)$ are ideals of N.

Proof. Clearly S is a subgroup of $(N, +)$. Since for all $s \in S, n \in N, sn = s$, we have $SN \subseteq S$. Take $n, n' \in N, s \in S$. Let $n = n_0 + n_c$, where $n_0 \in N_0, n_c \in N_c$. Now $n(n' + s) - nn' = (n_0 + n_c)(n' + s) - (n_0 + n_c)n' = n_0(n' + s) + n_c - n_0 n' - n_c = n_0(n' + s) - n_0 n' = n_0 n' + n_0 s - n_0 n' = n_0 s = 0 \in S$. Therefore S is an ideal of N. □

Theorem 6.10. Let N be a special Boolean near-ring and I a subgroup of N. Then the following conditions are equivalent
(a) I is a right ideal,
(b) I is a left ideal,
(c) I is an ideal.

Proof. (a) $\Longrightarrow$ (b). By (a) $IN \subseteq I$. Let $n, n' \in N, i \in I$. Let $n = n_0 + n_c, n' = n'_0 + n'_c, i = i_0 + i_c$, where $n_0, n'_0, i_0 \in N_0$ and $n_c, n'_c, i_c \in N_c$. Now $n(n' + i) - nn' = (n_0 + n_c)(n'_0 + n'_c + i_0 + i_c) - (n_0 + n_c)(n'_0 + n'_c) = n_0(n'_0 + n'_c + i_0 + i_c) + n_c - n_0(n'_0 + n'_c) - n_c = (n_0 n'_0 + n_0 n'_c + n_0 i_0 + n_0 i_c) - (n_0 n'_0 + n_0 n'_c) = n_0 n'_0 + n_0 i_0 - n_0 n'_0 = n_0 i_0 = i_0 n_0$.
Since $i = i_0 + i_c, i_0 = i_c \Longrightarrow i_c = i0 \in IN \subseteq I$. Thus $i_0 = i - i_c \in I$. Therefore $i_0 n_0 \in IN \subseteq I$.Thus $i_0 n_0 \in I$ and this implies that $n(n' + i) - nn' \in M$. Hence M is a left ideal.
(b) $\Longrightarrow$ (c) : By (b), I is a left ideal and hence $N_0 I \subseteq I$. Let $i \in I, n \in N$. Let $i = i_0 + i_c$, $n = n_0 + n_c$ where $i_0, n_0 \in N_0, i_c, n_c \in N_c$.Consider $in = (i_0 + i_c)(n_0 + n_c) = i_0 n_0 + i_c n_c = i_0 n_0 + i_c$. Since I is left ideal, $i_0 i \in N_0 I \subseteq I$. But $i_0 i = i_0(i_0 + i_c) = i_0 i_0 = i_0 \in I$. Therefore $i_c \in I$. Also since I is a left ideal, $n_0 i_0 \in N_0 I \subseteq I \Longrightarrow i_0 n_0 = n_0 i_0 \in I$. Hence $in \in I$. Thus I is a right ideal and hence it is an ideal.
(c) $\Longrightarrow$ (a) is trivial. □

Corollary 6.5. Every Special Boolean near-ring is a duo-near-ring.

Lemma 6.5. Let N be a Special Boolean near-ring. Then every N-subgroup of N is an invariant subnear-ring of N.

Proof. Let M be an N-subgroup of N. Then M is a subgroup of N such that $NM \subseteq M$. Clearly $N_c = N_c M \subseteq NM \subseteq M$. Let $m \in M, n \in N$. Let $m = m_0 + m_c, n = n_0 + n_c$, where $m_0, n_0 \in N_0, n_c, m_c \in N_c$. Now $mn = (m_0 + m_c)(n_0 + n_c) = m_0 n_0 + m_c n_c = m_0 n_0 + n_c = n_0 m_0 + m_c$.
Since $m = m_0 + m_c$ and $m, m_c \in M$, we have that $m_0 \in M$. This implies that $n_0 m_0 \in NM \subseteq M$. Therefore $n_0 m_0 + m_c \in M$, eventually, $mn \in M$. Therefore $MN \subseteq M$ and so M is invariant. □

Bibliography

1. Bh. Satyanarayana and K. Syam Prasad, "Near-rings, Fuzzy Ideals & Graph Theory", *CRC press, (Taylor & Francis group), New York/England*, 2013 (ISBN: 978-1-4398-7310-6).
2. Clay, James R. and Lawyer, Donald A., Boolean near-ring, *Canad. Math. Bull.* **12** (1969), 265–273.

3. C. V. L. N. Murty, On strongly regular near-rings, *Algebra and its Applications, Lecture notes in Pure and Appl. Math.* **91**, M. Dekkar, Newyork, (1984), 293–300.
4. G. Pilz, Near-rings, the Theory and its Applications, *North-Holland Pub. company*, 1983.
5. Hansen D. J. and Jiang Luh, Boolean near-rings and weak commutativity, *J. Aust. Math. Soc. (Series A)* (1989), 103–107.
6. K. Pushpalatha, On Special Boolean near-rings and Boolean-like near-rings, Ph.D. thesis, Acharya Nagarjuna University (2015).
7. K. Pushpalatha, 'One-sided ideals in Boolean near-rings' in *International Journal of Engineering Technology, Management and Applied Sciences(IJETMAS)*, November 2014, Vol. 2 (6).
8. K. Pushpalatha, On Subdirect product representation of Boolean near-rings, *International Journal of Scientific and Innovative Mathematical Research Special issue*, Vol. 2, issue-7(A), 77–81 (2015).
9. Subrahmanyam, N. V., Boolean semirings, *Math Annalen*, **148** (1962), 395–401.
10. Szasz, G., Introduction to lattice theory (3rd revised edition), *Academic Press, New York* (1963).
11. T. W. Hungerford, Algebra, Holt, *Rinehart and Winston Inc., New York* (1974).
12. Y. V. Reddy and C. V. L. N. Murty, 'On strongly regular near-rings', *Proc. Edinburgh Math. Soc.* **27** (1984), 61–64.

Recent results on the annihilator graph of a commutative ring: A survey

Ayman Badawi

Department of Mathematics & Statistics, American University of Sharjah,
P. O. Box 26666, Sharjah, United Arab Emirates,
E-mail: abadawi@aus.edu

Let R be a commutative ring with nonzero identity, $Z(R)$ be its set of zero-divisors, and if $a \in Z(R)$, then let $ann_R(a) = \{d \in R \mid da = 0\}$. The annihilator graph of R is the (undirected) graph $AG(R)$ with vertices $Z(R)^* = Z(R) \setminus \{0\}$, and two distinct vertices x and y are adjacent if and only if $ann_R(xy) \neq ann_R(x) \cup ann_R(y)$. It follows that each edge (path) of the zero-divisor graph $\Gamma(R)$ is an edge (path) of $AG(R)$. The extended zero-divisor graph of R is the undirected (simple) graph $EG(R)$ with the vertex set $Z(R)^*$, and two distinct vertices x and y are adjacent if and only if either $Rx \cap ann_R(y) \neq \{0\}$ or $Ry \cap ann_R(x) \neq \{0\}$. Hence it follows that the zero-divisor graph $\Gamma(R)$ is a subgraph of $EG(R)$. In this paper, we collect some properties (many are recent) of the two graphs $AG(R)$ and $EG(R)$.

Keywords: Zero-divisor graph; Annihilator ideal; Annihilator graph.

1. introduction

Let R be a commutative ring with nonzero identity, and let $Z(R)$ be its set of zero-divisors. Over the past several years, there has been considerable attention in the literature to associating graphs with commutative rings (and other algebraic structures) and studying the interplay between ring-theoretic and graph-theoretic properties; see the recent survey articles [10,47]. For example, as in Ref. 16, the *zero-divisor graph* of R is the (simple) graph $\Gamma(R)$ with vertices $Z(R) \setminus \{0\}$, and distinct vertices x and y are adjacent if and only if $xy = 0$. This concept is due to Beck[29], who let all the elements of R be vertices and was mainly interested in colorings. The zero-divisor graph of a ring R has been studied extensively by many authors, for example see Refs. 1–3, 11, 20, 21, 39–44, 48–54, 58. We recall from Ref. 12, the *total graph* of R, denoted by $T(\Gamma(R))$ is the (undirected) graph with all elements of R as vertices, and for distinct $x, y \in R$, the vertices x and y are adjacent if and only if $x + y \in Z(R)$. The total graph (as in Ref. 12) has been investigated in Refs. 5–8, 36, 47, 49, 52 and 56; and several variants of the total graph have been studied in Refs. 4, 13–15, 19, 28, 32–35, 37, 38 and 45. Let $a \in Z(R)$ and let $ann_R(a) = \{r \in R \mid ra = 0\}$. In 2014, Badawi[23] introduced the *annihilator graph of R*. We recall from Ref. 23 that the annihilator graph of R is the (undirected) graph $AG(R)$ with vertices $Z(R)^* = Z(R) \setminus \{0\}$, and two distinct vertices x and y are adjacent if and only if $ann_R(xy) \neq ann_R(x) \cup ann_R(y)$. It follows that each edge (path) of the classical zero-divisor of R is an edge (path) of $AG(R)$. For Further investigations of $AG(R)$, see Refs. 24, 25 and 31. The authors in Refs. 26 and 27 introduced the *extended zero-divisor graph of R*. We recall from Ref. 26 that the extended zero-divisor graph of R is the undirected (simple) graph $EG(R)$ with the vertex set $Z(R)^*$, and two distinct vertices x and y are adjacent if and only if either $Rx \cap ann_R(y) \neq \{0\}$ or $Ry \cap ann_R(x) \neq \{0\}$. Hence it follows that the zero-divisor graph $\Gamma(R)$ is a subgraph of $EG(R)$.

Let G be a (undirected) graph. We say that G is *connected* if there is a path between any two distinct vertices. For vertices x and y of G, we define $d(x,y)$ to be the length of a shortest path from x to y ($d(x,x)=0$ and $d(x,y)=\infty$ if there is no path). Then the *diameter* of G is $diam(G) = \sup\{ d(x,y) \mid x \text{ and } y \text{ are vertices of } G\}$. The *girth* of G, denoted by $gr(G)$, is the length of a shortest cycle in G ($gr(G)=\infty$ if G contains no cycles).

A graph G is *complete* if any two distinct vertices are adjacent. The complete graph with n vertices will be denoted by K^n (we allow n to be an infinite cardinal). A *complete bipartite graph* is a graph G which may be partitioned into two disjoint nonempty vertex sets A and B such that two distinct vertices are adjacent if and only if they are in distinct vertex sets. If one of the vertex sets is a singleton, then we call G a *star graph*. We denote the complete bipartite graph by $K^{m,n}$, where $|A|=m$ and $|B|=n$ (again, we allow m and n to be infinite cardinals); so a star graph is a $K^{1,n}$ and $K^{1,\infty}$ denotes a star graph with infinitely many vertices. By $\overline{G}$, we mean the complement graph of G. Let G_1, G_2 be two graphs. The join of G_1 and G_2, denoted by $G_1 \vee G_2$, is a graph with the vertex set $V(G_1 \cup G_2) = V(G_1) \cup V(G2)$ and edge set $E(G_1 \cup G_2) = E(G_1) \cup E(G_2) \cup \{u-v | u \in G_1, v \in G_2\}$. Finally, let $\overline{K}^{m,3}$ be the graph formed by joining $G_1 = K^{m,3}$ ($= A \cup B$ with $|A| = m$ and $|B| = 3$) to the star graph $G_2 = K^{1,m}$ by identifying the center of G_2 and a point of B.

Throughout, R will be a commutative ring with nonzero identity, $Z(R)$ its set of zero-divisors, $Nil(R)$ its set of nilpotent elements, $U(R)$ its group of units, $T(R)$ its total quotient ring, and $Min(R)$ its set of minimal prime ideals. For any $A \subseteq R$, let $A^* = A \setminus \{0\}$. We say that R is *reduced* if $Nil(R) = \{0\}$ and that R is *quasi-local* if R has a unique maximal ideal. A prime ideal P of R is called an associated prime ideal, if $ann_R(x) = P$, for some non-zero element $x \in R$. The set of all associated prime ideals of R is denoted by $Ass(R)$, and $\sum = \{ann_R(x) | 0 \neq x \in R\}$. The distance between two distinct vertices a, b of $\Gamma(R)$ is denoted by $d_{\Gamma(R)}(a,b)$. If $AG(R)$ is identical to $\Gamma(R)$, then we write $AG(R) = \Gamma(R)$; otherwise, we write $AG(R) \neq \Gamma(R)$. As usual, $\mathbb{Z}$ and $\mathbb{Z}_n$ will denote the integers and integers modulo n, respectively.

2. Basic properties of $AG(R)$

We recall the following basic results from Ref. 23.

Theorem 2.1. *(Ref. 23 Theorem 2.2) Let R be a commutative ring with $|Z(R)^*| \geq 2$. Then $AG(R)$ is connected and $diam(AG(R)) \leq 2$.*

Theorem 2.2. *(Ref. 23 Theorem 2.4) Let R be a commutative ring. Suppose that $x-y$ is an edge of $AG(R)$ that is not an edge of $\Gamma(R)$ for some distinct $x,y \in Z(R)^*$. If $xy^2 \neq 0$ and $x^2y \neq 0$, then there is a $w \in Z(R)^*$ such that $x-w-y$ is a path in $AG(R)$ that is not a path in $\Gamma(R)$, and hence $C: x-w-y-x$ is a cycle in $AG(R)$ of length three and each edge of C is not an edge of $\Gamma(R)$.*

In view of Theorem 2.2, we have the following result.

Corollary 2.3. *(Ref. 23 Corollary 2.5) Let R be a reduced commutative ring. Suppose that $x-y$ is an edge of $AG(R)$ that is not an edge of $\Gamma(R)$ for some distinct $x,y \in Z(R)^*$. Then* *there is a* $w \in ann_R(xy) \setminus \{x,y\}$ *such that* $x - w - y$ *is a path in* $AG(R)$ *that is not a path in* $\Gamma(R)$, *and hence* $C : x - w - y - x$ *is a cycle in* $AG(R)$ *of length three and each edge of* C *is not an edge of* $\Gamma(R)$.

In light of Corollary 2.3, the following result follows.

Theorem 2.4. *(Ref. 23 Theorem 2.6) Let* R *be a reduced commutative ring and suppose that* $AG(R) \neq \Gamma(R)$. *Then* $gr(AG(R)) = 3$. *Furthermore, there is a cycle* C *of length three in* $AG(R)$ *such that each edge of* C *is not an edge of* $\Gamma(R)$.

In view of Theorem 2.2, the following is an example of a non-reduced commutative ring R where $x - y$ is an edge of $AG(R)$ that is not an edge of $\Gamma(R)$ for some distinct $x, y \in Z(R)^*$, but every path in $AG(R)$ of length two from x to y is also a path in $\Gamma(R)$.

Example 2.5. Let $R = \mathbb{Z}_8$. Then $2 - 6$ is an edge of $AG(R)$ that is not an edge of $\Gamma(R)$. Now $2 - 4 - 6$ is the only path in $AG(R)$ of length two from 2 to 6 and it is also a path in $\Gamma(R)$. Note that $AG(R) = K^3$, $\Gamma(R) = K^{1,2}$, $gr(\Gamma(R)) = \infty$, $gr(AG(R)) = 3$, $diam(\Gamma(R)) = 2$, and $diam(AG(R)) = 1$.

The following is an example of a non-reduced commutative ring R such that $AG(R) \neq \Gamma(R)$ and if $x - y$ is an edge of $AG(R)$ that is not an edge of $\Gamma(R)$ for some distinct $x, y \in Z(R)^*$, then there is no path in $AG(R)$ of length two from x to y.

Example 2.6.

(1) Let $R = \mathbb{Z}_2 \times \mathbb{Z}_4$ and let $a = (0,1), b = (1,2)$, and $c = (0,3)$. Then $a - b$ and $c - b$ are the only two edges of $AG(R)$ that are not edges of $\Gamma(R)$, but there is no path in $AG(R)$ of length two from a to b and there is no path in $AG(R)$ of length two from c to b. Note that $AG(R) = K^{2,3}$, $\Gamma(R) = \overline{K}^{1,3}$, $gr(AG(R)) = 4$, $gr(\Gamma(R)) = \infty$, $diam(AG(R) = 2$, and $diam(\Gamma(R)) = 3$.

(2) Let $R = \mathbb{Z}_2 \times \mathbb{Z}_2[X]/(X^2)$. Let $x = X + (X^2) \in \mathbb{Z}_2[X]/(X^2)$, $a = (0,1), b = (1,x)$, and $c = (0, 1 + x)$. Then $a - b$ and $c - b$ are the only two edges of $AG(R)$ that are not edges of $\Gamma(R)$, but there is no path in $AG(R)$ of length two from a to b and there is no path in $AG(R)$ of length two from c to b. Again, note that $AG(R) = K^{2,3}$, $\Gamma(R) = \overline{K}^{1,3}$, $gr(AG(R)) = 4$, $gr(\Gamma(R)) = \infty$, $diam(AG(R) = 2$, and $diam(\Gamma(R)) = 3$.

If $AG(R) \neq \Gamma(R)$ and $gr(AG(R)) = 4$, then the following result characterize, up to isomorphism, all such rings.

Theorem 2.7. *(Ref. 23 Theorem 2.9) Let* R *be a commutative ring and suppose that* $AG(R) \neq \Gamma(R)$. *Then the following statements are equivalent:*

(1) $gr(AG(R)) = 4$;

(2) $gr(AG(R)) \neq 3$;

(3) If $x - y$ *is an edge of* $AG(R)$ *that is not an edge of* $\Gamma(R)$ *for some distinct* $x, y \in Z(R)^*$, *then there is no path in* $AG(R)$ *of length two from* x *to* y;

(4) There are some distinct $x, y \in Z(R)^*$ *such that* $x - y$ *is an edge of* $AG(R)$ *that is not an edge of* $\Gamma(R)$ *and there is no path in* $AG(R)$ *of length two from* x *to* y*;*
(5) R *is ring-isomorphic to either* $\mathbb{Z}_2 \times \mathbb{Z}_4$ *or* $\mathbb{Z}_2 \times \mathbb{Z}_2[X]/(X^2)$.

In view of Theorem 2.7, the following result follows

Corollary 2.8. *(Ref. 23 Corollary 2.10) Let* R *be a commutative ring such that* $AG(R) \neq \Gamma(R)$ *and assume that* R *is not ring-isomorphic to* $\mathbb{Z}_2 \times B$*, where* $B = \mathbb{Z}_4$ *or* $B = \mathbb{Z}_2[X]/(X^2)$*. If* E *is an edge of* $AG(R)$ *that is not an edge of* $\Gamma(R)$*, then* E *is an edge of a cycle of length three in* $AG(R)$*.*

A direct implication of Theorem 2.7 and Corollary 2.8 is the following result.

Corollary 2.9. *(Ref. 23 Corollary 2.11) Let* R *be a commutative ring such that* $AG(R) \neq \Gamma(R)$*. Then* $gr(AG(R)) \in \{3,4\}$*.*

Theorem 2.10. *(Ref. 24 Theorem 2.5) Let* R *be a non-reduced ring such that* R *is not ring-isomorphic to* $Z_2 \times B$*, where* $B = Z_4$ *or* $B = \frac{Z_2[X]}{(X^2)}$*. Then the following statements are equivalent:*

(1) $gr(AG(R)) = \infty$*;*
(2) $AG(R)$ *is a star graph;*
(3) $AG(R)$ *is a bipartite graph;*
(4) $AG(R)$ *is a complete bipartite graph;*
(5) $\Sigma^* = Ass(R) = \{ann_R(x), ann_R(y)\}$ *for some* $x, y \in Z(R)^*$*. Furthermore, if* $ann_R(x) = ann_R(y)$*, then* $|ann_R(x)| = |Z(R)| = 3$ *and if* $ann_R(x) \neq ann_R(y)$*, then* $\Sigma^* = \{Z(R), ann_R(Z(R))\}$ *and* $|ann_R(Z(R))^*| = 1$.

Theorem 2.11. *(Ref. 24 Corollary 2.3) Let* R *be a ring. Then* $AG(R)$ *is a complete bipartite graph if and only if one of the following statements holds:*

(1) $Nil(R) = \{0\}$ *and* $|Min(R)| = 2$*;*
(2) $Nil(R) \neq \{0\}$ *and either* $AG(R) = K^{1,n}$ *or* $AG(R) = K^{2,3}$*, where* $1 \leq n \leq \infty$.

Let x be a vertex of $AG(R)$. In the following result, the authors in Ref. 24 gave conditions under which x is adjacent to every vertex in $\Gamma(R)$.

Theorem 2.12. *(Ref. 24 Theorem 2.6) Let* R *be a ring and* x *be a vertex of* $AG(R)$*. Then the following statements are equivalent:*

(1) x *is adjacent to every other vertex of* $\Gamma(R)$*;*
(2) $ann_R(x)$ *is a maximal element of* Σ *and* x *is adjacent to every other vertex of* $AG(R)$*.*

Recall that a undirected simple graph $G = (V, E)$ is called an n-partite graph if $V = A_1 \cup A_2 \cup \cdots \cup A_n$ for some $n \geq 2$, where each $A_i \neq \phi$, $A_i \cap A_j = \phi$, $i \neq j$, $1 \leq i, j \leq n$, and $x, y \in A_i$ implies $x - y$ is not an edge of G.

The authors in Ref. 25 prove the following result.

Theorem 2.13. *(Ref. 25 Theorem 2.1) Let $R = D_1 \times \cdots \times D_n$, where $n \geq 2$ and D_i is an integral domain for every $1 \leq i \leq n$. Then the following statements hold:*

(1) $AG(R)$ is an $nC\lceil \frac{n}{2} \rceil$-partite graph (Recall that mCn (m choose n) = $\frac{m!}{n!(m-n)!}$.)
(2) $AG(R)$ is not an $nC\lceil \frac{n}{2} \rceil - 1$-partite graph.

3. When does $AG(R) = \Gamma(R)$?

It is natural to ask when does $AG(R) = \Gamma(R)$? For a reduced ring R that is not an integral domain, we have the following results.

3.1. *Case I: R is reduced*

Theorem 3.1. *(Ref. 23 Theorem 3.3) Let R be a reduced commutative ring that is not an integral domain. Then the following statements are equivalent:*
(1) $AG(R)$ is complete;
(2) $\Gamma(R)$ is complete (and hence $AG(R) = \Gamma(R)$);
(3) R is ring-isomorphic to $\mathbb{Z}_2 \times \mathbb{Z}_2$.

Theorem 3.2. *(Ref. 23 Theorem 3.4) Let R be a reduced commutative ring that is not an integral domain and assume that $Z(R)$ is an ideal of R. Then $AG(R) \neq \Gamma(R)$ and $gr(AG(R)) = 3$.*

Theorem 3.3. *(Ref. 23 Theorem 3.5) Let R be a reduced commutative ring with $|Min(R)| \geq 3$ (possibly $Min(R)$ is infinite). Then $AG(R) \neq \Gamma(R)$ and $gr(AG(R)) = 3$.*

Theorem 3.4. *(Ref. 23 Theorem 3.6) Let R be a reduced commutative ring that is not an integral domain. Then $AG(R) = \Gamma(R)$ if and only if $|Min(R)| = 2$.*

Theorem 3.5. *(Ref. 23 Theorem 3.7) Let R be a reduced commutative ring. Then the following statements are equivalent:*
(1) $gr(AG(R)) = 4$;
(2) $AG(R) = \Gamma(R)$ and $gr(\Gamma(R)) = 4$;
(3) $gr(\Gamma(R)) = 4$;
(4) $T(R)$ is ring-isomorphic to $K_1 \times K_2$, where each K_i is a field with $|K_i| \geq 3$;
(5) $|Min(R)| = 2$ and each minimal prime ideal of R has at least three distinct elements;
(6) $\Gamma(R) = K^{m,n}$ with $m, n \geq 2$;
(7) $AG(R) = K^{m,n}$ with $m, n \geq 2$.

Theorem 3.6. *(Ref. 23 Theorem 3.8) Let R be a reduced commutative ring that is not an integral domain. Then the following statements are equivalent:*
(1) $gr(AG(R)) = \infty$;
(2) $AG(R) = \Gamma(R)$ and $gr(AG(R)) = \infty$;
(3) $gr(\Gamma(R)) = \infty$;
(4) $T(R)$ is ring-isomorphic to $Z_2 \times K$, where K is a field;

(5) $|Min(R)| = 2$ and at least one minimal prime ideal ideal of R has exactly two distinct elements;

(6) $\Gamma(R) = K^{1,n}$ for some $n \geq 1$;

(7) $AG(R) = K^{1,n}$ for some $n \geq 1$.

In view of Theorem 3.5 and Theorem 3.6, we have the following result.

Corollary 3.7. *(Ref. 23 Corollary 3.9) Let R be a reduced commutative ring. Then $AG(R) = \Gamma(R)$ if and only if $gr(AG(R)) = gr(\Gamma(R)) \in \{4, \infty\}$.*

If R is non-reduced, then we have the following results.

3.2. *Case II: R is non-reduced*

Theorem 3.8. *(Ref. 24 Theorem 2.2) Let R be a ring such that for each edge of $AG(R)$, say $x - y$, either $ann_R(x) \in Ass(R)$ or $ann_R(y) \in Ass(R)$. Then $AG(R) = \Gamma(R)$. In particular, if $\sum^* = Ass(R)$, then $\Gamma(R) = AG(R)$.*

Theorem 3.9. *(Ref. 24 Theorem 2.3) Let R be a non-reduced ring. Then the following statements are equivalent:*

(1) $\Gamma(R) = AG(R) = K^n \vee \overline{K^m}$, where $n = |Nil(R)^*|$ and $m = |Z(R) \setminus Nil(R)|$.
(2) $ann_R(Z(R))$ is a prime ideal of R.
(3) $\sum^* = Ass(R)$ and $|\sum^*| \leq 2$.

Theorem 3.10. *(Ref. 23 Theorem 3.15) Let R be a non-reduced commutative ring such that $Z(R)$ is not an ideal of R. Then $AG(R) \neq \Gamma(R)$.*

Theorem 3.11. *(Ref. 23 Theorem 3.16) Let R be a non-reduced commutative ring. Then the following statements are equivalent:*

(1) $gr(AG(R)) = 4$;
(2) $AG(R) \neq \Gamma(R)$ and $gr(AG(R)) = 4$;
(3) R is ring-isomorphic to either $\mathbb{Z}_2 \times \mathbb{Z}_4$ or $\mathbb{Z}_2 \times \mathbb{Z}_2[X]/(X^2)$;
(4) $\Gamma(R) = \overline{K}^{1,3}$;
(5) $AG(R) = K^{2,3}$.

We observe that $gr(\Gamma(\mathbb{Z}_8)) = \infty$, but $gr(AG(\mathbb{Z}_8)) = 3$. We have the following result.

Theorem 3.12. *(Ref. 23 Theorem 3.17) Let R be a commutative ring such that $AG(R) \neq \Gamma(R)$. Then the following statements are equivalent:*

(1) $\Gamma(R)$ is a star graph;
(2) $\Gamma(R) = K^{1,2}$;
(3) $AG(R) = K^3$.

Theorem 3.13. *(Ref. 23 Theorem 3.18) Let R be a non-reduced commutative ring with $|Z(R)^*| \geq 2$. Then the following statements are equivalent:*

(1) $AG(R)$ is a star graph;

(2) $gr(AG(R)) = \infty$;

(3) $AG(R) = \Gamma(R)$ and $gr(\Gamma(R)) = \infty$;

(4) $Nil(R)$ is a prime ideal of R and either $Z(R) = Nil(R) = \{0, -w, w\}$ $(w \neq -w)$ for some nonzero $w \in R$ or $Z(R) \neq Nil(R)$ and $Nil(R) = \{0, w\}$ for some nonzero $w \in R$ (and hence $wZ(R) = \{0\}$);

(5) Either $AG(R) = K^{1,1}$ or $AG(R) = K^{1,\infty}$;

(6) Either $\Gamma(R) = K^{1,1}$ or $\Gamma(R) = K^{1,\infty}$.

Corollary 3.14. *(Ref. 23 Corollary 3.19) Let R be a non-reduced commutative ring with $|Z(R)^*| \geq 2$. Then $\Gamma(R)$ is a star graph if and only if $\Gamma(R) = K^{1,1}$, $\Gamma(R) = K^{1,2}$, or $\Gamma(R) = K^{1,\infty}$.*

Remark 3.15. In view of Theorem 2.10, the authors in Ref. 24 gave an alternative proof of Theorem 3.13 (see Ref. 24 Corollary 2.4).

In the following example, we construct two non-reduced commutative rings say R_1 and R_2, where $AG(R_1) = K^{1,1}$ and $AG(R_2) = K^{1,\infty}$.

Example 3.16.

(1) Let $R_1 = \mathbb{Z}_3[X]/(X^2)$ and let $x = X + (X^2) \in R_1$. Then $Z(R_1) = Nil(R_1) = \{0, -x, x\}$ and $AG(R_1) = \Gamma(R_1) = K^{1,1}$. Also note that $AG(\mathbb{Z}_9) = \Gamma(\mathbb{Z}_9) = K^{1,1}$.

(2) Let $R_2 = \mathbb{Z}_2[X,Y]/(XY, X^2)$. Then let $x = X + (XY + X^2)$ and $y = Y + (XY + X^2) \in R_2$. Then $Z(R_2) = (x,y)R_2$, $Nil(R_2) = \{0, x\}$, and $Z(R_2) \neq Nil(R_2)$. It is clear that $AG(R_2) = \Gamma(R_2) = K^{1,\infty}$.

Remark 3.17. Let R be a non-reduced commutative ring. In view of Theorem 3.10, Theorem 3.11, and Theorem 3.13, if $AG(R) = \Gamma(R)$, then $Z(R)$ is an ideal of R and $gr(AG(R)) = gr(\Gamma(R)) \in \{3, \infty\}$. The converse is true if $gr(AG(R) = gr(\Gamma(R)) = \infty$ (see Theorem 3.10 and 3.13). However, if $Z(R)$ is an ideal of R and $gr(AG(R)) = gr(\Gamma(R)) = 3$, then it is possible to have all the following cases:

(1) It is possible to have a commutative ring R such that $Z(R)$ is an ideal of R, $Z(R) \neq Nil(R)$, $AG(R) = \Gamma(R)$, and $gr(AG(R)) = 3$. See Example 3.18.

(2) It is possible to have a commutative ring R such that $Z(R)$ is an ideal of R, $Z(R) \neq Nil(R)$, $Nil(R)^2 = \{0\}$, $AG(R) \neq \Gamma(R)$, $diam(AG(R)) = diam(\Gamma(R)) = 2$, and $gr(AG(R)) = gr(\Gamma(R)) = 3$. See Example 3.19.

(3) It is possible to have a commutative ring R such that $Z(R)$ is an ideal of R, $Z(R) \neq Nil(R)$, $Nil(R)^2 = \{0\}$, $AG(R)$ is a complete graph (i.e., $diam(AG(R)) = 1$), $AG(R) \neq \Gamma(R)$, $diam(\Gamma(R)) = 2$, and $gr(AG(R)) = gr(\Gamma(R)) = 3$. See Theorem 3.20.

Example 3.18. Let $D = \mathbb{Z}_2[X,Y,W]$, $I = (X^2, Y^2, XY, XW)D$ is an ideal of D, and let $R = D/I$. Then let $x = X + I, y = Y + I$, and $w = W + I$ be elements of R. Then $Nil(R) = (x,y)R$ and $Z(R) = (x,y,w)R$ is an ideal of R. By construction, we have $Nil(R)^2 = \{0\}$, $AG(R) =$

$\Gamma(R)$, $diam(AG(R)) = diam(\Gamma(R)) = 2$, and $gr(AG(R)) = gr(\Gamma(R)) = 3$ (for example, $x - (x+y) - y - x$ is a cycle of length three).

Example 3.19. Let $D = \mathbb{Z}_2[X,Y,W]$, $I = (X^2,Y^2,XY,XW,YW^3)D$ is an ideal of D, and let $R = D/I$. Then let $x = X+I, y = Y+I$, and $w = W+I$ be elements of R. Then $Nil(R) = (x,y)R$ and $Z(R) = (x,y,w)R$ is an ideal of R. By construction, $Nil(R)^2 = \{0\}$, $diam(AG(R)) = diam(\Gamma(R)) = 2$, $gr(AG(R)) = gr(\Gamma(R)) = 3$. However, since $w^3 \neq 0$ and $y \in ann_R(w^3) \setminus (ann_R(w) \cup ann_R(w^2))$, we have $w - w^2$ is an edge of $AG(R)$ that is not an edge of $\Gamma(R)$, and hence $AG(R) \neq \Gamma(R)$.

Given a commutative ring R and an R-module M, the *idealization* of M is the ring $R(+)M = R \times M$ with addition defined by $(r,m)+(s,n) = (r+s, m+n)$ and multiplication defined by $(r,m)(s,n) = (rs, rn+sm)$ for all $r,s \in R$ and $m,n \in M$. Note that $\{0\}(+)M \subseteq Nil(R(+)M)$ since $(\{0\}(+)M)^2 = \{(0,0)\}$. We have the following result

Theorem 3.20. *(Ref. 23 Theorem 3.24) Let D be a principal ideal domain that is not a field with quotient field K (for example, let $D = \mathbb{Z}$ or $D = F[X]$ for some field F) and let $Q = (p)$ be a nonzero prime ideal of D for some prime (irreducible) element $p \in D$. Set $M = K/D_Q$ and $R = D(+)M$. Then $Z(R) \neq Nil(R)$, $AG(R)$ is a complete graph, $AG(R) \neq \Gamma(R)$, and $gr(AG(R)) = gr(\Gamma(R)) = 3$.*

The following example shows that the hypothesis "Q is principal" in the above Theorem is crucial.

Example 3.21. Let $D = \mathbb{Z}[X]$ with quotient field K and $Q = (2,X)D$. Then Q is a non-principal prime ideal of D. Set $M = K/D_Q$ and $R = D(+)M$. Then $Z(R) = Q(+)M$, $Nil(R) = \{0\}(+)M$, and $Nil(R)^2 = \{(0,0)\}$. Let $a = (2,0)$ and $b = (0, \frac{1}{X} + D_Q)$. Then $ab = (0, \frac{2}{X} + D_Q) \in Nil(R)^*$. Since $ann_R(ab) = ann_R(b)$, we have $a - b$ is not an edge of $AG(R)$. Thus $AG(R)$ is not a complete graph.

We terminate this section with the following open question.

(Open question,[24]): Let R be a non-reduced ring and $x - y$ be an edge of $AG(R)$. If $\Gamma(R) = AG(R)$, then is it true either $ann_R(x) \in Ass(R)$ or $ann_R(y) \in Ass(R)$?

4. Clique number and chromatic number of $AG(R)$

Let $G = (V,E)$ be a graph. The clique number of G, denoted by $w(G)$, is the largest positive integer n such that K_n is a subgraph of G. The chromatic number of G, denoted by $\chi(G)$, is the the minimal number of colors which can be assigned to the vertices of G in such a way that every two adjacent vertices have different colors. It should be clear that $w(G) \leq \chi(G)$. Again, recall that mCn (m choose n) $= \frac{m!}{n!(m-n)!}$.

Theorem 4.1. *(Ref. 25 Theorem 2.2) Assume that R is ring-isomorphic to $D_1 \times \cdots \times D_n$, where $n \geq 2$ and D_i is an integral domain for every $1 \leq i \leq n$. Then $w(AG(R)) = \chi(AG(R)) = nC\lceil \frac{n}{2} \rceil$. In particular, if R is an Artinian ring, then $w(AG(R)) = \chi(AG(R)) = |Max(R)|C\lceil \frac{|Max(R)|}{2} \rceil$.*

Theorem 4.2. *(Ref. 25 Theorem 2.3) Let R be a non-reduced ring. Then the following statements hold.*

(1) If $|Z(R)| < \infty$, then the following statements are equivalent:

(a) $w(AG(R)) = |Nil(R)|$.
(b) $\chi(AG(R)) = |Nil(R)|$.
(c) $AG(R) = K^{2,3}$.

(2) If $|Z(R)| = \infty$, $w(AG(R)) < \infty$ and $Z(R)$ is an ideal of R, then the following statements are equivalent:

(a) $W(AG(R)) = |Nil(R)$.
(b) $\chi(AG(R)) = |Nil(R)|$.
(c) $AG(R) = K_{|Nil(R)^*|} \vee \overline{K_\infty}$.
(d) $x-y$ is not an edge of $AG(R)$, for every $x, y \in Z(R) \setminus Nil(R)$.

It is well-known that if G is a bipartite graph, then $\chi(AG(R)) = 2$. In the following result, the authors in Ref. 25 classified all bipartite annihilator graphs of rings.

Theorem 4.3. *(Ref. 25 Theorem 2.4) Let R be a non-reduced ring. Then the following statements are equivalent:*

(1) $w(AG(R)) = 2$;
(2) $\chi(AG(R)) = 2$;
(3) $AG(R) = K_{2,3}$ or $AG(R) = K_2$ or $AG(R) = K_1 \vee \overline{K_\infty}$.

5. Genus of $AG(R)$

The genus of a graph G, denoted by $g(G)$, is the minimal integer n such that the graph can be embedded in S_n. Intuitively, G is embedded in a surface if it can be drawn in the surface so that its edges intersect only at their common vertices. A graph G with genus 0 is called a planar graph and a graph G with genus 1 is called as a toroidal graph. Note that if H is a subgraph of a graph G, then $g(H) \leq g(G)$. In the following result, the authors in Ref. 31 classified all quasi-local rings (up to isomorphism) that have planar annihilator graphs.

Theorem 5.1. *(Ref. 31 Theorem 15) Let R be a quasi-local ring. Then $AG(R)$ is a planar if and only if R is ring-isomorphic to one of the following rings: $Z_4, \frac{Z_2[X]}{(X^2)}, Z_9, \frac{Z_3[X]}{(X^3)}, Z_8, \frac{Z_2[X]}{(X^3)}, \frac{Z_4[X]}{(X^3, X^2-2)}, \frac{Z_2[X,Y]}{(X^2, XY, Y^2)}, \frac{Z_4[X]}{(2X, X^2)}, \frac{F_4[X]}{(X^2)}$ (where F_4 denotes a field with 4 elements), $\frac{Z_4[X]}{(X^2+X+1)}, Z_{25}$, or $\frac{Z_5[X]}{(X^2)}$.*

For a reduced finite ring, we have the following result.

Theorem 5.2. *(Ref. 31 Theorem 16) Let R be a reduced finite ring that is not a field, i.e., $R = F_1 \times \cdots \times F_n$, where each F_i is a finite field and $n \geq 2$. Then $AG(R)$ is planar if and only if R is ring-isomorphic to one of the following rings: $Z_2 \times F, Z_3 \times F, Z_2 \times Z_2 \times Z_2, Z_2 \times Z_2 \times Z_3$, where F is a finite field.*

If R is a non-reduced finite ring, then we have the following.

Theorem 5.3. *(Ref. 31 Theorem 17) Assume that R is ring-isomorphic to $R_1 \times \cdots \times R_n \times F_1 \times \cdots \times F_m$, where each R_i is a finite quasi-local ring that is not a field, each F_i is a finite field, and $n, m \geq 1$. Then $AG(R)$ is planar if and only if R is ring-isomorphic to one of the following rings: $Z_4 \times Z_2$, $\frac{Z_2[X]}{(X^2)} \times Z_2$.*

The following result classifies (up to isomorphism) all quasi-local rings that have genus one annihilator graphs.

Theorem 5.4. *(Ref. 31 Theorem 18) Let R be a quasi-local ring. Then $g(AG(R)) = 1$ if and only if R is ring-isomorphic to one of the following rings:*
Z_{16}, $\frac{Z_2[X]}{(X^4)}$, $\frac{Z_4[X]}{(X^4,X^2-2)}$, $\frac{Z_2[X]}{(X^3-2,X^4)}$, $\frac{Z_4[X]}{(X^3+X^2-2,X^4)}$, $\frac{Z_2[X]}{(X^3,X^2-2X)}$, $\frac{Z_2[X,Y]}{(X^3,XY,Y^2-X^2)}$, $\frac{Z_8[X]}{(X^2-4,2x)}$, $\frac{Z_4[X,Y]}{(X^3,XY,X^2-2,Y^2-2,Y^3)}$, $\frac{Z_4[X]}{(X^2)}$, $\frac{Z_4[X,y]}{(X^2,Y^2,XY-2))}$, $\frac{Z_2[X,Y]}{(X^2,Y^2)}$, $\frac{Z_2[X,Y]}{(X^2,Y^2,XY)}$, $\frac{Z_4[X]}{(X^3,2x)}$, $\frac{Z_4[X,Y]}{(X^3,X^2-2,XY,Y^2)}$, $\frac{Z_8[X]}{(X^2)}$, $\frac{F_8[X]}{(X^2)}$, $\frac{Z_4[X]}{(X^3+X+1)}$, $\frac{Z_4[X,Y]}{(2X,2Y,X^2,Y^2,XY)}$, $\frac{Z_2[X,Y,Z]}{(X,Y,Z)^2}$, Z_{49}, *or* $\frac{Z_7[X]}{(X^2)}$.

The following result classifies (up to isomorphism) all finite reduced rings that have genus one annihilator graphs.

Theorem 5.5. *(Ref. 31 Theorem 19) Let R be a reduced finite ring that is not a field, i.e.,R is ring-isomorphic to $F_1 \times \cdots \times F_n$, where each F_i is a finite field and $n \geq 2$. Then $g(AG(R)) = 1$ if and only if R is ring-isomorphic to one of the following rings: $F_4 \times F_4, F_4 \times Z_5, Z_5 \times Z_5$, or $F_4 \times Z_7$.*

If R is a non-reduced finite ring, then we have the following.

Theorem 5.6. *(Ref. 31 Theorem 20)Assume that R is ring-isomorphic to $R_1 \times \cdots \times R_n \times F_1 \times \cdots \times F_m$, where each R_i is a finite quasi-local ring that is not a field, each F_i is a finite field, and $n, m \geq 1$. Then $g(AG(R)) = 1$ if and only if R is ring-isomorphic to one of the following rings: $Z_4 \times Z_3$, or $\frac{Z_2[X]}{(X^2)} \times Z_3$.*

6. Extended zero-divisor graph of R: $EG(R)$

Recall[26] that the extended zero-divisor graph of R is the undirected (simple) graph $EG(R)$ with the vertex set $Z(R)^*$, and two distinct vertices x and y are adjacent if and only if either $Rx \cap ann_R(y) \neq \{0\}$ or $Ry \cap ann_R(x) \neq \{0\}$. Hence it follows that the zero-divisor graph $\Gamma(R)$ is a subgraph of $EG(R)$.

In the following result, we collect some basic properties of $EG(R)$.

Theorem 6.1.[26] *Let R be a ring. Then*

(1) (Ref. 26 Theorem 2.1) $EG(R)$ is connected and $diam(EG(R)) \leq 2$. Moreover, if $E(G)$ has a cycle, then $gr(EG(R)) \leq 4$.
(2) (Ref. 26 Theorem 2.2) If $EG(R)$ has a cycle, then $gr(E(G)) = 4$ if and only if R is reduced with $|Min(R)| = 2$.

(3) *(Ref. 26 Theorem 3.2) $EG(R)$ is a star graph if and only if one of the following statements holds:*

  (a) *R is ring-isomorphic to $Z_2 \times D$, where D is an integral domain.*
  (b) $|Z(R)| = 3$.
  (c) *$Nil(R)$ is a prime ideal of R and $|Nil(R)| = 2$.*

(4) *(Ref. 26 Theorem 3.3) Suppose that R is a non-reduced ring such that $EG(R)$ is a star graph. Then the following statements hold:*

  (a) *R is indecomposable.*
  (b) *Either $|Z(R)| = 3$ or $|Z(R)| = \infty$.*

(5) *Assume that R is ring-isomorphic to $D_1 \times \cdots \times D_n$, where $n \geq 2$ and each D_i is an integral domain. Then $EG(R)$ is a complete $(2^n - 2)$-partite graph.*

Let R be a ring and $x, y \in R$. The authors in Ref. 26 called an element x an Ry-regular element if $x \notin Z(Ry)$ and $RxRy \neq Ry$.

Theorem 6.2. *(Ref. 26 Theorem 3.5) Let R be a non-reduced ring. Then $EG(R)$ is complete if and only if R is indecomposable and either x is not Ry-regular or y is not Rx-regular, for every distinct $x, y \in Z(R)^*$.*

7. When does $EG(R) = \Gamma(R)$?

Since $\Gamma(R)$ is always an induced subgraph of $EG(R)$, it is natural to ask when does $EG(R) = \Gamma(R)$? First, we consider the case when R is reduced.

7.1. *Case I: R is reduced*

Theorem 7.1.[26] *Let R be a reduced ring that is not an integral domain.*

(1) *(Ref. 26 Theorem 4.1, Corollary 4.3) Assume $|Min(R)| = n$. The following statements are equivalent:*

  (a) $n = 2$;
  (b) $\Gamma(R) = EG(R)$;
  (c) $gr(EG(R)) = gr(\Gamma(R)) \in \{4, \infty\}$.

(2) *(Ref. 26 Corollary 4.1) The following statements are equivalent:*

  (a) $gr(EG) = \infty$;
  (b) *$EG(R) = \Gamma(R)$ and $gr(EG(R)) = \infty$;*
  (c) $gr(\Gamma(R)) = \infty$;
  (d) *$|Min(R)| = 2$ and at least one minimal prime ideal of R has exactly two distinct elements;*
  (e) *$\Gamma(R) = K_{1,n}$ for some $n \geq 1$.*
  (f) *$EG(R) = K_{1,n}$ for some $n \geq 1$.*

(3) *(Ref. 26 Corollary 4.2) The following statements are equivalent:*

(a) $gr(EG(R)) = 4$;
(b) $EG(R) = \Gamma(R)$ *and* $gr(\Gamma(R)) = 4$;
(c) $gr(EG(R)) = 4$;
(d) $|Min(R)| = 2$ *and each minimal prime ideal of R has at least three distinct elements;*
(e) $EG(R) = K_{m,n}$ *for some* $m, n \geq 2$;
(f) $\Gamma(R) = K_{m,n}$ *for some* $m, n \geq 2$.

Now we consider the case when R is non-reduced.

7.2. *Case II: R is non-reduced*

Theorem 7.2. *(Ref. 26 Theorem 4.3) Let R be a non-reduced ring. Then the following statements are equivalent:*

(1) $gr(EG(R)) = \infty$;
(2) $EG(R)$ *is a star graph;*
(3) $EG(R) = \Gamma(R)$ *and* $gr(\Gamma(R)) = \infty$;
(4) $ann_R(Z(R))$ *is a prime ideal of R and either* $|Z(R)| = |ann_R(Z(R))| = 3$ *or* $|ann_R(Z(R))| = 2$ *and* $|Z(R)| = \infty$;
(5) $EG(R) = K_{1,1}$ *or* $EG(R) = K_{1,\infty}$;
(6) $\Gamma(R) = K_{1,1}$ *or* $\Gamma(R) = K_{1,\infty}$.

8. When is $EG(R)$ planar?

Recall that a graph G is called a planar if it can be drawn in the plane so that the edges of G do not cross.

Theorem 8.1. *(Ref. 27 Theorem 3.2) Let R be a ring such that either R is ring-isomorphic to* $R_1 \times R_2 \times R_3$ *(for some rings* R_1, R_2, R_3*) or* $|Min(R)| \geq 3$ *and R is ring-isomorphic to* $R_1 \times R_2$*(for some rings* R_1, R_2*), then* $EG(R)$ *is not a planar.*

For a reduced ring R, we have the following result.

Theorem 8.2. *(Ref. 27 Theorem 3.3) Let R be a reduced ring. Then the following statements hold:*

(1) $EG(R)$ *is planar;*
(2) $|Min(R)| = 2$ *and one of the minimal prime ideals of R has at most three distinct elements.*

For a non-reduced ring R, we have the following result.

Theorem 8.3.[27] *Let R be a non-reduced ring. Then*

(1) *(Ref. 27 Theorem 3.4) Suppose that R is not ring-isomorphic to either* Z_4 *or* $\frac{Z_2[X]}{(X^2)}$. *Then*

(a) *Suppose that* $|Z(R)| < \infty$. *Then* $EG(R)$ *is planar if and only if* R *is ring-isomorphic to either* $Z_2 \times Z_4$ *or* $Z_2 \times \frac{Z_2[X]}{(X^2)}$.

(b) *Suppose that* $|Z(R)| = \infty$. *Then* $EG(R)$ *is planar if and only if* $ann_R(R)$ *is a prime ideal of* R.

(2) *(Ref. 27 Theorem 3.5) Suppose that* $|Nil(R)| = 3$. *Then* $ann_R(Z(R))$ *is a prime ideal of* R *if and only if* $EG(R)$ *is planar.*

(3) *(Ref. 27 Theorem 3.6) If* $|Nil(R)| \geq 6$, *then* $EG(R)$ *is not planar. If* $4 \leq |Nil(R)| \leq 5$, *then* $EG(R)$ *is planar if and only if* $Z(R) = Nil(R)$.

Bibliography

1. Akbari, S., Maimani, H. R., Yassemi, S., When a zero-divisor graph is planar or a complete r-partite graph, *J. Algebra* **270** (2003), 169–180.
2. Akbari, S., Mohammadian, A., On the zero-divisor graph of a commutative ring, *J. Algebra* **274** (2004), 847–855.
3. Anderson, D. D., Naseer, M., Beck's coloring of a commutative ring, *J. Algebra* **159** (1993), 500–514.
4. Abbasi, A., Habib, S., The total graph of a commutative ring with respect to proper ideals, *J. Korean Math. Soc.* **49** (2012), 85–98.
5. Akbari, S., Heydari, F., The regular graph of a non-commutative ring, *Bulletin of the Australian Mathematical Society* (2013), Doi: 10.1017/S0004972712001177.
6. Akbari, S., Aryapoor, M., Jamaali, M., Chromatic number and clique number of subgraphs of regular graph of matrix algebras, *Linear Algebra Appl.* **436** (2012), 2419–2424.
7. Akbari, S., Jamaali, M., Seyed Fakhari, S. A., The clique numbers of regular graphs of matrix algebras are finite, *Linear Algebra Appl.* **43** (2009), 1715-1718.
8. Akbari, S., Kiani, D., Mohammadi, F., Moradi, S., The total graph and regular graph of a commutative ring, *J. Pure Appl. Algebra* **213** (2009), 2224–2228.
9. Anderson, D. D., Winders, M., Idealization of a module, *J. Comm. Algebra* **1** (2009), 3–56.
10. Anderson, D. F., Axtell, M., Stickles, J., Zero-divisor graphs in commutative rings. In: Fontana, M., Kabbaj, S. E., Olberding, B., Swanson, I. (eds.) Commutative Algebra Noetherian and Non-Noetherian Perspectives, pp. 23–45, Springer-Verlag, New York (2010).
11. Anderson, D. F., Badawi, A., On the zero-divisor graph of a ring, *Comm. Algebra* **36** (2008), 3073–3092.
12. Anderson, D. F., Badawi, A., The total graph of a commutative ring, *J. Algebra* **320** (2008), 2706–2719.
13. Anderson, D. F., Badawi, A., The total graph of a commutative ring without the zero element, *J. Algebra Appl.* (2012), doi: 10.1142/S0219498812500740.
14. Anderson, D. F., Badawi, A., The generalized total graph of a commutative ring, *J. Algebra Appl.* (2013), doi: 10.1142/S021949881250212X.
15. Anderson, D. F., Fasteen, J., LaGrange, J. D., The subgroup graph of a group, *Arab. J. Math.* **1** (2012), 17–27.
16. Anderson, D. F., Livingston, P. S., The zero-divisor graph of a commutative ring, *J. Algebra* **217** (1999), 434–447.
17. Anderson, D. F., Mulay, S. B., On the diameter and girth of a zero-divisor graph, *J. Pure Appl. Algebra* **210** (2007), 543–550.
18. Afkhami, M., Khashyarmanesh, K. and Sakhdari, S. M., The annihilator graph of a commutative semigroup, *J. Algebra Appl.* **14** 1550015 (2015), [14 pages] DOI: 10.1142/S0219498815500152.

19. Atani, S. E., Habibi, S., The total torsion element graph of a module over a commutative ring, *An. Stiint. Univ. Ovidius Constanta Ser. Mat.* **19** (2011), 23–34.
20. Axtel, M., Coykendall, J. and Stickles, J., Zero-divisor graphs of polynomials and power series over commutative rings, *Comm. Algebra* **33** (2005), 2043–2050.
21. Axtel, M., Stickles, J., Zero-divisor graphs of idealizations, *J. Pure Appl. Algebra* **204** (2006), 235–243.
22. Badawi, A., On the dot product graph of a commutative ring, *Comm. Algebra* **43** (2015), 43–50.
23. Badawi, A., On the annihilator graph of a commutative ring, *Comm. Algebra*, Vol. (42)(1), 108–121 (2014), DOI: 10.1080/00927872.2012.707262.
24. Bakhtyiari, M., Nikandish, R., Nikmehr, M. J., More on the annihilator graph of a commutative ring, to appear in Hokkaido Mathematical Journal.
25. Bakhtyiari, M., Nikandish, R., Nikmehr, M. J., Coloring of the annihilator graph of a commutative ring, *J. Algebra Appl.* (2015), doi: 10.1142/S0219498816501243.
26. Bakhtyiari, M., Nikandish, R., Nikmehr, M. J., The Extended zero-Divisor graphof a commutative Ring I, to appear in Hokkaido Mathematical Journal.
27. Bakhtyiari, M., Nikandish, R., Nikmehr, M. J., The Extended zero-Divisor graphof a commutative Ring II, to appear in Hokkaido Mathematical Journal.
28. Barati, Z., Khashyarmanesh, K., Mohammadi, F., Nafar, K., On the associated graphs to a commutative ring, *J. Algebra Appl.* (2012), doi: 10.1142/S021949881105610.
29. Beck, I., Coloring of commutative rings, *J. Algebra* **116** (1988), 208–226.
30. Bollaboás, B., Graph Theory, An Introductory Course, Springer-Verlag, New York (1979).
31. Chelvam, T., Selvakumar, K., On the genus of the annhilator graph of a commutative ring, to appear in Algebra and Discrete Mathematics.
32. Chelvam, T., Asir, T., Domination in total graph on $\mathbb{Z}_n$, *Discrete Math. Algorithms Appl.* **3** (2011), 413–421.
33. Chelvam, T., Asir, T., Domination in the total graph of a commutative ring, *J. Combin. Math. Combin. Comput.*, to appear.
34. Chelvam, T., Asir, T., Intersection graph of gamma sets in the total graph, *Discuss. Math. Graph Theory* **32** (2012), 339–354.
35. Chelvam, T., Asir, T., On the Genus of the Total Graph of a Commutative Ring, *Comm. Algebra* **41** (2013), 142–153.
36. Chelvam, T., Asir, T., On the total graph and its complement of a commutative ring, *Comm. Algebra* (2013), doi:10.1080/00927872.2012.678956.
37. Chelvam, T., Asir, T., The intersection graph of gamma sets in the total graph I, *J. Algebra Appl.* (2013), doi: 10.1142/S0219498812501988.
38. Chelvam, T., Asir, T., The intersection graph of gamma sets in the total graph II, *J. Algebra Appl.* (2013), doi: 10.1142/S021949881250199X.
39. Chiang-Hsieh, H.-J., Smith, N. O., Wang, H.-J., Commutative rings with toroidal zerodivisor graphs, *Houston J. Math.* **36** (2010), 1–31.
40. DeMeyer, F., DeMeyer, L., Zero divisor graphs of semigroups, *J. Algebra* **283** (2005), 190–198.
41. DeMeyer, F., McKenzie, T., Schneider, K., The zero-divisor graph of a commutative semigroup, *Semigroup Forum* **65** (2002), 206–214.
42. DeMeyer, F., Schneider, K., Automorphisms and zero divisor graphs of commutative rings. In: Commutative rings. Hauppauge, NY: *Nova Sci. Publ.* 2002, pp. 25–37.
43. DeMeyer, L., DSa, M., Epstein, I., Geiser, A., Smith, K., Semigroups and the zero divisor graph, *Bull. Inst. Combin. Appl.* **57** (2009), 60–70.
44. DeMeyer, L., Greve, L., Sabbaghi, A., Wang, J., The zero-divisor graph associated to a semigroup, *Comm. Algebra* **38** (2010), 3370–3391.
45. Khashyarmanesh, K., Khorsandi, M. R., A generalization of the unit and unitary Cayley graphs of a commutative ring, *Acta Math. Hungar.* **137** (2012), 242–253.

46. LeVeque, W. L, *Fundamentals of number theory*, Addison-Wesley Publishing Company, Reading, Massachusetts, 1977.
47. Maimani, H. R., Pouranki, M. R., Tehranian, A., Yassemi, S., Graphs attached to rings revisited, *Arab. J. Sci. Eng.* **36** (2011), 997–1011.
48. Maimani, H. R., Pournaki, M. R., Yassemi, S., Zero-divisor graph with respect to an ideal, *Comm. Algebra* **34** (2006), 923–929.
49. Maimani, H. R., Wickham, C., Yassemi, S., Rings whose total graphs have genus at most one, *Rocky Mountain J. Math.* **42** (2012), 1551–1560.
50. Mojdeh, D. A., Rahimi, A. M., Domination sets of some graphs associated to commutative ring, *Comm. Algebra* **40** (2012), 3389–3396.
51. Mulay, S. B., Cycles and symmetries of zero-divisors, *Comm. Algebra* **30** (2002), 3533–3558.
52. Pucanović, Z., Petrović, Z., On the radius and the relation between the total graph of a commutative ring and its extensions, *Publ. Inst. Math.* (Beograd) (N.S.) **89** (2011), 1–9.
53. Redmond, S. P., An ideal-based zero-divisor graph of a commutative ring, *Comm. Algebra* **31** (2003), 4425–4443.
54. Smith, N. O., Planar zero-divisor graphs, *Comm. Algebra* **35** (2007), 171–180.
55. Sharma, P. K., Bhatwadekar, S. M., A note on graphical representations of rings, *J. Algebra* **176** (1995), 124–127.
56. Shekarriz, M. H., Shiradareh Haghighi, M. H., Sharif, H., On the total graph of a finite commutative ring, *Comm. Algebra* **40** (2012), 2798–2807.
57. Visweswaran, S., Patel, H. D., A graph associated with the set of all nonzero annihilating ideals of a commutative ring, *Discrete Math. Algorithm. Appl.* **06**, 1450047 (2014), [22 pages] DOI: 10.1142/S1793830914500475.
58. Wickham, C., Classification of rings with genus one zero-divisor graphs, *Comm. Algebra* **36** (2008), 325–345.

Different prime graphs of a nearring with respect to an ideal

Kedukodi Babushri Srinivas, Jagadeesha B., Kuncham Syam Prasad

Department of Mathematics, Manipal Institute of Technology, Manipal University, Manipal, Karnataka 576104, India

E-mail: babushrisrinivas.k@manipal.edu, jagadeesh1_bhat@yahoo.com, syamprasad.k@manipal.edu

Juglal Suresh

Department of Mathematics and Applied Mathematics, Nelson Mandela Metropolitan University, Port Elizabeth, 6031, South Africa

E-mail: suresh.juglal@nmmu.ac.za

Let I be an ideal of a nearring N. We introduce the notions of equiprime graph of N denoted by $EQ_I(N)$ and c-prime graph of N denoted by $C_I(N)$. We relate $EQ_I(N)$, $C_I(N)$ and the graph of a nearring with respect to an ideal, $G_I(N)$. We prove that $diam(EQ_I(N \setminus I)) \leq 3$ and $diam(C_I(N \setminus I)) \leq 3$ and deduce that the prime graphs are edge partitionable. It is well-known that the homomorphic image of a prime ideal need not be a prime ideal in general. We study graph homomorphisms and obtain conditions under which the primeness property of an ideal is preserved under nearring homomorphisms.

Keywords: graph, nearring, ideal, prime.

1. Introduction

Bhavanari, Kuncham and Kedukodi[13] introduced the graph of a nearring N with respect to an ideal I denoted by $G_I(N)$. The equiprime graph of a nearring N with respect to an ideal I is denoted by $EQ_I(N)$. We find that if I is an equiprime ideal of N then I is a vertex cover of $EQ_I(N)$. The c-prime graph of a nearring N with respect to an ideal I is denoted by $C_I(N)$. We find that if I is a c-prime ideal of N then I is a vertex cover of $C_I(N)$. We find bounds for the diameters of $EQ_I(N \setminus I)$ and $C_I(N \setminus I)$. For a commutative ring N, we prove that the zero divisor graph $\Gamma(N)$ is same as $C_{\{0\}}(N \setminus \{0\})$. We show that the graphs $EQ_I(N)$, $G_I(N)$ and $C_I(N)$ are edge partitionable. We prove that the graphs $G_I(N)$ and $EQ_I(N)$ coincide if I is an equiprime ideal of N or a totally reflexive ideal of N. We prove that $C_I(N)$ and $G_I(N)$ coincide if I is a c-prime ideal. If I is an equiprime ideal of a right permutable nearring N or a prime ideal of a commutative ring N then we prove that the graphs $G_I(N)$, $EQ_I(N)$ and $C_I(N)$ coincide. A graph homomorphism is a mapping between two graphs that preserves adjacency. We prove that an onto nearring homomorphism is a graph homomorphism between two prime graphs. The primeness property of an ideal need not be preserved under homomorphism. By using properties of graph homomorphisms and prime graphs we find conditions under which homomorphic images of different prime ideals are prime ideals.

Beck[10] introduced the zero divisor graph of a commutative ring R, denoted by $\Gamma(R)$. Anderson and Livingston[5] proposed a generalized version for the definition for $\Gamma(R)$. Redmond[38] found bounds on the size of a ring using the zero divisor graph of the ring. Anderson and Badawi[7] studied the zero divisor graph of a ring. Anderson, Mulay[6] studied zero divisor graphs of commutative rings and deduced properties of zero divisor graphs of polynomial rings and power series rings. Axtell and Stickles[9] studied diameter and girth of the zero divisor graph idealizations of the ring. Anderson, Levy and Shapiro[4] proved that the zero divisor graph of the total quotient ring of a commutative ring R and the zero divisor graph of R are isomorphic. Cannon, Neuerburg and Redmond[15] introduced the zero divisor graphs of nearrings and semigroups. Chelvam and Nitya[17] studied the zero divisor graph of the ideal of a nearring. Anderson and Badawi[3] defined the total graph of a commutative ring and studied induced subgraphs of total graphs. Anderson and Badawi[8] found properties of the total graphs of a commutative ring without zero element. Akbari, Kiani, Mohammadi and Moradi[2] obtained a relation between the total graph and the regular graph of a commutative ring. Chakrabarty, Ghosh, Mukherjee and Sen[16] introduced the intersection graph of nontrivial left ideals of a ring. Redmond[37] introduced ideal based zero divisor graphs of a nearring. Aichinger, Cannon, Ecker, Kabza and Neuerburg[1] studied relations between ideals of nearrings and Noetherian quotients. Groenewald, Juglal and Meyer[25] studiedprime ideals in group near-rings. Groenewald and Juglal[26] further studied primeness in the generalized group near-rings. Davvaz[18,19,22], Davvaz and Kim[20], Davvaz and Cristea[22] studied rough sets and hyperstructures. Veldsman[40] studied equiprime nearrings. Kedukodi, Kuncham and Bhavanari[31,32] introduced equiprime, 3-prime and c-prime fuzzy ideals and related the concept with rough sets. Jagadeesha, Kedukodi and Kuncham[29] introduced interval valued L-fuzzy ideals based on t-norms and t-conorms. Kuncham, Kedukodi and Jagadeesha[34] studied interval valued L-fuzzy cosets.

2. Definitions and Preliminaries

In this paper, N, N_1 and N_2 represent right nearrings and R represents a ring. We refer to Pilz[36], Ferrero and Ferrero[23] for nearrings, Harrary[27], Bhavanari and Kuncham[11,12] for topics on graph theory, Lafuerza–Guillén and Harikrishnan[35] for probabilistic normed spaces, Kuncham, Kedukodi, Harikrishnan and Bhavanari[33] for recent developments in nearrings.

(Booth, Groenewald and Veldsman[14], Groenewald[24], Veldsman[40]) An ideal I of N is called *equiprime* if for all $a,x,y \in N$ with $arx - ary \in I$ for all $r \in N$ implies $a \in I$ or $x - y \in I$. An ideal I of N is called *3-prime* if $a,b \in N$ and $anb \in I$ for all $n \in N$ implies $a \in I$ or $b \in I$. An ideal I of N is called *c-prime* if $a,b \in N$ and $ab \in I$ implies $a \in I$ or $b \in I$.

Definition 2.1. (Veldsman[39]) An ideal I of N is called *e-semiprime* if for $a,b \in N$ with $(a-b)ra - (a-b)rb \in I$ for all $r \in N$ implies $(a-b) \in I$.

(Pilz[36]) A nearring N is said to be *right permutable* if $abc = acb$ for all $a,b,c \in N$.

An ideal I of N is said to be *totally reflexive* if $aNb \subseteq I$ then $bNa \subseteq I$ for all $a,b \in N$.

An ideal I of N is called *3-semprime* if for $x \in N$ with $xNx \subseteq I$ implies $x \in I$.

An ideal I of N is said to have *insertion of factors property* (IFP) if $a,b \in N$ with $ab \in I$ implies $anb \in I$ for all $n \in N$.

(Harary[27]) A *graph* $G = (V,E)$ consists of a set of objects $V = \{v_1, v_2, ...\}$ called *vertices* (or points) and another set $E = \{e_1, e_2, ..\}$ whose elements are called *edges* such that each edge e is identified with an unordered pair (v_i, v_j) of vertices. Let $G = (V,E)$ be a graph. The degree of a vertex $v \in V$ is the number of edges having end point v denoted by $deg(v)$. A graph $G = (V,E)$ is said to be complete if $(u,v) \in E$ for all $u, v \in V, u \neq v$. A complete graph with n vertices is denoted by K_n. Let $G_1 = (V_1, E_1)$ and $G_2 = (V_2, E_2)$ be two graphs. Their union $G_1 \cup G_2$ is a graph $G = (V,E)$ having vertex set $V = V_1 \cup V_2$ and edges $E = E_1 \cup E_2$.

The *bipartite graph* G is a graph whose point set V is partitioned into two subsets V_1 and V_2 such that every edge of G joins V_1 with V_2. If G contains every edge joining V_1 and V_2 then G is called a *complete bipartite graph.*

If $|V_1| = m, |V_2| = n$, then G can be written as $K_{m,n}$. The graph $K_{1,n}$ is called a *star graph.* The graph $K_{1,n}$ denotes a star graph with $V_1 = \{v_1\}$ and here v_1 represents the *root vertex* of the star graph.

A graph G is said to be *connected* if there is an edge between every pair of vertices of G. An alternating sequence of points and edges of a graph G, beginning and ending with points is called *walk* of G. If all the points and edges of the walk are distinct then the walk is called a *path.* A path with n vertices is denoted by P_n. Let x and y be distinct vertices of graph G. Let $d(x,y)$ be the length of shortest path between x and y ($d(x,y) = \infty$ if there is no such path). The *diameter of* G is denoted by $diam(G)$, given by $diam(G) = sup\{d(x,y) \mid x$ and y are distinct vertices of $G\}$. A *vertex cover* of graph G is a subset K of V such that if (u,v) is an edge of G, then $u \in K$ or $v \in K$ or both $u \in K$ and $v \in K$. A maximal connected subgraph of G is called a *component* of G. If a component contains only one vertex then that component is called a *trivial component.*

Definition 2.2. (Hell[28]) Let $G_1 = (V_1, E_1)$ and $G_2 = (V_2, E_2)$ be two graphs. A *homomorphism* of G_1 to G_2 is a mapping $f : V_1 \to V_2$ such that $(f(u), f(v)) \in E_2$ whenever $(u,v) \in E_1$.

(Bhavanari, Kuncham and Kedukodi[13])
Let I be an ideal of N. The *graph of* N *with respect to* I is a graph with each element of N as vertex and two distinct vertices x and y are adjacent if and only if $xNy \subseteq I$ or $yNx \subseteq I$. The graph is denoted by $G_I(N)$. If we restrict the vertex set of $G_I(N)$ to $N \setminus I$ the graph thus obtained after deleting the isolated vertices if any is denoted by $G_I(N \setminus I)$. The graph $G_I(N)$ is said to be *ideal symmetric* if every pair of vertices $x, y \in G_I(N)$ with an edge between them, either $deg(x) = deg(0)$ or $deg(y) = deg(0)$.

Definition 2.3. (Anderson and Livingston[5]) Let R be a commutative ring and let $Z(R)$ be its set of zero divisors. Then the *zero divisor graph* of R denoted by $\Gamma(R)$ is a simple graph with vertices
$Z(R)^* = Z(R) - \{0\}$ (the set of nonzero zero divisors of R) and for distinct x and $y \in Z(R)^*$, the vertices x and y are adjacent if and only if $xy = 0$.

Lemma 2.1. *(Ferrero and Ferrero[23]) If I is an ideal of N which satisfies $n0 \in I$ for all $n \in N$ then $NI \subseteq I$.*

Remark 2.1. Let I be an ideal of N, $i \in I$, and $n_1, n_2 \in N$. By a property of the ideal,
(i) $n_1(n_2+i) - n_1n_2 \in I$. This implies $n_1(n_2+i) = i_1 + n_1n_2$ for some $i_1 \in I$.
(ii) $n_1(i+n_2) - n_1n_2 \in I$. This implies $n_1(i+n_2) = i_2 + n_1n_2$ for some $i_2 \in I$.
(iii) $n_1 + i - n_1 \in I$. This implies $n_1 + i = i_3 + n_1$ for some $i_3 \in I$.

3. Graphs $EQ_I(N)$ and $C_I(N)$

Notation 3.1. Let I be an ideal of N. We denote $H_I = \{p \in N \mid \text{for } x, y \in N, pr(x-y) - pr0 \in I \text{ for all } r \in N \Leftrightarrow prx - pry \in I \text{ for all } r \in N\}$.

Remark 3.1. Let I be an ideal of N. Then $0 \in H_I$. Therefore, $H_I \neq \emptyset$.

Example 3.1. Let $N = \{0, a, b, c\}$ be the nearring with addition $+$ and multiplication $\cdot$ defined as in Table 1.

$+$	0	a	b	c
0	0	a	b	c
a	a	0	c	b
b	b	c	0	a
c	c	b	a	0

$\cdot$	0	a	b	c
0	0	0	0	0
a	0	a	0	a
b	b	b	b	b
c	b	c	b	c

Table 1: Nearring for Example 3.1.

Ideals of N are $\{0\}, \{0,a\}, \{0,b\}, N$. We have $H_{\{0\}} = H_{\{0,a\}} = H_{\{0,b\}} = H_N = N$.

Proposition 3.1. *Let I be an ideal of N and ${}_IH = \{p \in N \mid prx_1 - pry_1 \in I \text{ for all } r \in N \Leftrightarrow prx_2 - pry_2 \in I \text{ for all } r \in N, \text{for every } x_1, x_2, y_1, y_2 \in N \text{ satisfying } x_1 - y_1 = x_2 - y_2\}$. Then ${}_IH = H_I$.*

Proof. Let I be an ideal of N and $p \in H_I$. Let $x_1, x_2, y_1, y_2 \in N$ be such that $x_1 - y_1 = x_2 - y_2$. Then for $r \in N$, we have $prx_1 - pry_1 \in I \Leftrightarrow pr(x_1 - y_1) - pr0 \in I \Leftrightarrow pr(x_2 - y_2) - pr0 \in I \Leftrightarrow prx_2 - pry_2 \in I$. Hence, we get $p \in {}_IH$. This proves $H_I \subseteq {}_IH$. To prove ${}_IH \subseteq H_I$, let $p \in {}_IH$. Fix $x, y \in N$. Then for $r \in N$, $prx - pry \in I \Leftrightarrow pr(x-y) - pr0 \in I$ (by taking $x_1 = x, y_1 = y, x_2 = x - y, y_2 = 0$ in the condition of ${}_IH$). This implies $p \in H_I$. Thus, ${}_IH = H_I$. □

Proposition 3.2. *Let I be an ideal of N. If $p \in N$ such that pr is a distributive element of N for all $r \in N$. Then $p \in H_I$.*

Proposition 3.3. *Let $a, x, y \in N$. If I is an ideal of N satisfying the condition, $arx - ary \in I$ for all $r \in N$ if and only if $xra - yra \in I$ for all $r \in N$, then $H_I = N$.*

Proposition 3.4. *Let N be a distributive nearring and I be an ideal of N. Then $H_I = N$.*

Proposition 3.5. *Let I be an ideal of N such that for every $x_1, y_1, x_2, y_2 \in N$ with $x_1 - y_1 = x_2 - y_2$ implies $-x_1 + x_2 \in I$ and $-y_1 + y_2 \in I$. Then $H_I = N$.*

Proof. Let $a \in N$ such that $akx_1 - aky_1 \in I$ for all $k \in N$. Fix $r \in N$.
Now, $arx_2 - ary_2 = ar(x_1 - y_1 + y_2) - ary_2$ because $x_1 - y_1 = x_2 - y_2$
$= ar(x_1 + (-y_1 + y_2)) - ary_2 = i_1 + arx_1 - ary_2$, for some $i_1 \in I$ because $-y_1 + y_2 \in I$
$= i_1 + arx_1 - ar(y_1 - x_1 + x_2)$ because $x_1 - y_1 = x_2 - y_2$
$= i_1 + arx_1 - ar((y_1 - x_1 + x_2) - y_1 + y_1) = i_1 + arx_1 - ar((y_1 + (-x_1 + x_2) - y_1) + y_1)$
$= i_1 + arx_1 - ar(i_2 + y_1)$ for some $i_2 \in I$ because $y_1 + (-x_1 + x_2) - y_1 \in I$
$= i_1 + arx_1 - ar((y_1 - y_1) + i_2 + y_1) = i_1 + arx_1 - ar(y_1 + (-y_1 + i_2 + y_1))$
$= i_1 + arx_1 - ar(y_1 + i_3)$ for some $i_3 \in I$ because $-y_1 + i_2 + y_1 \in I$
$= i_1 + arx_1 - (i_4 + ary_1)$, for some $i_4 \in I$ by Remark 2.1
$= i_1 + (arx_1 - ary_1) - i_4 \in I$ because $arx_1 - ary_1 \in I$.
Therefore, $arx_2 - ary_2 \in I$ for all $r \in N$. Thus, $H_I = N$. □

Definition 3.1. Let I be an ideal of N, $H_I = N$ and $p \in N$. Let $EQ_I^p(N)$ be the graph with vertex set N and the pair of distinct vertices p and $(x-y)$ are adjacent if and only if $prx - pry \in I$ for all $r \in N$ or $(x-y)rp - (x-y)r0 \in I$ for all $r \in N$. Then $\cup_{p \in N} EQ_I^p(N)$ is called the *equiprime graph of nearring N with respect to ideal I* denoted by $EQ_I(N)$.

Remark 3.2. Let I be an ideal of N. For $x, y \in N$, we have $0rx - 0ry = 0 \in I$ for all $r \in N$. Hence, $EQ_I^0(N)$ is a star graph with root vertex 0 for an ideal I.

Example 3.2. Consider the nearring in Example 3.1. For $I = \{0, b\}$ we have $H_I = N$. The graph $EQ_I^p(N)$ with $I = \{0, b\}$ and different $p \in N$ is shown in Figure 1 and $EQ_I(N)$ is shown in Figure 2.

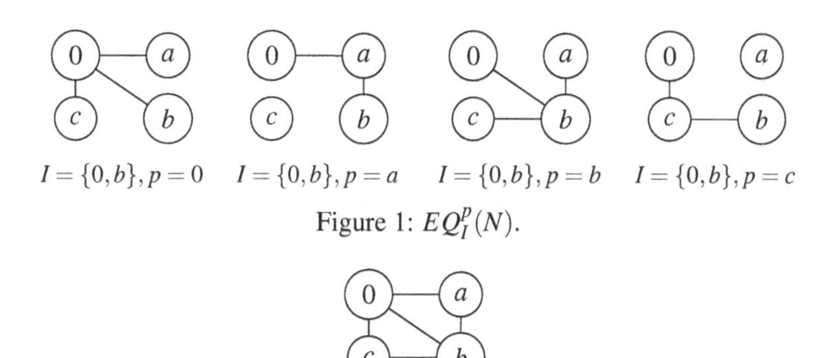

Figure 1: $EQ_I^p(N)$.

Figure 2: $EQ_I(N)$ when $I = \{0, b\}$.

Definition 3.2. Let I be an ideal of N. Let $C_I(N)$ be the graph with vertex set N in which the distinct pair of vertices x and y are adjacent if and only if $xy \in I$ or $yx \in I$. The graph $C_I(N)$ is called the *c-prime graph of N with respect to ideal I.*

Definition 3.3. The graph $C_I(N)$ is said to be *ideal symmetric* if for any pair of adjacent vertices $x, y \in C_I(N)$ either $deg(x) = deg(0)$ or $deg(y) = deg(0)$.

Example 3.3. Let $N = \{0, a, b, c\}$ be the nearring with addition $+$ and multiplication $\cdot$ defined as in Table 2.

$+$	0	a	b	c
0	0	a	b	c
a	a	0	c	b
b	b	c	0	a
c	c	b	a	0

$\cdot$	0	a	b	c
0	0	0	0	0
a	a	a	a	a
b	0	a	b	c
c	a	0	c	b

Table 2: Nearring for Example 3.3.

The graph $C_I(N)$ with respect to different ideals of N is shown in Figure 3.

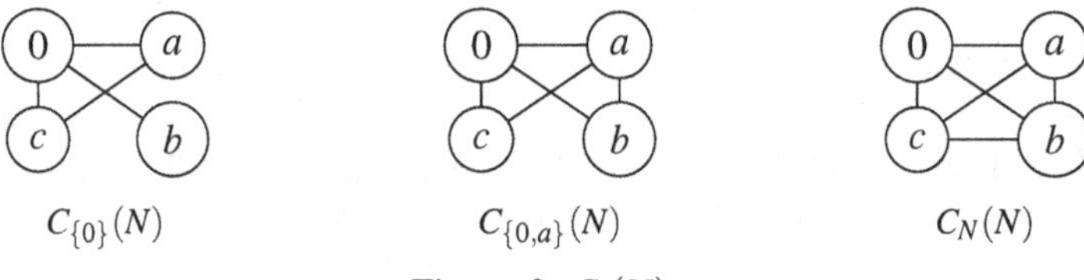

Figure 3: $C_I(N)$.

Proposition 3.6. *Let I be an ideal of N and $H_I = N$.*
(i) If $p \in I$ then $EQ_I^p(N)$ is a star graph.
(ii) If I is an equiprime ideal of N and $EQ_I^p(N)$ is a star graph with root vertex p then $p \in I$.
(iii) Let $x \in N$. If $x \in I$ then x is adjacent to all other vertices of N.

Proof. To prove (i), let $p \in I$ and $x \in N$. Then $prx \in I$ for all $r \in N$ and $pr0 \in I$ for all $r \in N \Rightarrow prx - pr0 \in I$ for all $r \in N$. Hence, p and $(x-0)$ are adjacent in $EQ_I^p(N)$. Therefore, p is adjacent to all other vertices of $EQ_I^p(N)$. Thus, $EQ_I^p(N)$ is a star graph with root vertex p.
To prove (ii), let $EQ_I^p(N)$ be a star graph with root vertex p. Then for $x \in N$, we get $prx - pr0 \in I$ for all $r \in N$ or $xrp - xr0 \in I$ for all $r \in N$. Let $prx - pr0 \in I$ for all $r \in N$. Suppose $I = N$, then $p \in I$. Let $I \subset N$. Choose $x \in N \setminus I$. As I is an equiprime ideal of N, we get $p \in I$. The proof is similar for $xrp - xr0 \in I$ for all $r \in N$. Thus, $EQ_I^p(N)$ is a star graph with root vertex p then $p \in I$.
To prove (iii), let x be a vertex in $EQ_I(N)$ and $x \in I$. Then by (i), we get $EQ_I^x(N)$ is a star graph with root vertex x. As $EQ_I(N) = \cup_{p \in N} EQ_I^p(N)$ we get, $deg(x) = deg(0)$ in $EQ_I(N)$. □

Remark 3.3. In Example 3.2, that $a \notin I$. Observe that $EQ_I^a(N)$ is not a star graph.

Remark 3.4. (i) If I is a totally reflexive ideal of N, then for all $a \in N$, $aN0 \subseteq I$.
(ii) We have $V(EQ_I(N)) = N = V(G_I(N))$.

Theorem 3.1. *Let I be an ideal of N and $H_I = N$. Then $G_I(N) = EQ_I(N)$ if any one of the following conditions is satisfied.*
(i) N is a zero-symmetric nearring.
(ii) N is a distributive nearring.
(iii) I is a totally reflexive ideal of N.
(iv) I is an equiprime ideal of N.

Proof. Let I be an ideal of N. To prove (i), suppose $x, y \in N$ such that $(x,y) \in E(G_I(N))$. Then $xNy \subseteq I$ or $yNx \subseteq I$. Let $xNy \subseteq I$, then $xny \in I$ for all $n \in N$. As N is zero-symmetric, $xny - xn0 \in I$. Then x and y are adjacent in $EQ_I^x(N)$. Hence, $(x,y) \in E(\cup_{p \in N} EQ_I^p(N)) = E(EQ_I(N))$. The proof is similar for $yNx \subseteq I$. Therefore, $E(G_I(N)) \subseteq E(EQ_I(N))$. Now suppose $a, b \in N$ such that $(a,b) \in E(EQ_I(N))$. Then $(a,b) \in E(\cup_{p \in N} EQ_I^p(N))$. Further, a and b are adjacent in $EQ_I^a(N)$ or a and b are adjacent in $EQ_I^b(N)$. Without loss of generality, assume a and b are adjacent in $EQ_I^a(N)$. Then $anb - an0 \in I$ for all $n \in N$ or $bna - bn0 \in I$ for all $n \in N$. As N is zero-symmetric, we get $an0 = 0, bn0 = 0$ for all $n \in N \Rightarrow anb \in I$ or $bna \in I$ for all $n \in N \Rightarrow aNb \subseteq I$ or $bNa \subseteq I \Rightarrow (a,b) \in E(G_I(N))$. Hence, $E(EQ_I(N)) \subseteq E(G_I(N))$. Therefore, $E(EQ_I(N)) = E(G_I(N))$. By Remark 3.4(ii), we have $V(EQ_I(N)) = V(G_I(N)) = N$. Thus, $G_I(N) = EQ_I(N)$. The proofs of (ii), (iii) and (iv) are similar to the proof of (i). □

We now provide examples to show that Theorem 3.1 is not true in general if conditions involved are excluded.

Example 3.4. Consider the nearring in Example 3.3. For $I = \{0\}$, we have $H_I = N$. The graph $EQ_I^p(N)$ with respect $I = \{0\}$ and different $p \in N$ is shown in Figure 4.

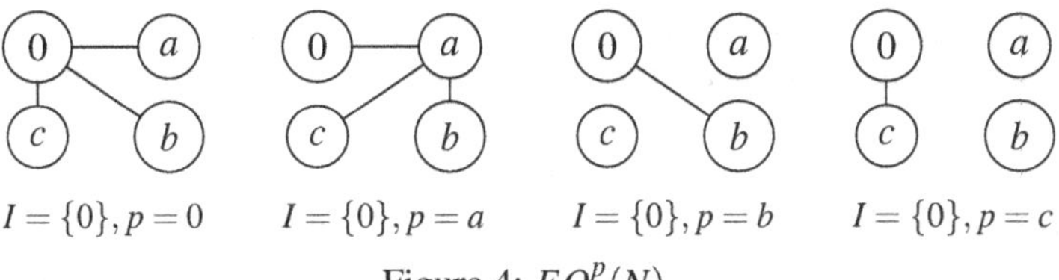

Figure 4: $EQ_I^p(N)$.

The graphs $G_I(N)$ and $EQ_I(N)$ are shown in Figure 5.

Figure 5: $G_I(N)$ and $EQ_I(N)$.

Note that N is not zero-symmetric. From Figure 5, observe that $G_I(N)$ is a proper subgraph of $EQ_I(N)$.

Example 3.5. Consider the nearring given in Example 3.1. For $I = \{0\}$ we get $H_I = N$. The graph $EQ_I^p(N)$ with respect to $I = \{0\}$ and different $p \in N$ is shown in Figure 6. The graph $G_I(N)$ and $EQ_I(N)$ are shown in the Figure 7.

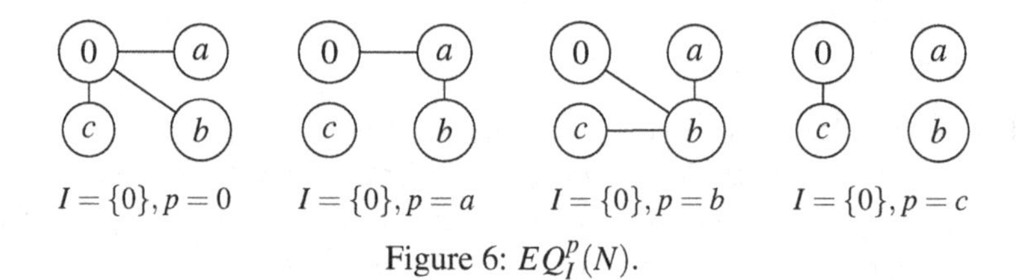

Figure 6: $EQ_I^p(N)$.

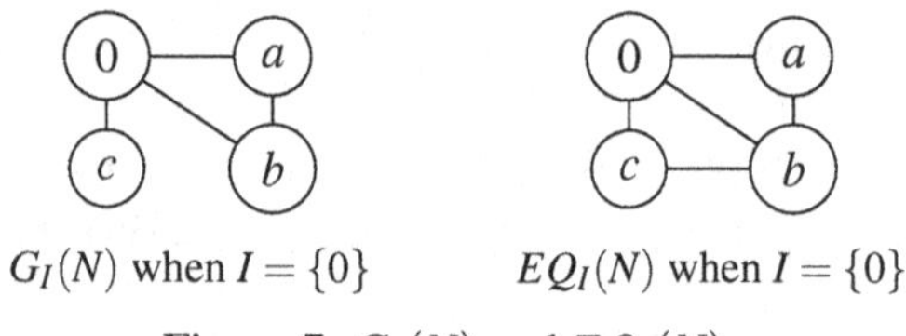

Figure 7: $G_I(N)$ and $EQ_I(N)$.

Note that $aNb = \{0\} \subseteq I$ however $bNa = \{b\} \nsubseteq I$, hence N is not totally reflexive. Observe that $G_I(N)$ is a proper subgraph of $EQ_I(N)$.

Remark 3.5. In Example 3.4, note that
(i) $I = \{0\}$ is not an equiprime ideal of N. Observe that $G_I(N)$ is a proper subgraph $EQ_I(N)$.
(ii) N is not distributive. Observe that $G_I(N)$ is a proper subgraph of $EQ_I(N)$.

Proposition 3.7. *Let I be an ideal of N and $H_I = N$. Suppose (i) I is a strong vertex cut of $G_I(N)$; (ii) $G_I(N) = EQ_I(N)$; and (iii) I is e-semiprime, then I is an equiprime ideal of N.*

Proof. Let $I = N$. Then I is an equiprime ideal of N. Let $I \subset N$. Let $x, y, p \in N$ such that $prx - pry \in I$ for all $r \in N$. Then p and $(x-y)$ are adjacent in $EQ_I(N)$. Suppose $p = 0$. Then $p \in I$. Suppose $x = y$. Then $(x-y) = 0 \in I$. Suppose $p = (x-y)$. Then $(x-y) \in I$ (I is e-semiprime). Let $p \neq 0$, $x \neq y$ and $p \neq (x-y)$. As $G_I(N) = EQ_I(N)$, we get p and $(x-y)$ are adjacent in $G_I(N)$. Then $pN(x-y) \subseteq I$ or $(x-y)Np \subseteq I$. Suppose $p \in N \setminus I$ and $(x-y) \in N \setminus I$. As I is a strong vertex cut of $G_I(N)$, we get $(p, (x-y)) \notin E(G_I(N))$. However we have $pN(x-y) \subseteq I$ or $(x-y)Np \subseteq I$, a contradiction. Hence, I is an equiprime ideal of N. □

Proposition 3.8. *Let I be an ideal of N and $H_I = N$. If I is a 3-prime ideal of N and $G_I(N) = EQ_I(N)$, then I is an equiprime ideal of N.*

Proposition 3.9. *Let I be an ideal of zerosymmetric nearring N and $H_I = N$. If (i) $G_I(N)$ is ideal symmetric, (ii) I is e-semiprime and (iii) for every $x \in N$ with $deg(x) = deg(0)$ in $G_I(N)$ implies $x \in I$, then I is an equiprime ideal of N.*

Proof. Let $p,x,y \in N$ be such that $prx - pry \in I$ for all $r \in N$. Suppose $p = 0$. Then $p \in I$. Suppose $x = y$. Then $(x-y) = 0 \in I$. Suppose $p = (x-y)$. Then $(x-y) \in I$ (I is e-semiprime). Let $p \neq 0$, $x \neq y$ and $p \neq (x-y)$. As $prx - pry \in I$, we get p and $(x-y)$ are adjacent in $EQ_I^p(N)$. As N is zerosymmetric by Theorem 3.1, we get $EQ_I(N) = G_I(N)$. Then p and $(x-y)$ are adjacent in $G_I(N)$. By ideal symmetry of $G_I(N)$, we get $deg(p) = deg(0)$ or $deg((x-y)) = deg(0)$ in $G_I(N)$. By assumption (iii), we get $p \in I$ or $(x-y) \in I$. Thus, I is an equiprime ideal of N. □

Proposition 3.10. *(i) Let x be a vertex in $C_I(N)$. If $x \in I$ then $deg(x) = deg(0)$.*
(ii) Let I be a c-prime ideal of N and x be a vertex in $C_I(N)$. If $deg(x) = deg(0)$ then $x \in I$.

Proposition 3.11. *If I is a 3-prime ideal of N then $G_I(N) \subseteq C_I(N)$.*

Proposition 3.12. *Let I be a c-prime ideal of N. Then $G_I(N) = C_I(N)$.*

Proof. We have $V(G_I(N)) = V(C_I(N)) = N$. Suppose I is a c-prime ideal of N. Then I is a 3-prime ideal of N. By Proposition 3.11, $E(G_I(N)) \subseteq E(C_I(N))$. Let $a,b \in N$ such that $(a,b) \in E(C_I(N))$. Then $ab \in I$ or $ba \in I$. Let $ab \in I$. By the property of c-prime ideal, we get $a \in I$ or $b \in I$. Now, suppose $a \in I$. Then $aN \subseteq I \Rightarrow aNb \subseteq Ib \subseteq I$. Hence, $aNb \subseteq I \Rightarrow (a,b) \subseteq E(G_I(N))$. Similarly for $b \in I$, we get $(b,a) \subseteq E(G_I(N))$. Therefore, $E(C_I(N)) \subseteq E(G_I(N))$. The proof is similar for $ba \in I$. Thus, $G_I(N) = C_I(N)$. □

Proposition 3.13. *Let N be a right permutable nearring and I be an ideal of N.*
(i) If I is a 3-prime ideal of N then I is a c-prime ideal of N.
(ii) If I is an equiprime ideal of N then I is a c-prime ideal of N.
(iii) If I is a 3-semiprime ideal of N then I is a c-semiprime ideal of N.

Proposition 3.14. *Let I be an ideal of N and $H_I = N$. Then $G_I(N) = C_I(N) = EQ_I(N)$ if any one of following conditions is satisfied.*
(i) I is an equiprime ideal of right permutable nearring N.
(ii) I is a prime ideal of commutative ring N.
(iii) I is an equiprime ideal of nearring N and I has IFP.

Proof. To prove (i), let I be an equiprime ideal of right permutable nearring N. By Proposition 3.13(ii), I is a c-prime ideal of N. By Proposition 3.12, we get $G_I(N) = C_I(N)$. As I is an equiprime ideal of N by Theorem 3.1(iv), $G_I(N) = EQ_I(N)$. Thus, $G_I(N) = C_I(N) = EQ_I(N)$. To prove (ii), in a commutative ring every prime ideal is an equiprime ideal and right permutable. Hence, the proof follows by (i). To prove (iii), suppose I is an equiprime ideal of nearring and I has IFP. Then by Lemma 2.17, in Kedukodi, Kuncham and Bhavanari[30], I is a c-prime ideal of N. Hence, by Proposition 3.12, $G_I(N) = C_I(N)$. As I is an equiprime ideal of N by Theorem 3.1(iv), we get $G_I(N) = EQ_I(N)$. Thus, $G_I(N) = C_I(N) = EQ_I(N)$. □

Proposition 3.15. *Let I be an ideal of N. If I is c-prime ideal of N then I is a strong vertex cut of $C_I(N)$. If I is a strong vertex cut of $C_I(N)$ and I is c-semiprime then I is c-prime.*

Proposition 3.16. *Let I be an ideal of N.*
(a) If I is an c-prime ideal of N, then $C_I(N)$ is ideal symmetric.
(b) Let (i) I be c-semiprime; (ii) $C_I(N)$ be ideal symmetric; and (iii) for every $x \in N$, $deg(x) = deg(0)$ in $C_I(N) \Rightarrow x \in I$, then I is c-prime and and I is a strong vertex cut of $C_I(N)$.

Remark 3.6. If I is an equiprime ideal of N then $G_I(N)$ is ideal symmetric.

Proposition 3.17. *(i) If I is a 3-prime ideal of N then I is a vertex cover of $G_I(N)$.*
(ii) If I is a c-prime ideal of N then I is a vertex cover of $C_I(N)$.
(iii) If I is an equiprime ideal of N and $H_I = N$ then I is a vertex cover of $EQ_I(N)$.

Proof. To prove (i), let $(x,y) \in E(G_I(N))$. Then $xNy \subseteq I$ or $yNx \subseteq I$. As I is a 3-prime ideal of N, we get $x \in I$ or $y \in I$. Thus, I is vertex cover of $G_I(N)$. To prove (ii), let $(x,y) \in E(C_I(N))$. Then $xy \in I$ or $yx \in I$. As I is a c-prime ideal of N, we get $x \in I$ or $y \in I$. Thus, I is vertex cover of $C_I(N)$. To prove (iii), let $(x,y) \in E(EQ_I(N)) = E(\cup_{p \in N} EQ_I^p(N))$. Then $(x,y) \in E(EQ_I^x(N))$ or $(x,y) \in E(EQ_I^y(N))$. Let $(x,y) \in E(EQ_I^x(N))$. Then $xry - xr0 \in I$ for all $r \in N$ or $yrx - yr0 \in I$ for all $r \in N$. As I is an equiprime ideal of N, we get $x \in I$ or $y \in I$. Thus, I is vertex cover of $EQ_I(N)$. □

Proposition 3.18. *Let I be an ideal of N. (i) Let x be a vertex in $C_I(N)$. If $x \in I$ then $\{x\}$ is a vertex cover of $C_I(N)$. (ii) Let N be a zero-symmetric nearring and x be a vertex in $G_I(N)$. If $x \in I$ then $\{x\}$ is a vertex cover of $G_I(N)$.*

Notation 3.2. If the vertex set of $EQ_I(N)$ is restricted to $N \setminus I$ then the non-trivial component of largest size of the resulting graph is denoted by $EQ_I(N \setminus I)$. If the vertex set of $C_I(N)$ is restricted to $N \setminus I$ then the non-trivial component of largest size of the resulting graph is denoted by $C_I(N \setminus I)$.

Proposition 3.19. *Let I be a proper ideal of N. If I is a c-prime ideal of N then $C_I(N \setminus I)$ is an empty graph.*

Proof. Let I be a c-prime ideal of N. Suppose there exists $x, y \in N \setminus I$ such that $\{x,y\} \in E(C_I(N \setminus I))$. Then $xy \subseteq I$ or $yx \subseteq I$. As I is c-prime we get $x \in I$ or $y \in I$, a contradiction. Hence, there are no non-trivial components in $C_I(N \setminus I)$. Therefore, $C_I(N \setminus I)$ is an empty graph. □

Proposition 3.20. *Let I be a proper ideal of N. If I is an equiprime ideal of N then $EQ_I(N \setminus I)$ is an empty graph.*

Proposition 3.21. *Let I be a proper ideal of N. If I is a 3-prime ideal of N then $G_I(N \setminus I)$ is an empty graph.*

Proposition 3.19, Proposition 3.20 and Proposition 3.21 are not true in general if conditions involved are excluded. We now provide an example to illustrate this fact.

Example 3.6. Consider the nearring in Example 3.3. For $I=\{0\}$, we have $H_I=N$. We have $G_I(N\setminus I)=\emptyset$. The graphs $C_I(N\setminus I)$ and $EQ_I(N\setminus I)$ are shown in Figure 8.

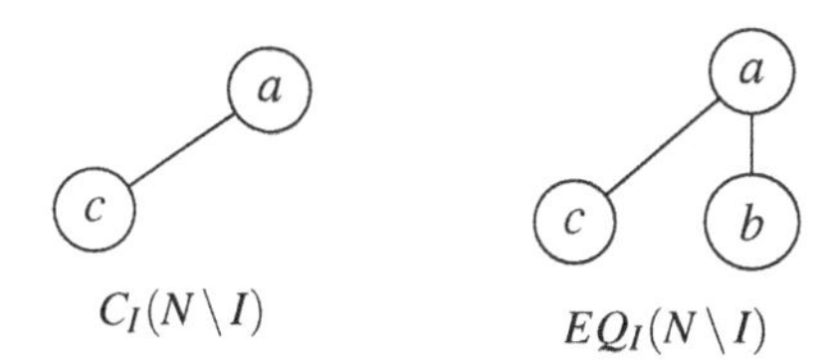

Figure 8: $C_I(N\setminus I)$ and $EQ_I(N\setminus I)$ are non-trivial, whereas $G_I(N\setminus I)=\emptyset$.

Note that I is neither a c-prime nor an equiprime ideal of N. Observe that $C_I(N\setminus I)$, $EQ_I(N\setminus I)$ and $G_I(N\setminus I)$ are distinct.

Proposition 3.22. *Let I be an ideal of N. If $EQ_I(N\setminus I)$ is not an empty graph, then $diam(EQ_I(N\setminus I))\leq 3$.*

Proof. Let $p,x\in(N\setminus I)$.

Case 1: Let $prx-pr0\in I$ for all $r\in N$. Then $(p,x)\in E(EQ_I(N\setminus I))$.

Case 2: Let $prx-pr0\notin I$ for some $r\in N$, $prp-pr0\in I$ and $xrx-xr0\in I$ for all $r\in N$. For $r\in N$, consider $pr(px)-pr0=(prp)x-pr0=(i_1+pr0)x-pr0$ (since $prp-pr0\in I$, we get $prp=i_1+pr0$ for some $i_1\in I$) $=i_2+pr0-pr0\in I$ ($i_1x=i_2$ for some $i_2\in I$). Hence, $(p,px)\in E(EQ_I(N\setminus I))$. Also, for $r\in N$ consider $(px)rx-(px)r0=p(xrx)-(px)r0=p(i_1+xr0)-(px)r0=i_2+pxr0-pxr0\in I$ (by Remark 2.1(ii)). Hence, $(px,x)\in E(EQ_I(N\setminus I))$. Therefore, p_px_x is a path in $EQ_I(N\setminus I)$.

Case 3: Let $prx-pr0\notin I$, $xrx-xr0\notin I$ for some $r\in N$ and $prp-pr0\in I$ for all $r\in N$. Then there exists $b\in(N\setminus(I\cup\{p,x\})$ such that $(b,x)\in E(EQ_I(N\setminus I))\Rightarrow brx-br0\in I$ or $xrb-xr0\in I$ for all $r\in N$. We get following subcases. (i) Let $brx-br0\in I$ for all $r\in N$. Suppose $brp-br0\in I$ for all $r\in N$. Then $(p,b)\in E(EQ_I(N\setminus I))$. Therefore, p_b_x is a path in $EQ_I(N\setminus I)$.

Now suppose $brp-br0\notin I$ for some $r\in N$. For $r\in N$, consider $(pb)rx-(pb)r0=p(brx)-(pb)r0=p(i_1+br0)-(pb)r0=i_2+pbr0-pbr0\in I$ for some $i_1,i_2\in I$. Hence, $(pb,x)\in E(EQ_I(N\setminus I))$. Also, for $r\in N$ consider $pr(pb)-pr0=(prp)b-pr0=(i_1+pr0)b-pr0=i_1b+pr0-pr0\in I$ for some $i_1\in I$. Hence, $(p,pb)\in E(EQ_I(N\setminus I))$. Therefore, p_pb_x is path in $EQ_I(N\setminus I)$. (ii) Let $xrb-xr0\in I$ for all $r\in N$. The proof is similar to (i).

Case 4: Let $prx-pr0\notin I$, $prp-pr0\notin I$ for some $r\in N$ and $xrx-xr0\in I$ for all $r\in N$. Then the proof is similar to Case (3).

Case 5: Let $prx-pr0\notin I$, $prp-pr0\notin I$ and $xrx-xr0\notin I$ for some $r\in N$. Then there exists $a,b\in(N\setminus(I\cup\{p,x\})$ such that $(a,p),(b,x)\in E(EQ_I(N\setminus I))\Rightarrow arp-ar0\in I$ or $pra-pr0\in I$, $brx-br0\in I$ or $xrb-xr0\in I$ for all $r\in N$. Suppose $a=b$. Then

$(a,p),(a,x) \in E(Q_I(N\setminus I))$. Hence, p_a_x is a path in $EQ_I(N\setminus I)$. Now suppose $a \neq b$ and $(a,b) \in E(EQ_I(N\setminus I))$. Then (a,p), (a,b), $(b,x) \in E(EQ_I(N\setminus I))$. Hence, $p\_a\_b\_x$ is a path in $EQ_I(N\setminus I)$. Suppose $a \neq b$ and $(a,b) \notin E(EQ_I(N\setminus I))$. Then we get following subcases.

(i) Suppose $arp - ar0 \in I$ for all $r \in N$ and $brx - br0 \in I$ for all $r \in N$. For $r \in N$, consider $ar(pb) - ar0 = (arp)b - ar0 = (i_1 + ar0)b - ar0 = i_2 + ar0 - ar0 \in I$ for some $i_1, i_2 \in I$. Hence, $(a, pb) \in E(EQ_I(N\setminus I))$. Similarly for $r \in N$, consider $(pb)rx - (pb)r0 = p(brx) - (pb)r0 \in I$. Hence, $(pb,x) \in E(EQ_I(N\setminus I))$. Therefore, $p\_a\_pb\_x$ is a path in $EQ_I(N\setminus I)$.

(ii) Suppose $arp - ar0 \in I$ and $xrb - xr0 \in I$ for all $r \in N$. For $r \in N$, consider $xr(ba) - xr0 = (xrb)a - xr0 \in I$ (similar to (i)). Hence, $(x, ba) \in E(EQ_I(N\setminus I))$. Similarly for $r \in N$, consider $(ba)rp - (ba)r0 = b(arp) - ar0 \in I$. Hence, $(ba, p) \in E(EQ_I(N\setminus I))$. Therefore, p_ba_x is a path in $EQ_I(N\setminus I)$.

(iii) Suppose $pra - pr0 \in I$ and $brx - br0 \in I$ for all $r \in N$. For $r \in N$, consider $pr(ab) - pr0 = (pra)b - pr0 \in I$ (similar to (i)). Hence, $(p, ab) \in E(EQ_I(N\setminus I))$. Similarly for $r \in N$, consider $(ab)rx - (ab)r0 = a(brx) - (ab)r0 \in I$. Hence, $(ab,x) \in E(EQ_I(N\setminus I))$. Therefore, p_ab_x is a path in $EQ_I(N\setminus I)$.

(iv) Suppose $pra - pr0 \in I$ and $xrb - xr0 \in I$ for all $r \in N$. Consider $xr(bp) - xr0 = (xrb)p - xr0 \in I$ (similar to (i)). Hence, $(x, bp) \in E(EQ_I(N\setminus I))$. Now for $r \in N$, consider $(bp)ra - (bp)r0 = b(pra) - (bp)r0 \in I$. Hence, $(bp, a) \in E(EQ_I(N\setminus I))$. Therefore, $p\_a\_bp\_x$ is a path in $EQ_I(N\setminus I)$. Thus, $diam(EQ_I(N\setminus I)) \leq 3$. □

Proposition 3.23. *Let I be an ideal of zero-symmetric nearring N. If $C_I(N\setminus I)$ is not an empty graph, then $diam(C_I(N\setminus I)) \leq 3$.*

Proposition 3.24. *If N is a commutative ring then $C_{\{0\}}(N\setminus\{0\})$ is the zero divisor graph of N.*

Proof. We have $V(C_{\{0\}}(N\setminus\{0\})) = Z(N) - \{0\} = V(\Gamma(N))$. Suppose $(x,y) \in E(C_{\{0\}}(N\setminus\{0\}))$. Then $x \neq 0, y \neq 0$ such that $xy = 0$ or $yx = 0$. Hence, $(x,y) \in E(\Gamma(N))$. Therefore, $E(C_{\{0\}}(N\setminus\{0\})) \subseteq E(\Gamma(N))$. Suppose $(x,y) \in E(\Gamma(N))$. Then $x \neq 0, y \neq 0$ such that $xy = 0$ or $yx = 0$. Hence, $(x,y) \in E(C_{\{0\}}(N\setminus\{0\}))$. Therefore, $E(\Gamma(N)) \subseteq E(C_{\{0\}}(N\setminus\{0\}))$. Thus, $C_{\{0\}}(N\setminus\{0\}) = \Gamma(N)$, the zero divisor graph of N. □

4. Nearring Homomorphism and Graph Homomorphism

Definition 4.1. Let $G = (V(G), E(G)$ be a graph. G is called *edge partitionable* if there exist subgraphs G_1 and G_2 of G such that (i) $V(G) = V(G_1) \cup V(G_2)$, (ii) $E(G) = E(G_1) \cup E(G_2)$ and $E(G_1) \cap E(G_2) = \emptyset$. By the notation $G = G_1 \oplus G_2$, we mean G is edge partitioned by subgraphs G_1 and G_2.

Definition 4.2. Let I be an ideal of N. Define $K(I,N)$ be the subgraph of a prime graph ($G_I(N)$ or $C_I(N)$ or $EQ_I(N)$) with vertex set equal to N and edge set formed in such a way that (i) each vertex of I is adjacent to every other vertex of N, and (ii) no two vertices of $N\setminus I$ are adjacent to each other.

Example 4.1. Let $N = Z_6$ be the ring of integers modulo 6. For $I = \{0\}$ the graphs $C_I(N), C_I(N \setminus I)$ and $K(I,N)$ are shown in Figure 11. Note that $C_{\{0\}}(N \setminus \{0\})$ coincides with the zero divisor graph of N.

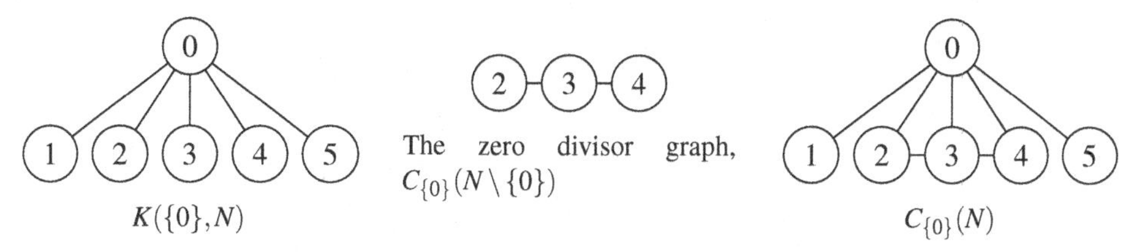

Figure 9: $C_{\{0\}}(N) = K(\{0\},N) \oplus C_{\{0\}}(N \setminus \{0\})$.

Theorem 4.1. *Let I be an ideal of N. Then*
(i) $G_I(N) = K(I,N) \oplus G_I(N \setminus I)$.
(ii) $C_I(N) = K(I,N) \oplus C_I(N \setminus I)$.
(iii) $EQ_I(N) = K(I,N) \oplus EQ_I(N \setminus I)$.

Proof. (i) Clearly $V(K(I,N)) \cup V(G_I(N \setminus I)) = N = V(G_I(N))$. Let $(x,y) \in E(G_I(N))$. Then $xNy \subseteq I$ or $yNx \subseteq I$. Without loss of generality, suppose $xNy \subseteq I$. Now we have only two possible cases. Case 1: $x \in I$ or $y \in I$. By Lemma 3.17 (ii) in Bhavanari, Kuncham, Kedukodi [13], we get $(x,y) \in E(K(I,N))$. Note that $(x,y) \notin E(G_I(N \setminus I))$. Case 2: $x \notin I$ and $y \notin I$. Then $(x,y) \in E(G_I(N \setminus I))$. By Definition 4.2, $(x,y) \notin E(K(I,N))$. Case 1 and Case 2 imply

$$E(G_I(N)) \subseteq E(K(I,N)) \cup E(G_I(N \setminus I)) \textit{ and } E(K(I,N)) \cap E(G_I(N \setminus I)) = \emptyset.$$

As $K(I,N)$ and $G_I(N \setminus I))$ are subgraphs of $G_I(N)$ we get $E(K(I,N)) \cup E(G_I(N \setminus I)) \subseteq E(G_I(N))$. Hence, $E(G_I(N)) = E(K(I,N)) \oplus E(G_I(N \setminus I))$. The proofs of (ii), (iii) are similar to the proof of (i). □

Corollary 4.1. *Let I be an ideal of* N_1 *and* $\theta : N_1 \to N_2$ *be an onto homomorphism. Then*
(i) $G_{\theta(I)}(N_2) = K(\theta(I), N_2) \oplus G_{\theta(I)}(N_2 \setminus \theta(I))$.
(ii) $C_{\theta(I)}(N_2) = K(\theta(I), N_2) \oplus C_I(N_2 \setminus \theta(I))$.
(iii) $EQ_{\theta(I)}(N_2) = K(\theta(I), N_2) \oplus EQ_I(N_2 \setminus \theta(I))$.

Proof. The result follows from Theorem 4.1 because of the fact that $\theta(I)$ is an ideal of N_2. □

Example 4.2. Let $N_1 = Z_2 \times Z_4$ and $N_2 = Z_4$, where Z_2 be the ring of integers modulo 2 and Z_4 be the ring of integers modulo 4. Define $\theta : N_1 \to N_2$ by $\theta(x,y) = y$. Then θ is an onto nearring homomorphism. Let $I = \{(0,0)\}$. Then $\theta(I) = \{0\}$. Figure 10 shows $G_I(N_1)$, $G_I(N_1 \setminus I)$, $K(I, N_1)$ and $G_{\theta(I)}(N_2)$.

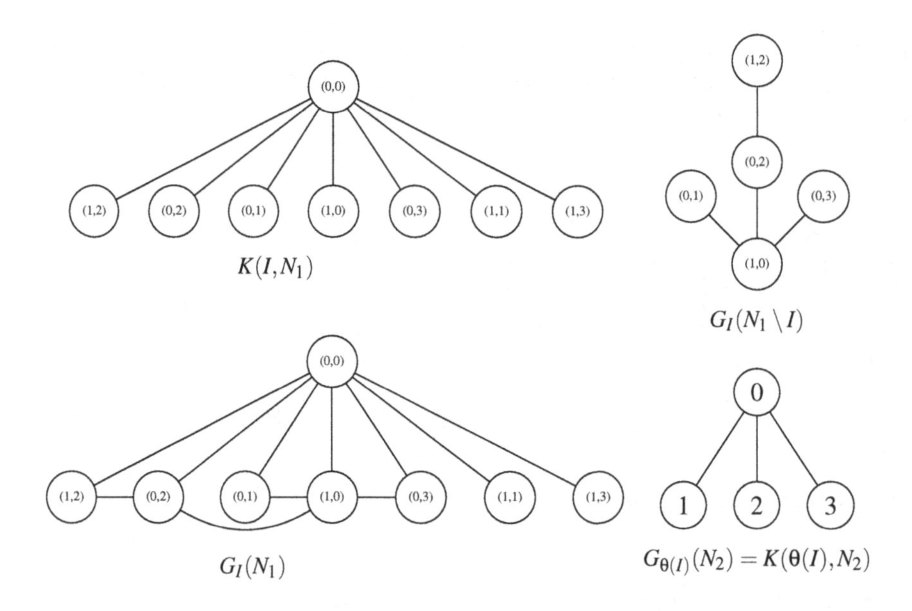

Figure 10: $K(I,N_1)$, $G_I(N_1 \setminus I)$, $G_I(N_1)$ and $G_{\theta(I)}(N_2)$.

In Figure 10, observe that $G_I(N_1) = K(I,N_1) \oplus G_I(N_1 \setminus I)$ and $G_{\theta(I)}(N_2) = K(\theta(I),N_2) \oplus G_{\theta(I)}(N_2 \setminus \theta(I))$ where $G_{\theta(I)}(N_2 \setminus \theta(I))$ is an empty graph.

Proposition 4.1. *Let $\theta : N_1 \to N_2$ be an onto nearring homomorphism. Let I be an ideal of N_1. Let $H_I = N_1$ and $H_{\theta(I)} = N_2$. Then*
(i) θ is a graph homomorphism from $G_I(N_1)$ to $G_{\theta(I)}(N_2)$,
(ii) θ is a graph homomorphism from $EQ_I(N_1)$ to $EQ_{\theta(I)}(N_2)$, and
(iii) θ is a graph homomorphism from $C_I(N_1)$ to $C_{\theta(I)}(N_2)$.

Proof. Let I be an ideal of N_1. Then $\theta(I)$ is an ideal of N_2. To prove (i), suppose $(x,y) \in E(G_I(N_1))$. Then $xN_1y \subseteq I$ or $yN_1x \subseteq I$. Without loss of generality, assume $xN_1y \subseteq I \Rightarrow \theta(xN_1y) \subseteq \theta(I) \Rightarrow \theta(x)\theta(N_1)\theta(y) \subseteq \theta(I)$. As θ is onto, we get $\theta(x)N_2\theta(y) \subseteq \theta(I) \Rightarrow (\theta(x),\theta(y)) \in E(G_{\theta(I)}(N_2))$. Therefore, θ is a graph homomorphism from $G_I(N_1)$ to $G_{\theta(I)}(N_2)$. To prove (ii), suppose $(x,y) \in E(EQ_I(N_1)) = \cup_{p \in N_1} EQ_I^p(N_1)$. Then $(x,y) \in EQ_I^x(N_1)$ or $(x,y) \in EQ_I^y(N_1)$. Let $(x,y) \in EQ_I^x(N_1) \Rightarrow (xry - xr(0_{N_1})) \in I$ for all $r \in N_1$ or $(yrx - yr(0_{N_1})) \in I$ for all $r \in N_1$. Without loss of generality, assume $(xry - xr(0_{N_1})) \in I$ for all $r \in N_1$. Then $\theta(xry - xr(0_{N_1})) \in \theta(I)$ for all $r \in N_1$. As θ is a nearring homomorphism, $\theta(x)\theta(r)\theta(y) - \theta(x)\theta(r)\theta((0_{N_1})) \in \theta(I)$ for all $r \in N_1$. As θ is onto, $\theta(x)r_2\theta(y) - \theta(x)r_2\theta((0_{N_1})) \in \theta(I)$ for all $r_2 \in N_2$. Also as θ is a nearring homomorphism, $\theta(0_{N_1}) = 0_{N_2}$. Hence, $\theta(x)r_2\theta(y) - \theta(x)r_2(0_{N_2})) \in \theta(I)$ for all $r_2 \in N_2 \Rightarrow (\theta(x),\theta(y)) \in E(EQ_{\theta(I)}^{\theta(x)}(N_2)) \subseteq E(EQ_{\theta(I)}(N_2))$. The proof is similar for $(x,y) \in EQ_I^y(N_1)$. Therefore, θ is a graph homomorphism. The proof of (iii) is similar to the proof of (i). $\square$

Lemma 4.1. *If I is an ideal of N and $x \in I$, then $deg(x) = deg(0)$ in $G_I(N)$.*

Proposition 4.2. *Let $\theta : N_1 \to N_2$ be an onto nearring homomorphism and I be an ideal of N_1.*
(i) If $x \in I$, then $deg(\theta(x)) = deg(0_{N_2})$ in $G_{\theta(I)}(N_2)$.
(ii) If $H_I = N_1$, $H_{\theta(I)} = N_2$ and $x \in I$, then $deg(\theta(x)) = deg(0_{N_2})$ in $EQ_{\theta(I)}(N_2)$.
(iii) If $x \in I$, then $deg(\theta(x)) = deg(0_{N_2})$ in $C_{\theta(I)}(N_2)$.

Proof. To prove (i), let $x \in I$. By Lemma 4.1, we get $(x,y) \in E(G_I(N_1))$ for all $y \in N_1 \Rightarrow xN_1y \subseteq I$ or $yN_1x \subseteq I$ for all $y \in N_1$. Without loss of generality, assume $xN_1y \subseteq I$ for all $y \in N_1 \Rightarrow \theta(x)N_2\theta(y) \subseteq \theta(I)$ for all $\theta(y)$ in N_2. Therefore, $deg(\theta(x)) = deg(0_{N_2})$ in $G_{\theta(I)}(N_2)$. The proofs of (ii), (iii) are similar to the proof of (i). □

Remark 4.1. It is well-known that the homomorphic image of a 3-prime (resp. c-prime, equiprime) ideal need not be a 3-prime (resp. c-prime, equiprime) ideal in general. For example, consider $\psi : Z \to Z_6$ (where Z is the ring of integers and Z_6 is the ring of integers modulo 6) defined by $\psi(x) = x \bmod 6$. Then ψ is an onto ring homomorphism. Note that $\{0\}$ is a 3-prime (resp. c-prime, equiprime) ideal of Z. However $\psi(I) = \{\overline{0}\}$ is not a 3-prime (resp. c-prime, equiprime) ideal of Z_6.

Lemma 4.2. *Let $\theta : N_1 \to N_2$ be an onto homomorphism and I be an ideal of N_1. Then $\theta(N_1 \setminus I) = N_2 \setminus \theta(I)$.*

Proposition 4.3. *Let I be an ideal of N_1 and $\theta : N_1 \to N_2$ be an onto homomorphism. Then*
(i) for every $x \in N_1$ with $deg(\theta(x)) = deg(0_{N_2})$ in $G_{\theta(I)}(N_2)$ implies $deg(x) = deg(0_{N_1})$ in $G_I(N_1)$.
(ii) for every $x \in N_1$ with $deg(\theta(x)) = deg(0_{N_2})$ in $C_{\theta(I)}(N_2)$ implies $deg(x) = deg(0_{N_1})$ in $C_I(N_1)$.
(iii) for every $x \in N_1$ with $deg(\theta(x)) = deg(0_{N_2})$ in $EQ_{\theta(I)}(N_2)$ implies $deg(x) = deg(0_{N_1})$ in $EQ_I(N_1)$.

Proof. Let $x \in N_1$ with $deg(\theta(x)) = deg(0_{N_2})$ in $G_{\theta(I)}(N_2)$. Suppose $deg(x) \neq deg(0_{N_1})$ in $G_I(N_1)$. Then there exists $y \in N_1$ such that $xN_1y \nsubseteq I$ and $yN_1x \nsubseteq I \Rightarrow xn_1y \notin I$ and $yn_2x \notin I$ for some $n_1, n_2 \in N_1$
$\Rightarrow xn_1y \in N_1 \setminus I$ and $yn_2x \in N_1 \setminus I \Rightarrow \theta(xn_1y) \in \theta(N_1 \setminus I)$ and $\theta(yn_2x) \in \theta(N_1 \setminus I)$
$\Rightarrow \theta(x)\theta(n_1)\theta(y) \in \theta(N_1 \setminus I)$ and $\theta(y)\theta(n_2)\theta(x) \in \theta(N_1 \setminus I)$. By Lemma 4.2, we get $\theta(x)\theta(n_1)\theta(y) \in N_2 \setminus \theta(I)$ and $\theta(y)\theta(n_2)\theta(x) \in N_2 \setminus \theta(I) \Rightarrow \theta(x)\theta(n_1)\theta(y) \notin \theta(I)$ and $\theta(y)\theta(n_2)\theta(x) \notin \theta(I)$
$\Rightarrow \theta(x)N_2\theta(y) \nsubseteq \theta(I)$ and $\theta(y)N_2\theta(x) \nsubseteq \theta(I) \Rightarrow \theta(y)$ is not adjacent with $\theta(x)$ in $G_{\theta(I)}(N_2)$. A contradiction to the fact that $deg(\theta(x)) = deg(0_{N_2})$ in $G_{\theta(I)}(N_2)$. The proofs of (ii) and (iii) are similar to that of (i). □

Example 4.3. Let $N_1 = \frac{Z}{8Z}$ and $N_2 = \frac{Z}{4Z}$. Then N_1 and N_2 are commutative rings. Define $\theta : N_1 \to N_2$ by $\theta(x+8Z) = x+4Z$. Then θ is an onto ring homomorphism and by Proposition 4.1(ii), θ is a graph homomorphism form $EQ_I(N_1)$ to $EQ_{\theta(I)}(N_2)$. Let $I = \{0+8Z, 4+8Z\}$. Then I is an ideal of N_1. We have $\theta(I) = \{0+4Z\}$. Graphs $EQ_I(N_1)$ and $EQ_{\theta(I)}(N_2)$ are shown in Figure 11.

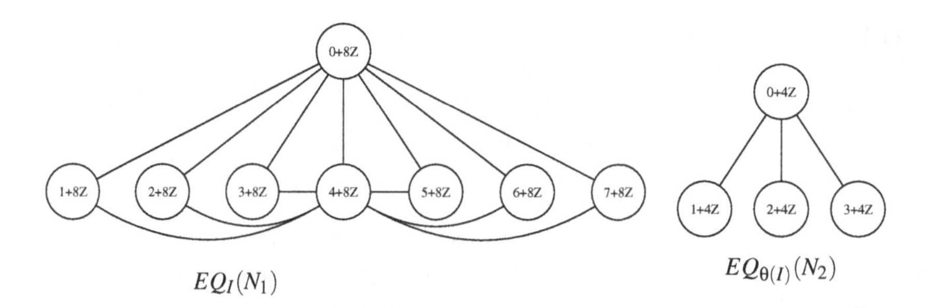

Figure 11: $EQ_I(N_1)$ and $EQ_{\theta(I)}(N_2)$.

Figure 11 shows that
(1) Proposition 4.2 (ii) is not true in general when the condition $x \in I$ is excluded. Note that $x = 5+8Z \notin I$. Observe that $deg(\theta(5+8Z)) = deg(1+4Z) = 1 < 3 = deg(0_{N_2})$ in $EQ_{\theta(I)}(N_2)$.
(2) Proposition 4.3 (iii) is not true in general when the condition $deg(\theta(x)) = deg(0_{N_2})$ in $EQ_{\theta(I)}(N_2)$ is excluded. For $x = 7+8Z$ we have $\theta(7+8Z) = 3+4Z$. Note that $deg(3+4Z) = 1 < 3 = deg(0_{N_2})$. Observe that $deg(7+8Z) = 2 < 7 = deg(0_{N_1})$ in $EQ_I(N_1)$.

Definition 4.3. Let I be an ideal of N_1 and $\theta : N_1 \to N_2$ be a homomorphism. Then
(i) θ is said to *preserve vertex covers* of 3-prime graphs if I is a vertex cover of $G_I(N_1) \Rightarrow \theta(I)$ is a vertex cover of $G_{\theta(I)}(N_2)$.
(ii) θ is said to *preserve vertex covers* of c-prime graphs if I is a vertex cover of $C_I(N_1) \Rightarrow \theta(I)$ is a vertex cover of $C_{\theta(I)}(N_2)$.
(iii) θ is said to *preserve vertex covers* of equiprime graphs if I is a vertex cover of $EQ_I(N_1) \Rightarrow \theta(I)$ is a vertex cover of $EQ_{\theta(I)}(N_2)$.

Proposition 4.4. *Let N_1 and N_2 be zero-symmetric nearrings. Let $\theta : N_1 \to N_2$ be an onto nearring homomorphism and I be a 3-prime ideal of N_1. Suppose*
(i) θ preserves vertex covers of 3-prime graphs and
(ii) $\theta(I)$ is 3-semiprime,
then $\theta(I)$ is a 3-prime ideal of N_2.

Proof. By Proposition 4.1(i), we get θ is a graph homomorphism from $G_I(N_1)$ to $G_{\theta(I)}(N_2)$. Suppose $I = N_1$. Then $\theta(I) = N_2$ is a 3-prime ideal of N_2. Let $I \subsetneq N_1$. As I is a 3-prime ideal of N_1, by Proposition 3.17(i), we get I is vertex cover of $G_I(N_1)$. As θ preserves vertex cover, we get $\theta(I)$ is vertex cover of $G_{\theta(I)}(N_2)$. Let $(x,y) \in E(G_{\theta(I)}(N_2))$. Further $x \in \theta(I)$ or $y \in \theta(I)$ ($\theta(I)$ vertex cover of $G_{\theta(I)}(N_2)$). By Lemma 3.17 (ii) of Bhavanari, Kuncham and Kedukodi[13], we get $deg(x) = deg(0_{N_2})$ or $deg(y) = deg(0_{N_2})$ in $G_{\theta(I)}(N_2)$. Hence, $G_{\theta(I)}(N_2)$ is ideal symmetric. Let $x \in N_2$ such that $deg(x) = deg(0_{N_2})$ in $G_{\theta(I)}(N_2)$. Then $xN_2y \subseteq \theta(I)$ for all $y \in N_2$. Suppose $x = y$. Then $xN_2x \subseteq \theta(I) \Rightarrow x \in \theta(I)$ ($\theta(I)$ is 3-semiprime). As θ is onto, we get $x = \theta(x_1)$ for some $x_1 \in N_1$. Now, let $deg(\theta(x_1)) = deg(0_{N_2})$ in $G_{\theta(I)}(N_2)$. Then by Proposition 4.3 (i), $deg(x_1) = deg(0_{N_1})$ in

$G_I(N_1)$. Also $x_1N_1y_1 \subseteq I$ for all $y_1 \in N_1$. Suppose $I = N_1$. Then $\theta(I) = N_2$ is an 3-prime ideal of N_2. Let $I \subsetneq N_1$. Choose $y_1 \in N_1 \setminus I$. As I is a 3-prime ideal of N_1, we get $x_1 \in I$. Hence, $\theta(x_1) \in \theta(I) \Rightarrow x \in \theta(I)$. Therefore, $G_{\theta(I)}(N_2)$ satisfies all conditions of Theorem 3.23 (b), given in Bhavanari, Kuncham and Kedukodi [13]. Thus, $\theta(I)$ is a 3-prime ideal of N_2. □

Proposition 4.5. *Let N_1 and N_2 are zerosymmetric nearrings. Let $\theta : N_1 \to N_2$ be an onto nearring homomorphism and I be an equiprime ideal of N_1. Let $H_I = N_1$ and $H_{\theta(I)} = N_2$. Suppose*
(i) θ preserves vertex covers of 3-prime graphs and
(ii) $\theta(I)$ is e-semiprime,
then $\theta(I)$ is an equiprime ideal of N_2.

Proof. By Proposition 4.1(i), we get θ is a graph homomorphism from $G_I(N_1)$ to $G_{\theta(I)}(N_2)$. As I is an equiprime ideal of N_1, then I is a 3-prime ideal of N_1. By Proposition 3.17(i), I is a vertex cover of $G_I(N_1)$. As θ preserves vertex cover of 3-prime graph, we get $\theta(I)$ is vertex cover of $G_{\theta(I)}(N_2)$. Let $(x,y) \in E(G_{\theta(I)}(N_2))$. Then $x \in \theta(I)$ or $y \in \theta(I)$ ($\theta(I)$ vertex cover of $G_{\theta(I)}(N_2)$). By Lemma 3.17 (ii) of Bhavanari, Kuncham and Kedukodi [13], we get $deg(x) = deg(0_{N_2})$ or $deg(y) = deg(0_{N_2})$ in $G_{\theta(I)}(N_2)$. Hence, $G_{\theta(I)}(N_2)$ is ideal symmetric. Let $x \in N_2$ such that $deg(x) = deg(0_{N_2})$ in $G_{\theta(I)}(N_2)$. Then $xN_2y \subseteq \theta(I)$ for all $y \in N_2$. Suppose $x = y$. Then $xN_2x \subseteq \theta(I) \Rightarrow xN_2x - xN_2(0_{N_2}) \subseteq \theta(I)$ (N_2 is zerosymmetric.) Then $x \in \theta(I)$ ($\theta(I)$ is e-semiprime). As θ is onto, we get $x = \theta(x_1)$ for some $x_1 \in N_1$. Now, let $deg(\theta(x_1)) = deg(0_{N_2})$ in $G_{\theta(I)}(N_2)$. Then by Proposition 4.3 (i), $deg(x_1) = deg(0_{N_1})$ in $G_I(N_1)$. Then $x_1N_1y_1 \subseteq I$ for all $y_1 \in N_1$. Suppose $I = N_1$. Then $\theta(I) = N_2$ is an equiprime ideal of N_2. Let $I \subsetneq N_1$. Choose $y_1 \in N_1 \setminus I$. As I is a 3-prime ideal of N_1, we get $x_1 \in I$. Therefore, $\theta(x_1) \in \theta(I) \Rightarrow x \in \theta(I)$. By Proposition 3.9, we get $\theta(I)$ is an equiprime ideal of N_2. □

Proposition 4.6. *Let $\theta : N_1 \to N_2$ be an onto nearring homomorphism and I be a c-prime ideal of N_1. Suppose*
(i) θ preserves vertex covers of c-prime graphs and
(ii) $\theta(I)$ is c-semiprime,
then $\theta(I)$ is a c-prime ideal of N_2.

Example 4.4. Consider the nearrings N_1, N_2 and the nearring homomorphism θ defined in Example 4.2.
Let $I = \{(0,0),(0,2),(1,0),(1,2)\}$, then $\theta(I) = \{0,2\}$. Figure 12 shows $C_I(N_1)$ and $C_{\theta(I)}(N_2)$.

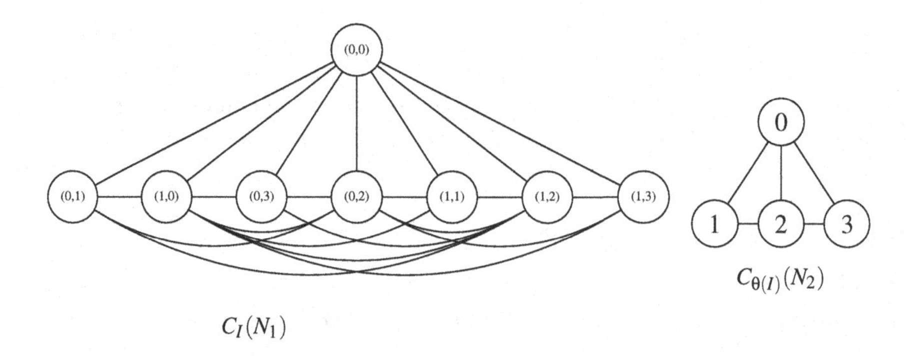

Figure 12: $C_I(N_1)$ and $C_{\theta(I)}(N_2)$ when $I = \{(0,0),(0,2),(1,0),(1,2)\}$.

Observe that I is a c-prime ideal of N_1 and θ satisfies all conditions of the Proposition 4.6. Note that $\theta(I)$ is a c-prime ideal of N_2 which is in agreement with the conclusion of Proposition 4.6.

Acknowledgments

The authors thank the reviewers for their valuable comments and suggestions. The authors thank MIT Manipal University, India; Nelson Mandella Metropolitan University, South Afrika; St. Joseph Engineering College Mangalore, India for thier encouragement and support.

Bibliography

1. E. Aichinger, G. A. Cannon, J. Ecker, L. Kabza, K. M. Neuerburg, Some nearrings in which all ideals are intersections of Noetherian quotients, *Rocky Mountain J. Math.* **38** (2008), 713–726.
2. S. Akbari, D. Kiani, F. Mohammadi, S. Moradi, The total graph and regular graph of a commutative ring, *J. Pure Appl. Algebra* **213** (2009), 2224–2228.
3. D. D. Anderson, A. Badawi, On total graph of a commutative ring, *J. Algebra* **320** (2008), 2706–2719.
4. D. D. Anderson, R. Levy, J. Shapiro, Zero-divisor graphs, von Neumann regular rings, and Boolean algebras, *J. Pure Appl. Algebra* **180** (2003), 221–241.
5. D. D. Anderson, P. S. Livingston, The zero divisor graph of a commutative ring, *J. Algebra* **217** (1999), 434–437.
6. D. F. Anderson, S. B. Mulay, On the diameter and girth of a zero divisor graph, *J. Pure Appl. Algebra* **210**(2) (2007), 543–550.
7. D. F. Anderson, A. Badawi, On the zero divisor graph of a ring, *Commun. Algebra* **36** (2008), 3073–3092.
8. D. F. Anderson, A. Badawi, On the total graph of a commutative ring without the zero element, *J. Algebra Appl.* **11**(4) (2012), 1250074-1-18.
9. M. Axtell, J. Stickles, Zero-divisor graphs of idealizations, *J. Pure Appl. Algebra* **204** (2006), 235–243.
10. I. Beck, Coloring of commutative rings, *J. Algebra* **116** (1998), 208–226.
11. S. Bhavanari, S. P. Kuncham, Nearrings, fuzzy ideals and graph theory, Chapman and Hall/ CRC Press, 2013.

12. S. Bhavanari, S. P. Kuncham, Discrete mathematics and graph theory, PHI Learning, ISBN: 978-81-203-4948-3, 2014.
13. S. Bhavanari, S. P. Kuncham, B. S. Kedukodi, Graph of a nearring with respect to an ideal, *Commun. Algebra* **38** (2010), 1957–1962.
14. G. L. Booth, N. J. Groenewald, S. Veldsman, A Kurosh-Amitsur prime radical for near-rings, *Commun. Algebra* **18**(9) (1990), 3111–3122.
15. G. A. Cannon, K. M. Neuerburg, S. P. Redmond, Zero divisor graphs of nearrings and semigroups, Nearrings and Nearfields, Springer Verlag, The Netherlands, (2005), 189–200.
16. I. Chakrabarty, S. Ghosh, T. K. Mukherjee, M. K. Sen Intersection graphs of ideals of rings, *Discrete Math.* **509** (2000), 5381–5392.
17. T. T. Chelvam, S. Nitya, Zero divisor graph of an ideal of a nearring, *Discrete Math. Algorithms and Appl.* **5** (2013), 1350007-1-11.
18. B. Davvaz, A study on the structure of H_v-near ring modules, *Indian J. Pure Appl. Math.* **34**(5) (2003), 693–700.
19. B. Davvaz, Roughness in rings, *Inform. Sci.* **164** (2004), 147–163.
20. B. Davvaz and K. H. Kim, Generalized hypernearring modules, *Southeast Asian Bull. Math.* **30** (2006), 621–630.
21. B. Davvaz, Roughness based on fuzzy ideals, *Inform. Sci.* **176** (2006), 2417–2437.
22. B. Davvaz and I. Cristea, Fuzzy Algebraic Hyperstructures–An Introduction, Springer, 2015.
23. C. C. Ferrero, G. Ferrero, Nearrings: Some developments Linked to Semigroups and Groups, Kluwer Acadamic Publishers, The Netherlands, 2002.
24. N. J. Groenewald, Different prime ideals in near-rings, *Commun. Algebra* **19** (1991), 2667–2675.
25. N. J. Groenewald, S. Juglal, J. H. Meyer, Prime ideals in group near-rings, *Algebra Colloq.* **15**(03) (2008), 501–510.
26. N. J. Groenewald, S. Juglal, On ideals and primeness in the generalised group near-ring, *Acta Math. Hungar.* **146**(1) (2015), 1–21.
27. F. Harary, Graph theory, Naraosa Publishing House, India, 2001.
28. P. Hell, J. Nesetril, Graphs and Homomorphisms, Oxford University Press New York, 2004.
29. B. Jagadeesha, B. S. Kedukodi, S. P. Kuncham, Interval valued L-fuzzy ideals based on t-norms and t-conorms, *J. Intell. Fuzzy Systems*, **28**(6) (2015), 2631–2641.
30. B. S. Kedukodi, S. P. Kucham, S. Bhavanari, C-prime fuzzy ideals of nearrings, *Soochow J. Math.* **33** (2007), 891–901.
31. B. S. Kedukodi, S. P. Kucham, S. Bhavanari, Equiprime, 3-prime, and c-prime ideals of nearrings, *Soft Comput.* **13** (2009), 933–944.
32. B. S. Kedukodi, S. P. Kucham, S. Bhavanari, Reference points and roughnesss, *Inform. Sci.* **180** (2010), 3348–3361.
33. S. P. Kuncham, B. S. Kedukodi, P. Harikrishnan, S. Bhavanari, Recent developments in nearrings with some applications, *Manipal Research Review* **2**(2) (2015), 1–21.
34. S. P. Kuncham, B. S. Kedukodi, B. Jagadeesha, Interval valued L-fuzzy cosets and isomorphism theorems, *Afr. Mat.* **27**(3) (2016), 393–408.
35. B. Lafuerza–Guillén, Panackal Harikrishnan, Probabilistic normed spaces, Imperial College Press, World Scientific, UK, London, 2014.
36. G. Pilz, Near-Rings, Revised edition, North Hollond, 1983.
37. S. P. Redmond, An ideal-based zero divisor graph of a commutative ring, *Commun. Algebra* **31** (2003), 4425–4443.
38. S. P. Redmond, On zero divisor graphs of small finite commutative rings, *Discrete Math.* **307** (2007), 1155–1166.
39. S. Veldsman, An overnilpotent radical theory for near-rings, *J. Algebra* **144** (1991), 248–265.
40. S. Veldsman, On equiprime near-rings, *Commun. Algebra* **20**(9) (1992), 2569–2587.

On probabilistic paranormed spaces and couplae

Panackal Harikrishnan

Department of Mathematics, Manipal Institute of Technology,
Manipal University, Manipal, Karnataka, India
E-mail: pk.harikrishnan@manipal.edu, pkharikrishnans@gmail.com

Bernardo Lafuerza–Guillén

Departamento de Matemática Aplicada y Estadística,
Universidad de Almería, Almería, Spain
E-mail: blafuerz@ual.es, blafuerza@gmail.com

Carlo Sempi

Dipartimento di Matematica "Ennio De Giorgi",
Università del Salento, 73100 Leece, Italy.
E-mail: carlo.sempi@unisalento.it

1. Introduction

The notion of probabilistic metric space was introduced by Karl Menger[18] in 1942 by intoducing the idea of statistical metric, i.e. of replacing the number $d(p,q)$, which gives the distance between two points p and q in a nonempty set S, by a distribution function F_{pq} whose value $F_{p,q}(t)$ at $t \in]0,+\infty]$ is interpreted as the probability that the distance between the points p and q is smaller than t. Karl Menger later introduced the concept probabilistic normed space (briefly PN spaces) as a natural consequence of the theory of Probabilistic Metric spaces. This theory of probabilistic normed spaces was later reframed into a new form by Alsina, Schweizer and Sklar and they gave a general definition of PN space based on the definition by Menger[18]. PN spaces may provide us a set of tools suitable to study the geometry of nuclear physics, for instance.

2. Probabilistic paranormed Spaces

We shall consider the space of all distance probability distribution functions (briefly, d.f.'s), namely the set of all left–continuous and non–decreasing functions from $\overline{\mathbb{R}}$ into $[0,1]$ such that $F(0) = 0$ and $F(+\infty) = 1$. Here, as usual, $\overline{\mathbb{R}} := \mathbb{R} \cup \{-\infty, +\infty\}$. The spaces of these functions will be denoted by Δ^+, while the subset $\mathcal{D}^+ \subseteq \Delta^+$ will denote the set of all proper distance d.f.'s, namely those for which $\ell^- F(+\infty) = 1$. Here $\ell^- f(x)$ denotes the left limit of the function f at the point x. The space Δ^+ is partially ordered by the usual pointwise ordering of functions i.e., $F \leq G$ if and only if $F(x) \leq G(x)$ for every x in $\mathbb{R}$. For each $a \geq 0$, ε_a is the d.f. given by

$$\varepsilon_a = \begin{cases} 0, & \text{if } x \leq a, \\ 1, & \text{if } x > a. \end{cases}$$

The space Δ^+ can be metrized in several ways[19], but we shall here adopt the Sibley metric d_S. If F, G are d.f.'s and h is in $]0,1[$, let $(F,G;h)$ denote the condition:

$$G(x) \leq F(x+h)+h \text{ for all} x \in \left]0, \frac{1}{h}\right[.$$

Then the Sibley metric[4] d_S is defined by

$$d_S(F,G) := \inf\{h \in]0,1[: \text{both } (F,G;h) \text{ and } (G,F;h) \text{ hold}\}.$$

In particular, under the usual pointwise ordering of functions, ε_0 is the maximal element of Δ^+. A triangle function is a binary operation on Δ^+, namely a function $\tau : \Delta^+ \times \Delta^+ \longrightarrow \Delta^+$ that is associative, commutative, nondecreasing in each place and has ε_0 as identity, that is, for all F, G and H in Δ^+:

(TF1) $\tau(\tau(F,G),H) = \tau(F,\tau(G,H))$,
(TF2) $\tau(F,G) = \tau(G,F)$,
(TF3) $F \leq G \Longrightarrow \tau(F,H) \leq \tau(G,H)$,
(TF4) $\tau(F,\varepsilon_0) = \tau(\varepsilon_0,F) = F$.

Moreover, a triangle function is said to be *continuous* if it is continuous in the metric space (Δ^+, d_S).

Typical continuous triangle functions are $\tau_T(F,G)(x) = \sup_{s+t=x} T(F(s),G(t))$, and $\tau_{T^*}(F,G) = \inf_{s+t=x} T^*(F(s),G(t))$. Here T is a continuous t–norm, i.e. a continuous binary operation on $[0,1]$ that is commutative, associative, nondecreasing in each variable and has 1 as identity; T^* is a continuous *t*–conorm, namely a continuous binary operation on $[0,1]$ which is related to the continuous t–norm T through $T^*(x,y) = 1 - T(1-x, 1-y)$. Three t–norms are particularly important:

$$M(x,y) := \min(x,y), \quad \Pi(x,y) := xy, \quad W(x,y) := \max\{x+y-1, 0\};$$

their conorms are given by

$$M^*(x,y) = \max\{x,y\}, \quad \Pi^*(x,y) = x+y-xy, \quad W^*(x,y) = \min\{x+y, 1\},$$

respectively.

A triangle function τ is said to be Archimedean if τ admits no idempotent other than ε_0 and ε_∞.

Let F be in Δ^+; then define $F^\wedge(t) := \sup\{u \in]0,+\infty[; F(u) < t\}$.

Definition 2.1.[4] A Probabilistic Normed Space, which will henceforth be called briefly a PN space, is a quadruple (V, ν, τ, τ^*), where V is a linear space, τ and τ^* are continuous triangle functions and the mapping $\nu : V \to \Delta^+$ satisfies, for all p and q in V, the conditions

(N1) $\nu_p = \varepsilon_0$ if, and only if, $p = \theta$ (θ is the null vector in V);
(N2) $\forall p \in V \quad \nu_{-p} = \nu_p$;
(N3) $\nu_{p+q} \geq \tau(\nu_p, \nu_q)$;
(N4) $\forall \alpha \in [0,1] \quad \nu_p \leq \tau^*\left(\nu_{\alpha p,}, \nu_{(1-\alpha)p}\right)$.

The function ν is called the *probabilistic norm.* If (V,ν,τ,τ^*) satisfies the condition, weaker than (N1),

$$\nu_\theta = \varepsilon_0\,,$$

then it is called a *Probabilistic Pseudo-Normed space* (briefly, a PPN space). If $\tau = \tau_T$ and $\tau^* = \tau_{T^*}$ for some continuous t–orm T and its t–conorm T^*, then $(V,\nu,\tau_T,\tau_{T^*})$ is denoted by (V,ν,T) and is called a *Menger PN space.*

Example 2.1.[4] Let $(V,\|\cdot\|)$ be a normed space and define $\nu_p := \varepsilon_{\|p\|}$. Let τ be a triangle function such that $\tau(\varepsilon_a,\varepsilon_b) = \varepsilon_{a+b}$ for all $a,b \geq 0$ and let τ^* be a triangle function with $\tau \leq \tau^*$. For instance, it suffices to take $\tau = \tau_T$ and $\tau^* = \tau_{T^*}$, where $T$is a continuous t-norm and T^* is its t–conorm. Then (V,ν,τ,τ^*) is a PN space.

Example 2.2.[4] Let $(V,\|\cdot\|)$ be a real normed space Define a map $\nu : V \to \Delta^+$ by

$$\nu_p(t) = \begin{cases} \dfrac{t}{t+\|p\|}, & \text{if } t \in\,]0,+\infty[\\ 1, & \text{if } t = +\infty \end{cases}$$

and triangle functions defined by

$$\Pi_\Pi(F,G)(t) := F(t)G(t) \quad \text{and} \quad \Pi_{\Pi^*}(F,G)(t) := F(t)+G(t)-F(t)G(t)\,;$$

then $(V,\nu,\Pi_\Pi,\Pi_{\Pi^*})$ is a PN space, called the *canonical* PN space associated with the normed space $(V,\|\cdot\|)$.

A PN space is called a Šerstnev space[4] if it satisfies (N1), (N3) and the following condition:

$$\nu_{\alpha p}(x) = \nu_p\left(\frac{x}{|\alpha|}\right),$$

holds for every $\alpha \neq 0 \in \mathbb{R}$ and $x > 0$. There is a natural topology in a PN space (V,ν,τ,τ^*), called *strong topology*[4]; it is defined for $p \in V$ and $t > 0$, by the neighborhoods

$$N_p(t) := \{q \in V; \nu_{q-p}(t) > 1-t\} = \{q \in V; d_s(\nu_{q-p},\varepsilon_0) < t\}.$$

We recall that a set A in a PN space (V,ν,τ,τ^*), is said to be $\mathcal{D}$*–bounded*[4], if its probabilistic radius R_A belongs to $\mathcal{D}^+$, where

$$R_A(x) = \begin{cases} \ell^-\inf\{\nu_p(x) : p \in A\}, & \text{if } x \in [0,+\infty[, \\ 1, & \text{if } x = +\infty. \end{cases}$$

Definition 2.2.[4] The class $\mathcal{L}$ is the set of all binary operations L on $\mathbb{R}^+$ that satisfy the following conditions:

(i) The range of L is $\mathbb{R}^+$, i.e., L is onto;
(ii) $L$is nondecreasing in each place;
(iii) $L$is continous on $\mathbb{R}^+ \times \mathbb{R}^+$, except possibly at the points $(0,\infty)$and $(\infty,0)$.
(iv) $u_1 < u_2$ and $v_1 < v_2$ imply $L(u_1,v_1) < L(u_2,v_2)$;
(v) $L(x,0) = x$.

Definition 2.3. Let T be a left–continuous t–norm, and let L belong to $\mathcal{L}$; then, for $F,G \in \Delta^+$ define the function $\tau_{T,L}$ as follows

$$\tau_{T,L}(F,G)(x) := \sup\{T(F(u),G(v) : L(u,v) = x)\}.$$

Theorem 2.1.[4] *Let T be a left–continuous t–norm, and let L belong to $\mathcal{L}$ and satisfy the following conditions:*

(a) *L is commutative;*
(b) *L is associative;*
(c) *$u_1 < u_2$ and $v_1 < v_2$ imply $L(u_1,v_1) < L(u_2,v_2)$;*
(d) *$L(x,0) = x$;*

then the function $\tau_{T,L}$ is a triangle function.

Lemma 2.1. *Let $L \in \mathcal{L}$ satisfy the assumptions of Theorem 1.1; then $L = \max$, if, and only if, $L(a,a) = a$ for every $a \in]0,+\infty[$.*

Definition 2.4.[4] A nonempty set Ain PN space (V,ν,τ,τ^*) is said to be

(a) certainly bounded, if $R_A(t_0) = 1$ for some $t_0 \in]0,+\infty[$;
(b) perhaps bounded, if one has $R_A(t) < 1$for every $t \in]0,+\infty[$, but

$$\lim_{t\to+\infty} R_A(t) = 1;$$

(c) perhaps unbounded, if $R_A(t_0) > 0$for some $t_0 \in]0,+\infty[$ and

$$\lim_{t\to+\infty} R_A(t) \in]0,1[;$$

(d) certainly unbounded, if $\lim_{t\to+\infty} R_A(t) = 0$, i.e., if $R_A = \varepsilon_\infty$.

Moreover, the set A will be said to be distributionally bounded (henceforth $\mathcal{D}$-bounded) if either (a) or (b) holds, i.e., if $R_A \in \mathcal{D}^+$; otherwise, i.e., if R_A belongs to $\Delta^+ \setminus \mathcal{D}^+$, A will be said to be *$\mathcal{D}$–unbounded.*

Definition 2.5.[4] Given a triangle function σ, the σ-product of the two PN spaces (V_1,ν_1) and (V_2,ν_2) is the pair $(V_1 \times V_2,\nu^\sigma)$ where $\nu^\sigma : V_1 \times V_2 \to \Delta^+$ is defined, for all $p_1 \in V_1$ and $p_2 \in V_2$, by

$$\nu^\sigma(p_1,p_2) := \sigma(\nu_1(p_1),\nu_2(p_2)).$$

The following theorem shows how the pair $(V_1 \times V_2,\nu^\sigma)$ can be endowed with a structure of PN space.

Theorem 2.2.[4] *Let (V_1,ν_1,τ,τ^*) and (V_2,ν_2,τ,τ^*) be two PN spaces under the same triangle functions τ and τ^*. If σ is a triangle function that dominates τ and is dominated by τ^*, i.e. $\sigma \gg \tau$ and $\tau^* \gg \sigma$, then their σ–product is a PN space under τ and τ^*.*

Example 2.3.[4] The Π_T–product of the PN spaces (V_1,ν_1,τ_T,Π_M) and (V_2,ν_2,τ_T,Π_M) is a PN space under $\tau_T$and Π_M.

If (V_1, ν_1, Π_M) and (V_2, ν_2, Π_M) are equilateral PN spaces with d.f.'s F and G respectively, then their Π_M–product is an equilateral space with d.f. given by $\Pi_M(F,G)$ if $F \neq G$, and by F if $F = G$. If $\tau^* \gg \tau$, both the τ^*–product and the τ–product of $(V_1, \nu_1, \tau, \tau^*)$ and $(V_2, \nu_2, \tau, \tau^*)$ are PN spaces under τ and τ^*. Moreover, if (V_1, ν_1, τ) and (V_2, ν_2, τ) are Šerstnev PN spaces and if $\tau_M \gg \tau$, then both their τ_M–product and their τ–product are Šerstnev PN spaces. Finally, if (V_1, ν_1, T) and (V_2, ν_2, T) are Menger PN spaces under the same continuous t–norm T, then their τ_M–product is a Menger PN space under T.

Definition 2.6. [12] A paranorm is a real function $p : V \to \mathbb{R}$ where $V$is a vector space and p satsfies conditions (i) through (v) for all vectors a and b in V:

(i) $p(\theta) = 0$;
(ii) $p(a) \geq 0$;
(iii) $p(-a) = p(a)$;
(iv) $p(a+b) \leq p(a) + p(b)$;
(v) If t_n is a sequence of scalars with $t_n \to t$ and (u_n) is a sequence of vectors with $u_n \to u$, then $p(t_n u_n - tu) \to 0$ (continuity of multiplication).

A paranorm p for which $p(a) = 0$ implies $a = \theta$ will be called total.

Definition 2.7. [12] A Probabilistic Total Paranormed space (briefly PTPN space) is a triple (V, ν, τ), where V is a real vector space, τ is a continuous triangle function and ν is a mapping (the probabilistic total paranorm) from V into Δ^+, such that for every choice of a and b in V the following axioms hold:

(P1) $\nu_a = \varepsilon_0$if, and only if, $a = \theta$ (the null vector in V);
(P2) $\nu_{-a} = \nu_a$;
(P3) $\nu_{a+b} \geq \tau(\nu_a, \nu_b)$;
(P4) If (u_n) and (α_n) are two sequences of vectors and scalars, respectively, with $(u_n) \to u$ with respect to the strong topology, and $(\alpha_n) \to \alpha$, then $\nu_{\alpha_n u_n - \alpha u} \to \varepsilon_0$.

A PTPN space is not generally a Šerstnev space as shown by the next example.

Example 2.4. [12] Let V be a normed vector space. Define $p(a) = \frac{\|a\|}{(1+\|a\|)}$ and $\nu_a = \varepsilon_{p(a)}$. Then (V, ν, τ) is a PTPN space for every triangle function τ such that $\tau(\varepsilon_a, \varepsilon_b) \leq \varepsilon_{a+b}$.

Definition 2.8. [13] An F–norm on a vector space V is a map $g : V \to \mathbb{R}_+$ that, for every $a \in V$ satisfies the following conditions:

(F1) $g(a) = 0$ if, and only if, $a = \theta$,
(F2) $g(\lambda a) \leq g(a)$ if $|\lambda| \leq 1$,
(F3) $g(a+b) \leq g(a) + g(b)$.

The pair (V, g) is called an F–normed space.

Theorem 2.3. *Let g be a map form V into $\mathbb{R}_+$ and define ν by $\nu_a = \varepsilon_{g(a)}$. Let τ and τ^* be two triangle functions. Then one has the following statements:*

(i) *If* $\tau(\varepsilon_a, \varepsilon_b) \geq \varepsilon_{a+b}$, *for all a and b in* $\mathbb{R}_+$ *and* (V, ν, τ, τ^*) *is a* PN *space, then g is an F–norm.*

(ii) *If* $\tau(\varepsilon_a, \varepsilon_b) \leq \varepsilon_{a+b}$, *for all a and b in* $\mathbb{R}_+$, *then g is a norm if, and only if,* (V, ν, τ, τ^*) *is a Šerstnev space.*

Proof. (F1) If $g(p) = 0$ then $\nu_p = \varepsilon_{g(p)} = \varepsilon_0 \to p = \theta$. Conversely if $p = \theta$, then $\nu_\theta = \varepsilon_{g(\theta)} = \varepsilon_0$ implies $g(\theta) = 0$.

(F2) One knows for that $\nu_{\lambda p} \geq \nu_p$ for $|\lambda| \leq 1$. Then $\varepsilon_{g(\lambda p)} \geq \varepsilon_{g(p)}$ and, hence, $g(\lambda p) \leq g(p)$.

(F3) From $\nu_{p+q} \geq \tau(\nu_p, \nu_q)$ one has

$$\varepsilon_{g(p+q)} \geq \tau(\varepsilon_{g(p)}, \varepsilon_{g(q)}) \geq \varepsilon_{g(p)+g(q)},$$

whence $g(p+q) \leq g(p) + g(q)$.

□

Theorem 2.4.[13] *If* (V, ν, τ) *is a* PTPN *space, then there exists an F–norm* $|\cdot|$ *such that* $(V, |\cdot|)$ *is an F–norm space.*

Theorem 2.5.[4] *Let* $E \subset V$ *where* (V, ν, τ) *is a Šerstnev space; then E is a* $\mathcal{D}$*–bounded set if, and only if, E is a bounded set.*

Proof. Let E be a $\mathcal{D}$–bounded set then by Theorem 6.2.1 in Ref. 6, there exists $G \in \mathcal{D}^+$ such that $\nu_p \geq G$ for every $p \in E$. E is bounded if, for each $t > 0$, there exists $s_t > 0$ such that $E \subset s_t N_\theta(t)$. For each $t \in]0, 1[$, there exists $x_t > 0$ such that $G(x_t) > 1 - t$. Otherwise, for every $x > 0$ one has $G(x) \leq 1 - t$ so that $\lim_{x \to \infty} G(x) < 1$, a contradiction.

Let $s_t := x_t / t$; then $G(s_t t) > 1 - t$ so that

$$\forall p \in E \quad \nu_p(s_t t) > 1 - t.$$

Then, for every $p \in E$, $\nu_{s_t^{-1} p}(t) > 1 - t$. Therefore for every $p \in E$, $s_t^{-1} p$ is in $N_\theta(t)$ and $E \subset s_t N_\theta(t)$.

Conversely, let E be a bounded set; then for every $t > 0$ there exists s_t such that $E \subset s_t N_\theta(t)$. Take $s_0 := 1$ and $s_t := +\infty$ whenever $t > 1$. Since $E \subset s_t N_\theta(t)$ one has $s_t^{-1} E \subset N_\theta(t)$ and, hence, $s_t^{-1} p \in N_\theta(t)$ for every $p \in E$ or, equivalently, $\nu_p(t) > 1 - \frac{t}{s_t}$ for every $p \in E$.

Now let $S_t = \min\{s_t : E \subset s_t N_\theta(t)\}$ and $G(t) := 1 - \frac{1}{S_t}$; then $\nu_p \geq G$ for every $p \in E$. The proof is complete if we show that $G \in \mathcal{D}^+$.

One has $G(0) = 0$ and $G(+\infty) = 1$. If G is non-decreasing, then it belongs to $\mathcal{D}^+$. Let $t_1 < t_2$, so that $\nu_p(t_1) < \nu_p(t_2)$ and $N_\theta(t_1) \subset N_\theta(t_2)$. Therefore, if $s_t^{-1} E \subset N_\theta(t_1)$, then $s_t^{-1} E \subset N_\theta(t_2)$. Thus $S_{t_2} < S_{t_1}$ and $G(t_1) < G(t_2)$, which concludes the proof. □

Corollary 2.1.[13] *A* PN *space* (V, ν, τ, τ^*) *is a* PTPN *space if, and only if,* (V, ν, τ) *is a* PTPN *space.*

Theorem 2.6.[13] *If* (V, ν, τ) *is a* PTPN *space then there exists an F–norm* $|\cdot|$ *such that* $(V, |\cdot|)$ *is an F–normed space.*

Proof. Let (V,ν,τ) be a PTPN space then (V,ν,τ) is a topological vector space and the set $\{N_\theta(1/n) : n \in \mathbb{N}\}$ is a countable neighborhood base of θ. Let $\{N_\theta(1/n_i) : i \in \mathbb{N}\}$ be a base of circled neighborhoods of θ satisfying

$$N_\theta(1/n_{i+1}) + N_\theta(1/n_{i+1}) \subset N_\theta(1/n_i) \qquad (i \in \mathbb{N})\,. \tag{1}$$

For each non–empty finite subset K of $\mathbb{N}$, define the circled neighborhood V_K by $V_K = \sum_{n\in K} N_\theta(1/n_k)$ and the positive number $P_K := \sum_{n\in K} 2^{-n}$. It follows from (2.1) by induction on the number of elements of K that the following implications hold:

$$p_K < 2^{-n} \Longrightarrow n < K \Longrightarrow V_K \subset N_\theta(1/n_k)\,, \tag{2}$$

where $n < K$ means that $n < k$ for every $k \in K$. We define the real–valued function $x \to |x|$ on V by $|x| = 1$ if x is not contained in any V_K, and by

$$|x| = \inf_K \{P_K : x \in V_K\}$$

otherwise; the range of this function is contained in the unit interval. The function $|\cdot|$ is the desired F–norm. □

For every $\varepsilon > 0$, let $B_\varepsilon = \{x \in V : |x| \le \varepsilon\}$; then

$$B_{2^{-k-1}} \subset N_\theta(1/n_k) \subset B_{2^{-k}} \qquad (n \in \mathbb{N})\,. \tag{3}$$

The inclusion $N_\theta(1/n_k) \subset B_{-k}$ is obvious since $x \in N_\theta(1/n_k)$ implies $|x| \le 2^{-k}$. On the other hand, if $|x| \le 2^{-n-1}$, then there exists K such that $x \in V_K$ and $p_K < 2^{-k}$; hence (4.2) implies $x \in N_\theta(1/n_k)$

The following Theorem and Corollary are geometrical forms of the Hahn–Banach Theorem in PTPN spaces.

Theorem 2.7.[13] *Let (V,ν,τ) be a* PTPN *space, A an open convex set in V and L a vector subspace disjoint from A; then there exists a continuous function f on V such that $f(L) = 0$ and $f(A) \ne 0$.*

Proof. (V,ν,τ) is a PTPN space then V is a topological vector space when endowed with the strong topology. Since A is convex, it may be represented by an equation $P(x - x_0) < 1$, where x_0 is a point of V and P is subadditive and positively–homogeneous function on V, and since it is open P is continuous. Since $L \cap A = \emptyset$, $P(y - x_0) \ge 1$ for $y \in L$. Define the linear function f_0 on $L + \mathbb{R}x_0$ by $f(y - \alpha x_0) = \alpha$ for $y \in L$ and $\alpha \in \mathbb{R}$. $f_0 \le P$ at every point of $L + \mathbb{R}x_0$. Since $P$is a positive gauge function on E, the Hahn–Banach theorem shows that f_0 may be extended to a linear functional f on V such that $f \le P$ on V.

Moreover $H = f^{-1}(0)$ is a closed hyperplane in V which contains $L = (L + \mathbb{R}x_0) \cap f^{-1}(0)$. $f(x) = 0$ for $x \in H$; as a consequence, $0 = f(x) = f(x - x_0) + f(x_0) = f(x - x_0) + f_0(x_0) = f(x - x_0) - 1 \le P(x - x_0) - 1$, which shows that $p(x - x - x_0) \ge 1$ for $x \in H$, namely that A dos not meet H. Since $L \subset H$, $f(L) = 0$. Since H does not meet A and A is convex, $f(A)$ must be a real interval not containing 0. Therefore $f(A) > 0$. □

The next Corollary is an immediate consequence of Theorem.

Corollary 2.2.[13] *Let (V, ν, τ) be a PTPN space, A an open convex set and B a convex set in the strong topology of (V, ν, τ); moreover suppose that $A \cap B = \emptyset$. Then there exists a continuous linear form $f \neq 0$ on V and a real number α such that*

$$f(A) < \alpha, \qquad f(B) \geq \alpha.$$

3. Copulæ

Definition 3.1.[4] A copula is a function $C : [0,1]^2 \to [0,1]$ that satisfies the following conditions:

(C1) for every $t \in [0,1]$, $C(0,t) = C(t,0) = 0$and $C(1,t) = C(t,1) = t$;
(C2) C is 2–increasing, i.e., for all s, s', t and t' in $[0,1]$, with $s \leq s'$ and $t \leq t'$,

$$C(s',t') - C(s',t) - C(s,t') + C(s,t) \geq 0. \tag{4}$$

It follows from Definition 3.1 that every copula C is increasing in each place. Moreover for any copula C one has $W \leq C \leq M$.

Definition 3.2.[4] Let φ be a continuous, strictly decreasing function from $\mathbb{I} = [0,1]$ to $[0,+\infty]$ such that $\varphi(1) = 0$. The pseudo–inverse of φ is the function $\varphi^{[-1]}$ with $\mathrm{Dom}\left(\varphi^{[-1]}\right) = [0,+\infty]$ and $\mathrm{Ran}\left(\varphi^{[-1]}\right) = \mathbb{I}$ defined by

$$\varphi^{[-1]}(t) := \begin{cases} \varphi^{-1}(t), & 0 \leq t \leq \varphi(0), \\ 0, & \varphi(0) \leq t \leq +\infty. \end{cases}$$

Lemma 3.1.[4] *Let φ and its pseudo-inverse be as in Definition 3.2 and let $C_\varphi : \mathbb{I}^2 \to \mathbb{I}$ be defined by*

$$C_\varphi(u,v) = \varphi^{[-1]}\left(\varphi(u) + \varphi(v)\right). \tag{5}$$

Then C satisfies the boundary conditions (C1). *Moreover C is 2–increasing if, and only if, for every $v \in \mathbb{I}$,*

$$C_\varphi(u_2,v) - C_\varphi(u_1,v) \leq u_2 - u_1, \tag{6}$$

whenever $u_1 \leq u_2$,

Theorem 3.1.[4] *Let φ and its pseudo–inverse be as in Definition 3.2. Then the function $C_\varphi : \mathbb{I}^2 \to \mathbb{I}$ defined by* (5) *is a copula if, and only if, φ is convex.*

Copulæ of the form (5) are called *Archimedean*. An Archimedean copula is associative and a t–norm.

Bibliography

1. C. Alsina, B. Schweizer and A. Sklar, Continuity properties of probabilistic norms, *J. Math. Anal. Appl.*, **208** (1997), 446–452.
2. C. Alsina, B. Schweizer and A. Sklar, On the definition of a probabilistic normed space, *Aequationes Math.*, **46** (1993), 91–98.
3. F. Durante and C. Sempi, Principles of copula theory, Chapman & Hall/CRC, Boca Raton FL, 2015.
4. B. Lafuerza–Guillén, Panackal Harikrishnan, Probabilistic normed spaces, Imperial College Press, World Scientific, UK, London, 2014.
5. B. Lafuerza–Guillén, Finite products of probabilistic normed spaces, Radovi Matematički, **13** (2004), 111–117.
6. B. Lafuerza–Guillén, A. Rodríguez Lallena and C. Sempi, A study of boundedness in probabilistic normed spaces, *J. Math. Anal. Appl.*, **232** (1999), 183–196.
7. B. Lafuerza–Guillén, $\mathcal{D}$-bounded sets in probabilistic normed spaces and their products, *Rend. Mat. Ser. VII*, **21** (2001), 17–28.
8. B. Lafuerza–Guillén, C. Sempi and Gaoxun Zhang, A Study of Boundedness in Probabilistic Normed Spaces, *Nonlinear Analysis*, **73** (2010), 1127–1135.
9. B. Lafuerza–Guillén, J. A. Rodríguez Lallena and C. Sempi, Normability of probabilistic normed spaces, *Note Mat.*, **29** (2008), 99–111.
10. B. Lafuerza-Guillén, J. A. Rodríguez Lallena and C. Sempi, Some classes of Probabilistic Normed Spaces, *Rend. Mat.*, **17** (1997), 237–252.
11. Gaoxun Zhang and Minxian Zhang, On the normability, *J. Math. Anal. Appl.*, **340** (2008), 1000–1011.
12. M. B. Ghaemi, B. Lafuerza–Guillén and S. Saiedinezhad, Some properties of continuous linear operators in topological vector PN spaces, *Int. J. Nonlinear Anal. Appl.* **1** (2010) No. 1, 58–64.
13. M. B. Ghaemi, B. Lafuerza–Guillén, Probabilistic Total Paranorms, F-norms and PN Spaces, *Int. J. Mathematics and Statistics*, Vol. 6, No. A10, Autumn (2010).
14. P. K. Harikrishnan, B. Lafuerza–Guillén, K. T. Ravindran, Compactness and $\mathcal{D}$–boundedness in Menger's 2-Probabilistic Normed Spaces (accepted in FILOMAT).
15. P. K. Harikrishnan, K. T. Ravindran, Some Results Of Accretive Operators and Convex Sets in 2-Probabilistic Normed Space, *Journal of Prime Research in Mathematics*, **8** (2012), 76–84.
16. B. Jagadeesha, B. S. Kedukodi, S. P. Kuncham, Interval valued L-fuzzy ideals based on t-norms and t-conorms, *J. Intell. Fuzzy Systems*, **28**(6) (2015), 2631–2641.
17. S. P. Kuncham, B. S. Kedukodi, P. Harikrishnan, S. Bhavanari, Recent developments in nearrings with some applications, *Manipal Research Review*, **2**(2) (2015), 1–21.
18. K. Menger, Statistical Metrics, *Proc. Nat. Acad. Sci. USA*, **28** (1942), 535–537.
19. B. Schweizer and A. Sklar, Probabilistic Metric Spaces, Elsevier, North-Holland, New York, 1983; reprinted with additions, Dover, Mineola NY, 2004.

Structural results on nearrings with some applications

Gerhard Wendt

Institut für Algebra, Johannes Kepler Universität Linz
Email: Gerhard.Wendt@jku.at
http://www.algebra.uni-linz.ac.at

Tim Boykett

Institut für Algebra, Johannes Kepler Universität Linz
Email: Tim.Boykett@jku.at http://www.algebra.uni-linz.ac.at
and Time's Up Research
Email: tim@timesup.org http://timesup.org

We look at structure results in nearring theory, especially with regard to Jacobson style primitivity. We look at some applications in state automata and cellular automata, where we find interesting structures. We then look at generalising planarity and nearfields, some of the best structured *nearrings*, and find that Jacobson style primitivity plays an important role.

The paper explores some of our recent work and presents several ways forward that we think might be of value. While we hope to answer some of the questions that we raise here, it is likely that other researchers will find simpler ways forward that we are unable to see. We look forward to some surprises!

Keywords: Nearrings, structure theory, Jacobson, automata, cellular automata, planar nearrings, units, generators, endomorphisms, fixedpointfree.

1. Introduction

This paper reports on various recent results to advance the structural theory for nearrings. We take two mutually supportive approaches to this, description and generation. The first part of the paper introduces the results that have been found so far, the second part looks at some research questions that we consider worth investigating.

One of the topics algebra is concerned with is computing with objects as if they were numbers. If these objects can be added and subtracted we speak of a *group*. If these objects can also be multiplied as we are used to with the numbers, we speak of a *ring*, for example the ring of integers with the two distributive laws we know. If we only have one distributive law w.r.t. multiplication, that means that we can multiply out brackets only from the right hand or only from the left hand side, then we speak of a *nearring*. Nearrings arise naturally when studying functions mapping from a group into itself, where addition is the usual addition of functions and multiplication is function composition. We see nearrings as the nonlinear generalisation of rings. Considering nearrings as sets of functions of a group can be extremely helpful because we can apply well-developed methods of group theory, for example knowledge about the endomorphisms of a group which then model the nearring multiplication, to study such nearrings.

Ring theory has the central density theorem of Jacobson, which introduces primitivity. In nearring theory, this concept splits into different concepts and satisfying results akin to

Jacobson's exist only partially. Betsch's famous density results for 2-primitive nearrings with identity is perhaps the most well known. Recent work by one of the authors has extended these ideas for 1-primitive and 2-primitive nearrings with and without identity, as well as finding partial results for 0-primitivity. We hope it will b epossible to close this gap and to investigate other forms of Jacobson-style primitivity, as well as investigating related properties such as primeness. We will be using techniques such as double centralizers which have been used significantly in the Jacobson theory of rings, as well as annihilators, sandwich multiplication and left ideals, which have been important tools within nearring theory.

Near-fields and planar nearrings are two of the best understood parts of nearring theory, with these classes of nearrings providing important structural information about nearrings. We will look at nearrings in which the units act fixedpointfreely and/or are additively closed (with 0), two generalisations of planarity and near-fields.

In previous related work we have found near-vector spaces, a near-field based generalisation of vector spaces introduced in order to investigate some geometrical constructions, to be of value. We will investigate the ways that near-vector spaces can be used to construct important classes of nearrings.

These structural efforts are closely related to the properties of the semigroup of right multiplicative mappings induced by a nearring upon its additive group. These are endomorphisms, by the right distributive law. However they are not, in general, additively closed, unless the nearring is a ring. One appraoch is to investigate the properties of this endomorphism semigroup, both as a centralizer and as a generating set for a nearring. Both of these constructions have special meanings in ring theory. The first has been extensively investigated for nearrings and continues to be a productive technique. The second has been used only in special cases.

One ongoing project in the nearring community is the development of a theory akin to Jacobson's work in ring theory, generalised for nearrings. Where rings have several concepts that fall together for Jacobson's analysis, these concepts differ in the theory of nearrings, leading to various classes of elementary nearrings. Rather than *primitive* rings we have several classes of 0-, 1- and 2-primitive nearrings. Some of these classes have been successfully classified, while classification of others, for example 0-primitive nearrings, is still open. As a consequence of these considerations, which go deeply into the structure theory of nearrings, we will be able to further study fundamental questions of nearring theory, for example studying various types of primeness in nearrings.

Complementary to the description of nearrings is the generation of nearrings. It is of value to find (in some sense) "nice" generators of nearrings. In this line of discussion we will focus on *units*, these are the elements which are invertible w.r.t. multiplication, which generate nearrings, and on *distributive* elements which generate a nearring.

Endomorphisms of the additive group, which arise naturally from the multiplicative structure of a nearring, will play a central role in our discussion. These endomorphisms serve as a tool to represent a nearring as a nearring of functions on a group and also they can be used to directly generate nearrings.

We will further investigate nearrings whose units are in some sense well behaved, we can see these nearrings as a generalisation of planar nearrings and near-fields. Planar nearrings and near-fields are rich in applications, for instance in the design of efficient statistical experiments as well as in group theory and geometry. Therefore we hope for results triggering applications not only in structure theory of nearrings but also in other areas of mathematics, for example the theory of automata and further links to geometry and group theory.

2. Background

A *right nearring* is an algebra $(N.+,*)$ such that

- $(N,+)$ is a group with identity 0
- $(N,*)$ is a semigroup
- for all $a,b,c \in N$, $(a+b)*c = a*c+b*c$.

We know that $0*x = 0$ for all x. If $x*0 = 0$ for all x, we say N is *zero symmetric*. Let $N^* = N \setminus \{0\}$. If $(N^*,*)$ is a group, N is a *nearfield*. If there is no risk of confusion we will omit the symbol $*$. If both distributive laws hold, we call the nearring *distributive*. If the additive group is abelian, then we call the nearring *abelian*. So a ring is an abelian, distributive nearring. We refer to the books[9,10,20,23] as standard references.

Every ring is a zero symmetric nearring. Every field is a nearfield.

Let $(\Gamma,+)$ be a group. Define $M_0(\Gamma) = \{f : \Gamma \to \Gamma | f(0) = 0\}$. Then $(M_0(\Gamma,+,\circ)$ is a nearring, the nearring of *zero symmetric mappings* on Γ.

Let $S \leq End(\Gamma)$ be a semigroup of endomorphisms of Γ. Then $M_S(\Gamma) = \{f : \Gamma \to \Gamma | f(s(\gamma) = s(f(\gamma)) \forall \gamma \in \Gamma, s \in S\}$ is the *centralizer nearing* on Γ with respect to S.

Let $A \subseteq N$ be a normal subgroup of $(N,+)$.

A is a *left ideal* if $n(m+a) - nm \in A$ for all $n,m \in N$, $a \in A$. A is a *right ideal* if $an \in A$ for all $n \in N$, $a \in A$.

A is an *ideal* if it is a left ideal and a right ideal. Ideals are the kernels of nearring homomorphisms.

A nearring satisfies *DCCL* (descending chain condition on left ideals) if every strictly descending chain of left ideals terminates. Similarly DCCI applies to ideals.

2.1. *Actions*

In this section we introduce the concepts relating to the action of a nearring on a group, generalising the idea of ring modules.

Let $(N,+,*)$ be a nearring. An $N-group$ is a group Γ such that there is a mapping $\alpha : N \times \Gamma \to \Gamma$ such that

- $\alpha((n+m),\gamma) = \alpha(n,\gamma) + \alpha(m,\gamma)$
- $\alpha((nm),\gamma) = \alpha(n,\alpha(m,\gamma))$.

We often write $\alpha(n,\gamma)$ as $n\gamma$, so we rewrite these as

- $(n+m)\gamma = n\gamma + m\gamma$
- $(nm)\gamma = n(m\gamma)$.

If N has an identity 1, then Γ is *unital* if $1\gamma = \gamma$ for all $\gamma \in \Gamma$. Γ is *monogenic* if there exists some $\gamma \in \Gamma$ such that $\Gamma = N\gamma$. Let $U \subseteq \Gamma$. Then $(0:U) = \{n \in N | nU = \{0\}\}$ is a left ideal of N, called the *annihilator* of U. Γ is *faithful* if $n\Gamma = \{0\}$ iff $n = 0$, i.e. $(0:\Gamma) = \{0\}$. $U \leq \Gamma$ is a *N-subgroup* if $NU \subseteq U$. If U is a normal subgroup of Γ and $n(\gamma+u) - n\gamma \in U$ then U is a *N-ideal.*

N is a natural N-group.

N satisfies DCCN if every descending chain of N-subgroups in N terminates.

The concept of primitivity in rings generalises in nearring theory. Let Γ be an N-group. Then we say that:

- Γ is type 0 if it is monogenic with no nontrivial N-ideals.
- Γ is type 1 if it is type 0 and for all $\gamma \in \Gamma$, $N\gamma = N$ or $N\gamma = \{0\}$.
- Γ is type 2 if $N\Gamma \neq \{0\}$ and there are no nontrivial N-subgroups in Γ.

We define $J_\nu = \bigcap_{\Gamma \text{ of type } \nu} (0:\Gamma)$ to be the *Jacobson* ν*-radical* of N. N is ν-primitive if it has a faithful N-group of type ν. That is:

- N is *0-primitive* if it has a faithful monogenic N-group Γ with no nontrivial N-ideals.
- N is *1-primitive* if it is 0-primitive and for all $\gamma \in \Gamma$, $N\gamma = N$ or $N\gamma = \{0\}$.
- N is *2-primitive* if it has a faithful N-group Γ with $N\Gamma \neq \{0\}$ and there are no nontrivial N-subgroups in Γ.

If N is a ring, then these three definitions are equivalent.

Let N be a right nearring. For $u \in N$ let $\psi_u : N \longrightarrow N, n \mapsto n * u$ be the right translation map induced by u. Note that ψ_u is an endomorphism of $(N,+)$. For a subsemigroup $S \subseteq N$ of the multiplicative semigroup of the nearring we define $\Psi_S := \{\psi_s \mid s \in S\}$. Therefore, $(\Psi_S, \circ)$ is a semigroup of endomorphisms of $(N,+)$. In particular, we can let $S = N$. In case N has an identity element, one can take the semigroup of endomorphisms Ψ_N and then show that N is isomorphic to a centralizer nearring.

Theorem 2.1 (Ref. 20, Theorem 2.8). *Every zero symmetric nearring N with identity is (isomorphic to) a centralizer nearring $M_{\Psi_N}(N)$.*

In case N has no identity element the construction of Theorem 2.1 can be generalised to get so called sandwich centralizer nearrings (see Section 7.1). One can use[35] this generalised construction to describe all zero symmetric 1- and 2-primitive nearrings with an efficient construction method. Again in this construction the semigroup Ψ_N of a right nearring plays an important role. We will come to the details later in Section 7.1.

First we note that this semigroup of endomorphisms also naturally shows up in a ring. Let R be a ring and Ψ_R the semigroup of all right translation maps. R is a ring so it follows from the left distributive law, which we do not have in a right nearring, that the sum of two right translation maps is again a right translation map. Hence, one can equip Ψ_R with the operation of pointwise function addition and we get a ring $(\Psi_R, +, \circ)$ which we will abbreviate by R_E, emphasizing that R_E is a subring of the endomorphism ring of $(R, +)$. It is elementary to see that $\phi : R \to R_E, r \mapsto \psi_r$ is a ring anti-epimorphism with kernel $A := \{a \in R | ra = 0 \, \forall r \in R\}$. In case R has an identity we see that $A = \{0\}$ and R and R_E are anti-isomorphic rings.

The situation drastically changes when we only have a nearring N. In this case, Ψ_N cannot be equipped with a group operation. However one can use the semigroup Ψ_N of the nearring N to generate a distributively generated nearring N_{dg}. In case $(N, +)$ is abelian, N_{dg} will be a ring. There has been some studies concerning the structure of this ring, see Ref. 25 for references. In case $(N, +)$ is not abelian, there is little knowledge about the nearring N_{dg}.

We see that the semigroup of endomorphisms Ψ_N has a number of important properties. It defines the nearring itself as a centralizer nearring, in case N has an identity and if the nearring is a ring with identity, it generates the opposite ring. It remains open as to the structure and properties of the distributively generated nearring that Ψ_N generates for an arbitrary nearring. The rest of this paper looks at the ways in which we propose to use information about the right endomorphisms in order to tackle several problems in nearring theory.

3. Sandwich centralizers

We use a special construction in order to create nearrings with special properties.

Definition 3.1. Let $(\Gamma, +)$ be a group, $X \subseteq \Gamma$ a subset of Γ containing the zero 0 of $(\Gamma, +)$ and $\phi : \Gamma \longrightarrow X$ a map such that $\phi(0) = 0$. Let $S \subseteq \mathrm{End}(\Gamma, +)$, S not empty, be such that $\forall s \in S, \forall \gamma \in \Gamma :\ \phi(s(\gamma)) = s(\phi(\gamma))$ and such that $S(X) \subseteq X$. Let

$$M_0(X, \Gamma, \phi, S) := \{f : X \longrightarrow \Gamma \mid f(0) = 0, f(s(x)) = s(f(x)) \forall s \in S, x \in X\}.$$

Then $(M_0(X, \Gamma, \phi, S), +, *)$ with pointwise addition and multiplication $f * g = f \circ \phi \circ g$ is a zero symmetric nearring, which we call a *sandwich centralizer nearring.*

This is a generalisation of centralizer nearrings: $M_0(\Gamma, \Gamma, id, S)$ is $M_S(\Gamma)$.

Density is a way to talk about approximation or interpolation.

Definition 3.2. F is a *dense* subnearring of $M_0(X, \Gamma, \phi, S)$ if and only if $\forall s \in \mathbb{N}\, \forall x_1, \ldots, x_s \in X\ \forall g \in M_0(X, \Gamma, \phi, S)\ \exists f \in F$: $f(x_i) = g(x_i)$ for all $i \in \{1, \ldots, s\}$.

The following definition will be very important in our considerations.

Definition 3.3. Let S be a group of automorphisms of a group Γ. Let $M \subseteq \Gamma \setminus \{0\}$ such that $S(M) \subseteq M$. S is called *fixed point free* on M if for any $s \in S$ and $m \in M$, $s(m) = m$ implies s is the identity function. S is called a fixedpointfree automorphism group of Γ if it acts fixedpointfree on $\Gamma \setminus \{0\}$.

The following is the classic result about primitivity.

Theorem 3.1 (Betsch). *Let N be a zero symmetric nearring with identity which is not a ring. Then N is 2-primitive iff there exist*

(1) *a group* $(\Gamma,+)$,
(2) $S \leq \mathrm{Aut}(\Gamma,+)$, *with S acting without fixed points on* Γ,

such that N is isomorphic to a dense subnearring M_S of $M_0(\Gamma,\Gamma,id,S)$, which is $M_S(\Gamma)$.

It has been generalised to the case that the nearring has no identity.

Theorem 3.2 (Ref. 35). *Let N be a zero symmetric nearring which is not a ring. Then N is 2-primitive iff there exist*

(1) *a group* $(\Gamma,+)$,
(2) *a set* $X = \{0\} \cup X_1 \subseteq \Gamma$, $X_1 \neq \emptyset$, $0 \notin X_1$ *and 0 being the zero of* Γ,
(3) $S \leq \mathrm{Aut}(\Gamma,+)$, *with $S(X) \subseteq X$ and S acting without fixed points on* X_1,
(4) *a function* $\phi : \Gamma \to X$ *with* $\phi|_X = id$, $\phi(0) = 0$ *and such that* $\forall \gamma \in \Gamma\, \forall s \in S : \phi(s(\gamma)) = s(\phi(\gamma))$,

such that N is isomorphic to a dense subnearring M_S of $M_0(X,\Gamma,\phi,S)$ where $\Gamma_0 := \{\gamma \in \Gamma | \phi(\gamma) = 0\}$ does not contain any non-trivial subgroups of Γ.

The following very technical property has been defined in order to get a good description of 1-primitive nearrings. If X,Γ,ϕ,S for a sandwich centraliser nearring additionally satisfy the following property, we say they satisfy property P:

Let $\Gamma_0 := \{\gamma \in \Gamma | \phi(\gamma) = 0\}$ and $C := \{I \lhd \Gamma | I \subseteq \Gamma_0$ and $\Gamma_0 = \cup_{\delta \in \Gamma_0} \delta + I$ and $\forall \gamma \in \Gamma \setminus \Gamma_0 \forall i \in I : S(\phi(\gamma + i)) = S(\phi(\gamma))\}$. Then $I \in C \Rightarrow (I = \{0\}$ or $\exists i \in I \exists \gamma_1 \in \Gamma \setminus \Gamma_0 \exists s \in S \exists \gamma \in \Gamma : \phi(\gamma_1 + i) = s(\phi(\gamma_1))$ and $s(\gamma) - \gamma \notin I)$.

This allows us to make the following statement.

Theorem 3.3 (Ref. 35). *Let N be a zero symmetric nearring which is not a ring. Then N is 1-primitive iff there exist*

(1) *a group* $(\Gamma,+)$,
(2) *a set* $X = \{0\} \cup X_1 \subseteq \Gamma$, $X_1 \neq \emptyset$, $0 \notin X_1$ *and 0 being the zero of* Γ,
(3) $S \leq \mathrm{Aut}(\Gamma,+)$, *with $S(X) \subseteq X$ and S acting fpf on* X_1,
(4) *a function* $\phi : \Gamma \to X$ *with* $\phi|_X = id$, $\phi(0) = 0$ *and* $\forall s \in S : \phi s = s\phi$,

such that N is isomorphic to a dense subnearring M_S of $M_0(X,\Gamma,\phi,S)$, where X,Γ,ϕ,S satisfy property P.

We find that there is an interesting connection between subdirect irreducibility and 0-primitivity. If G is a subdirectly irreducible non simple group, then the *heart* of G is the unique minimal nontrivial normal subgroup.

Theorem 3.4 (Ref. 36). *Let N be a zero symmetric nearring containing a minimal ideal H. Suppose that N is 0-primitive on the N-group Γ. Then N is a subdirectly irreducible nearring with heart H.*

Theorem 3.5 (Ref. 36). *Let N be a zero symmetric and subdirectly irreducible nearring with heart H, which is not a ring. Suppose that H satisfies the decending chain condition on left ideals of N contained in H and suppose H contains a minimal left ideal L of N such that $L^2 \neq \{0\}$. Suppose that L does not contain N-subgroups properly contained in L and being N-isomorphic to L. Then, N is 0-primitive on L, $H = (0 : \{l \in L | Nl \neq L\})$ and H is a finite direct sum of left ideals isomorphic to L.*

By combining the previous two results, we re-prove an older result of Kaarli's.

Theorem 3.6 (Refs. 15, 36). *Let N be a zero symmetric nearring with DCCN. Then the following are equivalent:*

(1) N is 0-primitive.
(2) N is subdirectly irreducible with non-nilpotent heart.

We see that there are many interesting results about the internal structure of nearrings, but we are still missing some conclusive results about 0-primitive 0-symmetric nearrings.

4. State Automata

In this section, we report on some recent work on state automata over groups, that can thus be described as nearrings. In particular, we are interested the way we can find the structures of primitivity. There have been several papers investigting connections between nearring theory and automata theory, for instance Refs. 3, 13, 22.

If we imagine that an infinite sequence of symbols can be seen as the behaviour of a system over time, then a mapping that models the behaviour of a machine should not be able to look into the future to decide what to do now. The following could be called the "no time travel" condition.

Definition 4.1. $n \in M_0(G^{\mathbb{N}})$ is *prefix preserving* if

$$\forall k \in \mathbb{N}, x, y \in G^{\mathbb{N}} : x_i = y_i \forall i < k \Rightarrow (nx)_i = (ny)_i \forall i < k.$$

Prefix preserving maps form a subnearring of $M_0(G^{\mathbb{N}})$, we will call it $PP(G)$. Prefix preserving maps in $M(G)$, which are not necessarily zero symmetric, will be called $PP_c(G)$.

Given a group $(G, +)$, it is possible to define the nearring of state automata or state machines over the alphabet G. If we fix the input-output alphabet G, a *state automaton* is defined as (Q, t, f, s), where

- Q is a set of states,
- $t : Q \times G \to Q$ is a state transition map,
- $f : Q \times G \to G$ is the output map and
- $s \in Q$ is the start state.

We can add and multiply state automata. Let $\Sigma_1 = (Q_1, t_1, f_1, s_1)$ and $\Sigma_2 = (Q_2, t_2, f_2, s_2)$.

Define $\Sigma_1 + \Sigma_2 = (Q_1 \times Q_2, t_1 \times t_2, f_1 + f_2, (s_1, s_2))$ with

$$\begin{aligned} t_1 \times t_2 : (Q_1 \times Q_2) \times G &\to Q_1 \times Q_2 \\ ((q_1, q_2), g) &\mapsto (t_1(q_1, g), t_2(q_2, g)) \\ f_1 + f_2 : (Q_1 \times Q_2) \times G &\to G \\ ((q_1, q_2), g) &\mapsto f_1(q_1, g) + f_2(q_2, g). \end{aligned}$$

Define $\Sigma_1 * \Sigma_2 = (Q_1 \times Q_2, t_1 \otimes t_2, f_1 * f_2, (s_1, s_2))$ with

$$\begin{aligned} t_1 \otimes t_2 : (Q_1 \times Q_2) \times G &\to Q_1 \times Q_2 \\ ((q_1, q_2), g) &\mapsto (t_1(q_1, f_2(q_2, g)), t_2(q_2, g)) \\ f_1 * f_2 : (Q_1 \times Q_2) \times G &\to G \\ ((q_1, q_2), g) &\mapsto f_1(q_1, f_2(q_2, g)). \end{aligned}$$

The following result shows that state machines and prefix preserving maps are equivalent. It is a specific version of a more general result from computer science.

Theorem 4.1 (Ref. 5). *Let G be a group. $PP_c(G)$ is isomorphic to $SM(G)$.*

A state $r \in Q$ is called *reachable* if there is an input sequence $x \in G^{\mathbb{N}}$ such that $r = q_i$ for some q_i in the state sequence q. It is clear that unreachable states do not affect the properties of state machines. A state $r \in Q$ is *0-reachable* if for some i, $r = q_i$ in the state sequence induced by the zero sequence $(0, 0, \ldots)$.

Lemma 4.1 (Ref. 5). *A state machine is zero symmetric iff the state output maps $f_q : g \mapsto f(q, g)$ are 0-symmetric for every state that is 0-reachable.*

The following result indicates that the nearring of prefix preserving maps is complex and complicated. Let $\mathbb{V}(G)$ be the variety generated by the group G.

Theorem 4.2 (Ref. 5). *Let K be a finite group. Then for all finite groups $G \in \mathbb{V}(K)$, $M_0(G) \hookrightarrow PP(K)$.*

This means that all of these groups can be embedded into the nearring of state machines.

Definition 4.2. The map $\alpha : PP(G) \to PP(G)$ defined by: $\forall n \in PP(G)$, $\forall x \in G^{\mathbb{N}}$

$$((\alpha n)x)_i = (n(0, 0, \ldots, 0, x_i, 0, \ldots))_i$$

is called the *amnesiac map*.

Lemma 4.2. *The amnesiac map is a nearring homomorphism.*

Definition 4.3. $n \in PP(G)$ is *delaying* if

$$\forall k \in \mathbb{N} : x_i = y_i \forall i < k \Rightarrow (nx)_i = (ny)_i \forall i \leq k.$$

We write $D(G)$ for this set of state machines.

Lemma 4.3. *$D(G)$ is an N-subgroup.*

Theorem 4.3 (Ref. 5). *Let G be a group. Then $D(G) \leq J_2(PP(G)) \leq \ker\alpha$.*

Definition 4.4. A group $(G,+)$ has *property X* if there is an element $k \in G$ and a function $f : G \to G$ such that $f(x+k) - f(x) = x$ for all $x \in G$.

Theorem 4.4 (Ref. 5). *Finite abelian groups have property X iff they are of odd order.*

Vinay Madhusudanan[17] has found several examples of nonabelian finite groups with this property and is developing further descriptions, working towards a classification of all finite groups with this property. This would be of interest, as it allows us to describe the Jacobson 2-radical of the nearring of state automata.

Theorem 4.5 (Ref. 5). *Let $(G,+)$ be a group with property X. Then $\mathcal{J}_2(PP(G)) = \ker\alpha$.*

This is an interesting result, as it says that the radical, i.e. the "difficult" part of the full nearring of state automata, is the part that remembers what has happened.

We hope that further work will be able to determine the radical in other cases, and that the 0-radical might be able to be determined.

5. Cellular Automata

Cellular Automata have been developed as a technique for modelling many types of systems, as well as being relevant for theoretical computer science in their own right. We find that they offer stimulating connections to nearring theory. The results in this section are to be found in Ref. 8.

Let S be a set of *states*, A a finitely generated group. We are interested in S^A, the mappings from A to S. For some $x \in S^A$, $a \in A$, we will usually write x_a for $x(a)$. We call elements of S^A *configurations*.

The set S^A has an induced metric topology. Let $x,y \in S^A$ then define the distance $d(x,y) = 2^{-i}$ where i is the shortest element (in terms of a set of generators) of A for which $x_i \neq y_i$.

The set S^A has a group of *shift maps* isomorphic to A. For $x \in S^A$, $a,b \in A$, $(\sigma_a x)_b = x_{(b-a)}$. A acts upon S^A by these shift maps.

Theorem 5.1. *A function $f : S^A \to S^A$ commutes with σ_a for all $a \in A$ iff there exists a well ordered index set $I \subseteq A$ and a function $F : S^I \mapsto S$ such that for all $x \in S^A$, $f(x)_a = F(x_{a+\sigma_i})$. Moreover f is continuous iff I is finite.*

We are interested in structured sets S. Let $(S,+)$ be a group.

The *support* of an element $x \in S^A$ is $supp(x) = \{a \in A : x_a \neq 0\}$.

Let $S^A_{fin} = \{x \in S^A | supp(x) \text{ is finite}\}$.

Theorem 5.2. *Let A be torsion free. Then $M_A(S^A_{fin})$ and $M^c_A(S^A_{fin})$ are 2-primitive nearrings.*

Let $H \leq A$ be a subgroup, $\gamma \in S$. Define

$$c_{H,\gamma}(x) = \begin{cases} \gamma \text{ if } x \in H \\ 0 \text{ otherwise.} \end{cases}$$

Theorem 5.3. *Let A be a finite cyclic group, S a group.*

If $S \neq \mathbb{Z}_2$, then $\mathcal{J}_2(M_A(S^A)) = \{f | f(c_{A,\gamma}) = 0 \forall \gamma \in S\}$.

If $S = \mathbb{Z}_2$, $\mathcal{J}_2(M_A(S^A)) = \{f | f(c_{A,\gamma}) = 0, f(c_{H,\gamma}) \in \{c_{A,0}, c_{A,1}\} \forall \gamma \in S, \forall H \leq A$ of index $2\}$.

A group is *virtually free* if there is a subgroup H of finite index such that H is free. A Baumslag-Solitar group $BS(n,m)$ has presentation of the form $< a, b : ba^n b^{-1} = a^m >$ for some integers m, n. $BS(1,1)$ is isomorphic to $\mathbb{Z}^2$.

The following result is interesting, as it shows that there are undecidable questions embedded within nearring theory.

Theorem 5.4. *The units in $M_A^c(S_{fin}^A)$ are decidable if A is virtually free. They are undecidable if A is $\mathbb{Z}^n$ for $n > 1$ or A a Baumslag-Solitar group.*

Lemma 5.1. *Let A be torsion free. Then $M_A^c(S_{fin}^A)$ and $M_A(S_{fin}^A)$ do not have DCCL.*

We see that nearrings of cellular automata offer a range of nearrings with interesting and complex properties from very simple ingredients.

6. Units generalising Planar Nearrings and Nearfields

In this section we report on some work generalising planar nearrings and nearfields, reported in Ref. 6. Recall the definition of Ψ_S above. A finite nearring is *planar* iff Ψ_{N^*} is fixed point free and $|\Psi_N| \geq 3$. A finite nearring is a *nearfield* iff it is planar and $1 \in N$ (unless N of order 2).

Planar nearrings are a very interesting class of nearrings, see Refs. 9, 16, 23, 30, 32–34. They have many applications. On one hand, they have real life applications because they can be used to construct highly efficient balanced incomplete block designs (BIBDs)[2,16,23]. These have applications in designing optimal statistical experiments. On the other hand, they have a strong impact on the structure theory of nearrings[30,34].

In Ref. 6, we showed that there exist nearrings with identity which are not near-fields and were the units U of the nearring induce a group of fixedpointfree automorphisms on N. Thus, Ψ_U is a fixedpointfree automorphism group of the additive group of the nearring. Such a nearring is called an f-nearring and can be seen as the mutual generalisation of a near-field and a planar nearring. We also showed that an f-nearring is simple iff it is a near-field and we could give a construction method for f-nearrings of all possible orders using near-vector spaces, which generalise vector spaces, see e.g. Ref. 14 and the references therein.

Definition 6.1. An *f-nearring N* is a zero symmetric nearring with identity and set of units U where $(\Psi_U, \circ)$ acts as a fixedpointfree automorphism group on the additive group of the nearring. An *a-nearring* N is a zero symmetric nearring with identity and set of units U such that $U_0 = U \cup \{0\}$ forms a subnearfield of the nearring w.r.t. the nearring operations. Nearrings with both properties will be called *af-nearrings*.

These generalisations are closely related and show similar structures, but as we see, they are distinct properties. These generalisations are not very interesting for rings.

Theorem 6.1. *Let $(R,+,*)$ be a ring with identity and group of units U such that $|U| \geq 2$. Then the following are equivalent:*

(1) R is an a-nearring with DCCL,
(2) R is an f-nearring with DCCL,
(3) R is a field.

We are able to determine the structure of f-nearrings and a-nearrings explicitly, in terms of a generating structure.

Theorem 6.2. *The following are equivalent:*

(1) $(N,+,\cdot)$ is an f-nearring.
(2) There exists

(a) a group $(M,+)$
(b) a semigroup $S = G \cup E$ of group endomorphisms of $(M,+)$ where G is a group of fpf automorphisms of $(M,+)$ and E is a semigroup of non-bijective endomorphisms containing the zero map $\overline{0}$
(c) an element $m \in M$ such that $S(m) = M$ and for all $n \in M$ there is a unique $s_n \in S$ with $n = s_n(m)$

such that $(N,+,\cdot) \cong (M,+,*)$, where $a * b = s_b(a)$ $(a,b \in M)$.

We thus obtain the following.

Theorem 6.3. *Let $(N,+,*)$ be a finite f-nearring. Then, $(N,+)$ is an elementary abelian group.*

Theorem 6.4. *The following are equivalent:*

(1) $(N,+,\cdot)$ is an a-nearring.
(2) There exists

(a) a group $(M,+)$ with a subgroup $H \subseteq M$
(b) a semigroup $S = G \cup E$ of group endomorphisms of $(M,+)$ where G is a group of automorphisms and E is a semigroup of non-bijective endomorphisms containing the zero map $\overline{0}$
(c) an element $m \in M$ such that $S(m) = M$, $G(m) = H \setminus \{0\}$ and for all $n \in M$ there is a unique $s_n \in S$ with $n = s_n(m)$

such that $(N,+,\cdot) \cong (M,+,*)$, where $a * b = s_b(a)$ $(a,b \in M)$.

A similar result to above holds, although slightly different.

Theorem 6.5. *Let $(N,+,*)$ be a finite a-nearring. Then, $(N,+)$ is a group of exponent p, $ord(1) = p$, p a prime number.*

For nearrings that satisfy both properties, we are able to find a specific construction, based upon the construction of nearvector spaces[7,14].

Theorem 6.6. *Let $(F,+,*)$ be a near-field and $(V,+) = (F^k,+)$ the k dimensional direct sum of $(F,+)$. For $i \in \{1,\ldots,k\}$ let $\alpha_i : F \to F$ be zero preserving maps which are multiplicative automorphisms of the group $(F \setminus \{0\},*)$ where α_k is the identity map. For $\underline{a},\underline{x} \in V$ define the operation $*_1$ in V in the following way:*

$$\underline{x} *_1 \underline{a} := (x_1 * \alpha_1(a_k),\ldots,x_{k-1} * \alpha_{k-1}(a_k), x_k * a_k) \text{ if } a_1 = a_2 = \ldots = a_{k-1} = 0$$
$$\underline{x} *_1 \underline{a} := (x_k * a_1,\ldots,x_k * a_k) \text{ otherwise.}$$

Then $(V,+,*_1)$ is an af-nearring with units $U = \{(0,\ldots,0,a) | a \in F, a \neq 0\}$, identity $(0,\ldots,0,1)$, and $(U \cup \{0\},+,*_1)$ isomorphic to $(F,+,*)$.

We are able to find some significant results about the structure of these nearrings.

Theorem 6.7. *Suppose that N is an f-nearring or a-nearring with DCCN which is not a ring. If $J_2(N) = \{0\}$ then N is a near-field or $N \cong M_0(\mathbb{Z}_3)$.*

We are also able to use this result to find all examples of order p^2 for a prime p.

Theorem 6.8. *Let $(N,+,*)$ be an f-nearring of order p^2 which is not a near-field. Then N is also an* a-nearring. *Moreover, there exists a group $(V,+)$ isomorphic to $(N,+)$ and a map $\alpha : U_0 \to U_0$ with $\alpha(0) = 0$ and α an automorphism of the multiplicative group $(U,*)$ such that $(N,+,*)$ is isomorphic to the nearring $(V,+,*_1)$, where the nearring operations are defined above.*

There is every indication that these techniques can be extended for higher prime powers.

7. Ongoing research questions

In the following, we will outline some questions that have arisen in our work on structural issues in nearrings and their applications, along with motivations and necessary background.

7.1. *Sandwich centralizer nearrings, 0-primitivity and structure*

The paper[35] gives a precise construction method for zero symmetric 1- and 2-primitive nearrings using sandwich nearrings. This method of construction was also used in Refs. 32 and 33 to show that any nearring with a right identity element can be constructed combining the ideas of a sandwich nearring and a centralizer nearring. This was used in Ref. 33 to classify 1-primitive nearrings with a right identity element, and then was generalized in Ref.

35 to nearrings not necessarily containing a multiplicative right identity element. These nearrings generalize e.g. Ref. 19. We listed the definitions and one of the main theorems (e.g. Theorem 3.2) of Ref. 35 above to give some feeling for the nature of the construction involved.

With these tools, one could look into the following problems.

(1) In the proof of the results, we see that S is nothing other than Ψ_N. In Ref. 35, there is an explicit construction method for the sandwich function ϕ. Using the main theorems of Ref. 35 and the method of constructing ϕ it would be possible to systematically investigate primitive nearrings. One could obtain more concrete examples of primitive nearrings with special type of S and ϕ. Also of interest is the question, whether the methods of construction can also be applied (in a modified fashion) to construct 0-primitive nearrings. Moreover, it would be of value to classify nearrings which have properties close to being 1- or 2-primitive using the sandwich multiplication construction, see (4) below.

(2) Sandwich centralizer nearrings of the type $M_0(X,\Gamma,\phi,S)$ where S is a group and acts without fixpoints on Γ and X is an orbit with zero of the group action of S on Γ are planar nearrings. It was shown in Refs. 30 and 32 that all planar nearrings are of this type. Planar nearrings are a class of nearrings which are rich in applications. Natural questions in this line of discussion would be to investigate the properties of $M_0(X,\Gamma,\phi,S)$ when:

    (a) X is a union of two or a special discrete number of orbits with zero of the action of S, S a group.
    (b) S is not necessarily fixpointfree but another special type of group.
    (c) S is a special type of semigroup.
    (d) X is a subgroup of Γ.

In the paper[36] the structure of 0-primitive nearrings could be described to some satisfying extent, however not completely. 0-primitive nearrings are of fundamental importance in the structure theory because when we study nearrings with a suitable chain condition, the 0-primitive nearrings are precisely the subdirectly irreducible nearrings with non-nilpotent heart as in Theorem 3.4.

It is straightforward to see that when N is a ring with *DCCN* then subdirect irreducibility of N with non-nilpotent heart is equivalent to N being a primitive ring. These are completely classified by Jacobson's Density Theorem. In case of nearrings we end up with 0-primitive nearrings and these need more classification.

This is a special case of a more general structure theory developed in Ref. 36. We know that N, when considered as an N-group, is the direct sum $(0:\theta_1)+(0:\theta_0)$. Here $(0:\theta_1)$ is the annihilator of all the generators of the N-group (which would be zero in case N is 1-primitive) and $(0:\theta_0)$ is the annihilator of the non-generators. $(0:\theta_0)$ is an ideal of the nearring and in case N is 1-primitive, $(0:\theta_0)=N$. Indeed, $(0:\theta_0)$ shares a lot of properties a 1-primitive nearring has, for example it is the direct sum of minimal left ideals. This gives powerful tools to look into the following problems.

(3) When describing 0-primitive nearrings we need knowledge about $(0:\theta_1)$. One could take the methods of problem (1) above to obtain a construction method for certain classes of 0-primitive nearrings where we know something about $(0:\theta_1)$, for example $(0:\theta_1)$ having zero multiplication, being nilpotent or being a direct sum of minimal left ideals.

(4) One could use the methods of Ref. 35 to describe nearrings N which act faithfully and strongly monogenic on a subdirectly irreducible N-group. The topic of subdirect irreducibility comes into play also when studying 0-primitive nearrings, as we saw in Theorem 3.6 above.

7.2. Ψ_N *as generating set*

In the foregoing section we have seen in the discussion that given a nearring N then taking Ψ_N as a semigroup of endomorphisms gives us the possibility to form sandwich centralizer nearrings which are isomorphic to the original nearring N. Another way to use the semigroup of endomorphisms Ψ_N is to use it as generators of a nearring of functions on the group $(N,+)$. This nearring will be distributively generated and in case $(N,+)$ is abelian it will be a ring. This fact was already pointed out above and we have seen that the nearring N_{dg} distributively generated by Ψ_N of a nearring N will lead to interesting results in nearring theory.

(5) Study N_{dg} for primitive nearrings and hope for contributions in the structure theory of these nearrings. Probably the simplest primitive nearrings are the near-fields, so start investigating N_{dg} when N is a near-field. This builds upon work of Williams[37] and Smith[25]. Since the additive group of a near-field is abelian, N_{dg} will be a ring.

(6) In general, these nearrings will not be rings. Still, distributively generated nearrings are more well behaved than general nearrings. What information flows from N to N_{dg}, what is lost? In which cases can we reconstruct N from N_{dg}? When is $(N_{dg})_{dg}=N$, as we see in rings?

(7) Study N_{dg} when Ψ_N is a group of fixedpointfree automorphisms, a generalisation of the situation for (finite) near-fields in the question above. This includes planar nearrings as a class.

(8) From ring theory we know that for describing a primitive ring R with a minimal left ideal L one uses the so called double centralizer construction. $D:=End_R(L)$ is a skew field and, in case L is finite dimensional over D, R is isomorphic to $M_n(D)$, a finite dimensional matrix ring. A similar construction is not known for nearrings. It is known (see Ref. 34) that a minimal left ideal of a 2-primitive nearring is planar. Investigate to which degree the nearring N_{dg} arising from a minimal left ideal in a 2-primitive nearring has significant connections to the original primitive nearring. For example, study when taking a matrix nearring over N_{dg} yields a similar construction as the centralizer construction in ring theory. Use the results obtained in Ref. 18 and build upon them.

(9) Let N be a nearring and Γ an N-group. The semigroup of endomorphisms $S_\Gamma(N)=\{s\in End(\Gamma)|n(s(\gamma))=s(n(\gamma)\forall n\in N,\forall\gamma\in\Gamma\}$ arises naturally. The func-

tions $N \mapsto S_\Gamma(N)$ and $S \mapsto M_S(\Gamma)$ form a Galois connection. Taking $\Gamma = N$ as a group, we obtain $\Psi_N \subseteq S_\Gamma(N)$, reflecting Theorem 2.1. The Galois connection and these bounds would help in developing results about the relation between Ψ_N as a generating set and as a centralizer set. This is again closely related to the double centralizer construction in ring theory. In ring theory, the endomorphisms are additively closed, whereas here we can only work with centralizer nearrings and induced endomorphisms. Investigate the structure of the Galois closed nearrings and the corresponding endomorphism semigroups. Investigate the development of density type results in these cases and transfer techniques from ring theory to nearring theory as appropriate.

When looking for possible applications similar to those of planar nearrings and near-fields, it will pay to learn more about f-nearrings:

(10) What is the precise connection between near-vector spaces and f-nearrings? Of course, when we have a near-vector space, then this near-vector space is not an f-nearring, because we do not have non-units in a near-vector space. A near-vector space leads naturally (see Ref. 29) to a centralizer nearring of mappings that commute with the action of the scalars. Which f-nearrings are constructable from a near-vector space, either through the centralizer construction or otherwise? When are they isomorphic?

(11) Since we have a fixedpointfree action of the units, similar to the situation in planar nearrings and in near-fields, an f-nearring also gives rise to a BIBD. f-nearrings that are not near-fields are non simple, so we have an ideal in the nearring, so Ψ_U acts without fixedpoints on the ideal. Hence we get a second BIBD. Other geometrical structures which can be obtained from planar nearrings and near-fields are also constructable from f-nearrings. Investigate these for possible applications.

(12) We know that f-nearrings have elementary abelian additive group. Consequently, the distributively generated nearring N_{dg} generated by Ψ_U will be a ring R. What is this ring and what can the structure of this ring tell us about the structure of N, or give us some construction method for constructing f-nearrings? In Ref. 6, we were able to construct all f-nearrings with identity of order p^2. Develop construction methods for higher orders. For doing so we need more knowledge on the structure of f-nearrings, as our previous techniques are not immediately generalisable.

(13) In a similar fashion we can consider a-nearrings introduced above where we found similar but different results. The same questions as for f-nearrings, like questions of geometrical and structural nature, can be asked for a-nearrings, as well as for nearrings that satisfy both properties.

(14) When studying N_{dg} generated by Ψ_U when N is an f-nearring we have a special case of a nearring which is generated by (some of) its units. Address the question as to when a nearring with identity is generated by units. For rings this is known for a broad class of examples and is an ongoing line of discussion, see Ref. 11

or the survey paper[26] for references. For nearrings there has only been investigation of special situations. For example, it is known, see Ref. 24 for example, when $M_0(G)$ is generated by its units, as well as much about the generation by a single unit. The full transformation nearring on a group is an example of a simple nearring with identity. What can be said in general about simple nearrings with identity? These are closely related to 0-primitive nearrings. When are they generated by their units?

Looking once again at applications, we are interested in the ways in which structure theory can be applied to applications in automata and cellular automata.

(15) In the above, we showed that the entire variety of groups generated by a given group could be embedded into the nearring of state machines. In which way does the equational theory of the variety get embedded into the equational theory of the nearring?
(16) We have only been able to determine the Jacobson 2-radical for the full zerosymmetric nearring of state machines over certain groups. Vinay Madhusudanan[17] has found some further examples where we can determine the radical of the full zerosymmetric nearring of state automata. Are we able to extend these results to larger, or more importantly, smaller nearrings of automata?
(17) In Ref. 3, we introduced a class of state machines that correspond to perturbations of systems, a natural definition for state machines on groups. Are we able to determine the structure of these state machines in a similar way to the determination in Ref. 5?
(18) Can we completely determine the structure of group cellular automata over torsion, in particular, finite space groups? It seems that these have very large radicals, can we determine these explicitly?

8. Conclusion

We see a wide variety of results and thus emerging questions dealing with the building blocks of nearring theory. With roots that reach back into the very early developments in nearrings, we see that these questions retain their usefulness for guiding research. They also give a wide area for application in various related areas of mathematics and computer science. We have not touched upon the applications in design theory or polynomial theory, which are also quite relevant.

We have chosen to share this collection of questions that have arisen in recent research work with the community, hoping that it will inspire some new developments, building upon what we have found.

Bibliography

1. E. Aichinger, F. Binder, J. Ecker, P. Mayr and C. Nöbauer, SONATA - system of nearrings and their applications, GAP package, Version 2; 2003. (http://www.algebra.uni-linz.ac.at/Sonata/)

2. M. Bäck, H. Köppl, G. Pilz, G. Wendt, EinfluΣ verschiedener Parameter auf den Mykotoxingehalt von Winterweizen Versuchsdurchführung mit Hilfe eines neuen statistischen Modells, Proceedings, 63th ALVA-Tagung, Raumberg, Austria, 2008.
3. T. Boykett, An algebraic perturbation theory for state automata. In *Contributions to general algebra, 12 (Vienna, 1999)*, pp. 109–119, Heyn, Klagenfurt, 2000.
4. T. Boykett, Efficient exhaustive listings of reversible one dimensional cellular automata, *Theoret. Comput. Sci.*, **325**(2):215–247, 2004.
5. T. Boykett and G. Wendt, J_2 radical in automata nearrings. *Internat. J. Found. Comput. Sci.*, **25**(5):585–595, 2014.
6. T. Boykett and G. Wendt, Units in nearrings. *Comm. Alg.*, to appear, 2015.
7. T. Boykett and K.-T. Howell, The multiplicative automorphism group of a nearfield, with an application. *Comm. Alg.*, to appear, 2016.
8. T. Boykett, Nearrings of Cellular Automata, submitted, 2016.
9. J. R. Clay, *Nearrings*, Oxford University Press, 1992.
10. C. Cotti Ferrero and G. Ferrero, *Nearrings: Some Developments Linked to Semigroups and Groups*, Springer, 2002.
11. H. K. Grover *et al.*, Sums of Units in Rings, *J. Algebra Appl.* **13**, DOI: 10.1142/S0219498813500722, 2014.
12. J. F. T. Hartney and A. M. Matlala, Structure theorems for the socle-ideal of a nearring, *Comm. Algebra*, Vol. 36, No. 3, pp. 1140–1152, 2008.
13. G. Hofer, Left ideals and reachability in machines, *Theor. Computer Science*, **68**:49–56, 1989.
14. K.-T. Howell and J. H. Meyer, Finite-dimensional near-vector spaces over fields of prime order, *Comm. Algebra* **38** (2010), no. 1, 86–93.
15. K. Kaarli, On non-zerosymmetric near-rings with minimum condition, *Nearrings, nearfields and K-loops*, (Hamburg, 1995) *Math. Appl.*, 426, Kluwer Acad. Publ., Dordrecht, 21–33, 1997.
16. W. F. Ke, G. Pilz, Abstract Algebra in Statistics, *Journal of Algebraic Statistics* **1** (2010), 6–12.
17. V. Madhusudanan, personal communication, 2016.
18. W. F. Ke, J. H. Meyer, J. H. Meyer, G. Wendt, Matrix maps over planar nearrings, *Proceedings of the Royal Society of Edinburgh: Section A Mathematics*, Vol. 140A, 83–99, 2010.
19. K. D. Magill, Jr., Isomorphisms of sandwich nearrings of continuous functions, *Boll. Un. Mat. Ital. B (6)*, **5**(1):209–222, 1986.
20. J. D. P. Meldrum, *Near-rings and their links with groups*, Pitman Advanced Publishing Programm, 1985.
21. G. Peterson, S. Scott, Units in compatible nearrings, III. *Monatshefte für Mathematik*, Vol. 171, Issue 1, pp. 103–124, 2012.
22. G. Pilz, Strictly connected group automata, *BProc, Roy. Irish Acad.*, **86A**:115–118, 1986.
23. G. Pilz, *Near-Rings*, Revised edition, North Holland, 1983.
24. S. Scott, The p-gen nature of $M_0(V)$ (I), *Algebra and Discrete Mathematics*, Vol. 15, No. 2, pp. 237–268, 2013.
25. K. C. Smith, A ring associated with a nearring, *J. Algebra*, **182**(1):329–339, 1996.
26. A. K. Srivastava, A survey of rings generated by units, *Ann. Fac. Sci. Toulouse Math.* **6**(19) (2010), Fascicule Spécial, 203–213.
27. S. Veldsman, On equiprime nearrings, *Comm. Algebra*, **20**(9):2569–2587, 1992.
28. H. Wähling, *Theorie der Fastkörper*, Thales Verlag, 1987.
29. A. P. J. van der Walt, Matrix nearrings contained in 2-primitive nearrings with minimal subgroups, *J. Algebra*, **148**(2):296–304, 1992.
30. G. Wendt, Planarity in Near-rings, Dissertation, Universität Linz, 2004.
31. G. Wendt, On the multiplicative semigroup of nearrings, *Math. Pannonica* **15**(2) (2004), 209–220.

32. G. Wendt, Planar nearrings, sandwich nearrings and nearrings with right identity, *Nearrings and Nearfields*, 277–291, Springer, 2005.
33. G. Wendt, Primitive Near-rings, *Algebra Colloquium*, **14**:3, 417–424, 2007.
34. G. Wendt, Minimal Left Ideals of Near-rings, *Acta Math. Hungar.*, **127** (2010), 52–63, Doi: 10.1007/s10474-010-9090-1.
35. G. Wendt, 1-Primitive Near-rings, *Math. Pannonica*, **24**(2), pp. 269–287, 2013.
36. G. Wendt, 0-Primitive Near-rings, Minimal Ideals and simple Near-rings, *Taiwanese Journal of Mathematics*, Vol. 19, No. 3, pp. 875–905, June 2015, DOI: 10.11650/tjm.19.2015.5077.
37. R. E. Williams, Simple Near-rings and Their Associated Rings. PhD Thesis, University of Missouri, 1965.

Lie rings and Lie algebra bundles

B. S. Kiranagi

PI, SERB/DST Project,
Department of Mathematics,
University of Mysore, Mysore-570006, India
E-mail: bskiranagi@gmail.com

We show that a Lie ring P over $C(X)$, the ring of real valued continuous functions on X, is isomorphic to a Lie ring of the set of all sections of a Lie algebra bundle ξ denoted by $\Gamma(\xi)$, if and only if P is finitely generated projective module over $C(X)$, where X is a compact Hausdorff space. The above results are extended to arbitrary topological space X for bundles of finite type. This result gives rise to a common intuition that a study of rings which are also projective modules over commutative rings are like the study of algebra bundles over Topological spaces.

1. Introduction

In 1955, J. P. Serre[37] has shown that there is a one to one correspondence between algebraic vector bundles over an affine variety and finitely generated projective modules over its coordinate ring.

In 1962, Richard G. Swan[40] has shown that a similar correspondence exists between topological vector bundles over a compact Hausdorff space X and finitely generated projective modules over $C(X)$, the ring of real valued continuous functions on X. Swan theorem relates the geometric notation of vector bundles to the algebraic concept of projective modules. Serre theorem is more algebraic in nature, and concerns vector bundles on an algebraic variety over an algebraically closed field of any characteristic. Where as Swan theorem is more analytic and concerns (real, complex or quarternionic) vector bundles on a compact Hausdroff space or smooth manifold.

Later in 1984 Goodearl[9] observed that the equivalence holds in the more general cases of Paracompact Hausdorff space X if one restricts to the bundles of finite type (i.e. if there exists a finite open covering $\Im$ of X such that the restriction of the bundle to each $U \in \Im$ is trivial). This restriction excludes vector bundles of unbounded dimension which cannot come from the category of finitely generated $C(X)$ modules.

In 1986, Vaserstein[41] has extended this result to an arbitrary topological space X, with an appropriate definition of finite type. According to Vaserstein, the bundle over an arbitrary space X is of finite type[41] if there is a finite partition S of 1 on X (that is a finite set S of nonnegative continuous function on X whose sum is 1) such that the restriction of the bundle to the set $\{x \in X | f(x) \neq 0\}$ is trivial for each f in S.

A vector bundle over a compact Hausdorff space X is of finite type. The definitions of finite type of Vaserstein and Goodearl are equivalent if a base space X of a bundle is normal.

Here we show that a similar correspondence exist between Lie algebra bundles over a space X and Lie rings which are finitely generated projective modules over $C(X)$.

2. Lie Algebra Bundle

An eminent mathematician J. P. Serre who was awarded Fields Medal, the highest award in mathematics, in 1954 at Amsterdam during International congress of mathematicians and the first recipient of the Abel Laureate in 2002 for the most valuable contributions to the development of mathematics, posed the question: For every given Lie algebra bundle does there exist a Hausdorff Lie group bundle whose Lie algebra bundle is isomorphic to a given Lie algebra bundle?

In 1966, A. Douady and M. Lazard[8] defined a family of Lie algebras; a (analytic) family of Lie algebras parametrised by X is a (separable) fiber space ξ over a finite dimensional (analytic manifold) X, whose fiber ξ_x at each point $x \in X$ is a finite dimensional vector space, together with an (analytic) anti symmetric, bilinear fiber map $[\quad]: \quad \xi \times \xi :\rightarrow \xi$ giving each fiber ξ, the structure of a Lie algebra.

A family of Lie groups; a nonseparable (analytic) family of Lie groups parametrized by X is a non separable (i.e., not necessarily separable) (analytic) manifold G together with a surjective submersion $\pi : G \rightarrow X$, a section $e : X \rightarrow G$ and a morphism $\Theta : G \times G \rightarrow G$, such that for each x belonging to X the fiber $\pi^{-1}(x) = G_x$ has a Lie group structure given by multiplication $\Theta|_{G_x \times G_x}$ and the identity $(e(x)$.

The base space X and the fibers G_x are all separable manifolds; only the total space G is allowed to be non separable (non-Hausdorff). Further, Douady and Lazard constructed a (family) Lie group bundle G (not necessarily Hausdorff) for a given Lie algebra bundle ξ such that the Lie algebra bundle of a Lie group bundle G is isomorphic to a given Lie algebra bundle ξ in their remarkable paper[8]. They had left the following problem open[8] in an analytic case. That is, if all the spaces of a bundle are analytic manifolds and the corresponding morphisms are analytic. They ask whether an analogue theorem still holds locally (around each point of X) for a given analytic family of Lie algebras, if one requires $G(\xi)$ to be separable. Don Coppersmith has given an example which provides a negative answer for an analytic family of Lie algebras[7].

Problem: Does there exist for each s in X, an open set S of s in X and a Hausdorff Lie group bundle over S and whose Lie algebra bundle is isomorphic to Lie algebra bundle $\xi_S = p^{-1}(S)$. They also suggested to tackle this question from the point of view of algebraic geometry[8][p.151].

We have answered[15] the open problem partially by proving a fundamental result in algebraic geometry: The real orbit of a real point is open in the real part of its complex orbit.

There is a result in complex algebraic geometry known as the closed orbit lemma due to Chevalley[2]: The (complex) orbit is open in its closure. But this result is not true in the real case.

The following results are true for Associative rings, Alternative rings, Jordan rings and corresponding algebra bundles, but however, here we prove for Lie rings and Lie bundles.

To define Lie algebra bundles, we need the following definitions.

3. Vector Bundle — motivation

For the motivation we follow the beautiful book by *Gerd Rudolph and Matthias Schmidt*[11]. A fibre bundle is a natural and useful generalization of the product of two manifolds. One way to interpret a product manifold is to place a copy of M_1 at each point of M_2. For example take $M_1 = S^1$ and M_2 the line segment $(0,1)$. By placing a line segment at each point of a circle or placing a circle at each point of the line segment. The product topology here gives us a piece of cylinder. This idea leads to more general concept of a fibre bundle.

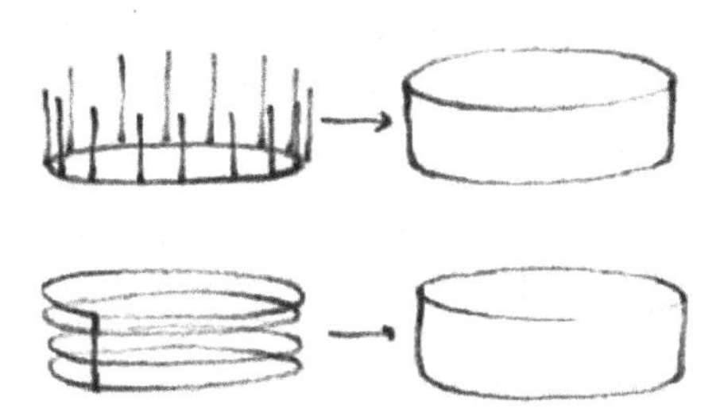

Globally, however a fibre bundle is not a product manifold in general. Möbius strip is an example of a non-trivial fibre bundle. Similarly to the case with the cylinder, we can create Möbius strip by starting with the circle and at each point on the circle attaching a bunch of parallel line segments (intervals of same length) to the circle, our intervals perform a 180deg twist as we go round the circle. This manifold gives much more interesting geometry. Note that it is locally indistinguishable from the cylinder. The twist in the Möbius strip is entirely a global property but not located at any particular point of the strip. This example leads to generalise the language of product spaces, to include objects like the Möbius strip which are only locally product space. This generalization led to the concept of fibre bundle and then to the special cases like vector bundles.

Definition 3.1. A triple $\xi = (\xi, p, X)$ is said to be a *vector bundle* if
(i) A map $p : \xi \to X$ is a continuous map from a topological space ξ onto a topological space X, p is called the projection map and for each x in $X, p^{-1}(x) = \xi_x$ is called the fibre over x.
(ii) For each x in X, there is an open set U of x in X, a vector space V and a homeomorphism $\phi : U \times V \to p^{-1}(U)$ such that for each x in $U, \phi_x : V \to p^{-1}(x)$ is a vector space isomorphism, (U, ϕ) is called a local chart. We generally denote the vector bundle $\xi = (\xi, p, X)$ by ξ itself.

Definition 3.2. Let $\xi = (\xi, p_1, X)$ and $\eta = (\eta, p_2, X)$ be two vector bundles. *Whitney sum* of two vector bundles ξ and η is again a vector bundle $\xi \oplus \eta = (\cup_{x\in X}(\xi_x \times \eta_x), q, X)$ where $\cup \xi_x \times \eta_x \subset \xi \times \eta$, with the induced topology. For, $q : \xi \oplus \eta \to X, q(u,v) = p_1(u) = p_2(v)$ is continuous being p_1 and p_2 continuous. Let $\phi_1 : U \times V \to p^{-1}(U)$ and $\phi_2 : U \times W \to p^{-1}(U)$ be two local charts of ξ and η, respectively. Define $(\phi_1 \times \phi_2) : U \times V \times W \to p^{-1}(U)$ by $(\phi_1 \times \phi_2)(x,v,w) = (\phi_1(x,v), \phi_2(x,w))$, then $\phi_1 \times \phi_2$ is homeomorphic and $(\phi_1 \times \phi_2)_x : V \times W \to p^{-1}(x)$ is linear for all x in U. Thus $(U, \phi_1 \times \phi_2)$ is a local chart of $\xi \oplus \eta$.

Definition 3.3. Let ξ, η be two vector bundles over the same base space X then a continuous map $f : \xi \to \eta$ is said to be a *morphism* if for each x in X, $f_x : \xi_x \to \eta_x$ is linear.

Definition 3.4. (Marius Sophus Lie (1842–99))
A vector space L over a field F, with an operation $[\ ]: L \times L \to L$ is called a Lie algebra over F if the following axioms are satisfied

(1) The bracket operation [ ] is bilinear.
(2) $[x,x] = 0$ for all x in L.
(3) $[x,[y,z]] + [y,[z,x]] + [z,[x,y]] = 0$ for all x,y,z in L. Axiom 3 is called the Jacobi identity.

Remark 3.1. (i) $[x,x] = 0 \Rightarrow [x,y] = -[y,x]$ since $[x+y, x+y] = 0$.
(ii) $[x,y] = -[x,y] \Rightarrow [x,x] = 0$, if $\text{Ch}F \neq 2$.

Example 3.1.

(1) Any associative algebra with the Lie product $[x,y] = xy - yx$, is a Lie algebra where xy is the associative product of x and y.
(2) Set of all vectors in R^3 with cross product $A \times B = (|A||B| sin\theta)u$, is Lie algebra, where u is a unit vector perpendicular to the plane containing the vectors A and B. For Lie algebras refer Refs. 13, 14.

Definition 3.5. A *Lie algebra bundle*[8], for short a Lie bundle, is a vector bundle ξ over a topological space X, together with a morphism $\theta : \xi \oplus \xi \to \xi$ which induces a Lie algebra structure on each fibre ξ_x.

A locally trivial Lie algebra bundle is found in Ref. 10.

Definition 3.6. A *locally trivial Lie bundle* ξ is a vector bundle in which each fibre is Lie algebra and further there exists an open cover $\{U_\alpha\}$ of X, Lie algebras $\{L_\alpha\}$ and a homeomorphisms
$\phi_\alpha : U_\alpha \times L_\alpha \to p^{-1}(U_\alpha)$ such that for each x in U_α,
$\phi_{\alpha,x} : L_\alpha \to p^{-1}(x)$ is a Lie algebra isomorphism.

Remark 3.2. Locally trivial Lie bundle is a Lie bundle, but the converse need not be true in general.

Definition 3.7. A *Locally trivial Lie bundle is a Lie bundle.* For, let ξ be a locally trivial Lie bundle. Define $\theta : \xi \oplus \xi \to \xi$ by $\theta(u,v) = \theta_x(u,v) = [u,v]_x$, if $u,v \in \xi_x$, where $[,]_x$ is a Lie product on ξ_x. Continuity being a local problem, consider $\phi : U \times L \to p^{-1}(U)$, where $\phi_x : L \to P^{-1}(x)$ is a Lie isomorphism for each $x \in U$.

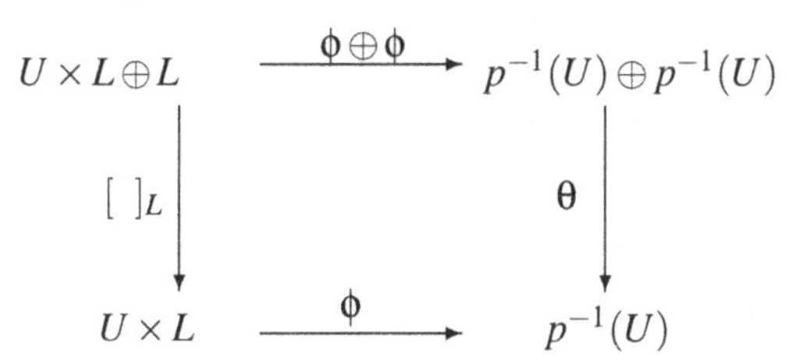

Obviously $\theta(\phi \oplus \phi) = \phi[,]_L$. Then $\theta(\phi \oplus \phi)$ is continuous, since the composition $\phi[,]_L$ is continuous being both bilinear. Then θ is continuous since the composition $\theta(\phi \oplus \phi)$ is continuous and $\phi \oplus \phi$ is homeomorphism. Thus ξ is a Lie bundle.

Example 3.2. Example-Lie bundle which is not locally trivial [18].
We construct an example of a Lie bundle which is not a locally trivial Lie bundle. Let L be a simple Lie algebra over the field R of real numbers. Then $(I \times L, p, I)$ is a trivial vector bundle over the closed interval $[0,1] = I$, with the morphism $I \times L \to I, (t,l) \to t$. Now we define a map $\theta : I \times L \oplus L \to I \times L$ by $\theta(t,(x,y)) = (t, \theta_t(x,y))$, where $\theta_t(x,y) = t[x,y]$ and [ ] is the given Lie product of L. It induces a Lie algebra structure θ_t on each fibre L_t and θ is continuous. For, let β be the set of all bilinear maps from $L \times L$ to L. Then β is a vector space over a real field R. Therefore the scalar product $R \times \beta \to \beta$ is continuous. Then $I \to I[,], t \to t[,]$, is continuous. Thus $\xi = (I \times L, p, I, \theta)$ is a Lie bundle but not locally trivial. For, $\xi_t = L_t$ is Abelian for $t = 0$ and $t \neq 0, \xi_t$ is non-abelian being isomorphic to the simple Lie algebra L. For, $\phi : L \to L_t, \phi(v) = t^{-1}v$ is isomorphism. Consider the Lie bracket θ_t on $L_t, \theta_t(\phi(\mu), \phi(v)) = t[\phi(u), \phi(v)]$=$t[t^{-1}u, t^{-1}v] = tt^{-1}t^{-1}[u,v] = t^{-1}[u,v] = \phi[u,v]$.

Hence ϕ is homomorphism and obviously bijective.
Thus $\xi = (I \times L, p, I, \theta)$ is a Lie bundle but not locally trivial.

Remark 3.3.

(1) Surprising to know that it is an example of a simple bundle (no proper ideal bundles) but not semi simple Lie bundle (each fibre is semi simple).
(2) It is an example of the jump phenomenon of "Deformation theories". These types of examples are important in quantum theory because it connects the simplicity and abelian which are opposite to each other.

Refer Ref. 22 for another important example in deformation theory which is semi simple over a quotient field and abelian over a residue field.

For further information on Lie algebra bundles, Jordan algebra bundles and associative algebra bundles refer Refs. 6, 22, 34, 35.

4. Lie Rings

Definition 4.1. A *Lie ring* is defined as a nonassociative ring with multiplication that is anticommutative and satisfies the Jocabi identity. More specifically we can define a Lie ring L to be an abelian group with an operation $[\cdot,\cdot]$ that has the following properties:

(1) Bilinearity:
$[x+y,z] = [x,z]+[y,z]$

$[z,x+y] = [z,x]+[z,y], \forall\, x,y,z \in L.$

(2) The Jacobi identity

$$[x,[y,z]]+[y,[z,x]]+[z,[x,y]] = 0$$

$x,y,z \in L.$

(3) Anticommutative : For all $x \in L$.

$$[x,x] = 0$$

$[x,y]+[y,x] = 0.$

Example 4.1.

(1) Any Lie algebra over a general ring instead of a field is an example of a Lie ring. Lie rings are not Lie groups under addition, despite the name.
(2) Any associative ring can be made into a Lie ring by defining a bracket operator $[x,y] = xy-yx$.
(3) For an example of a Lie ring arising from the study of groups, let G be a group with $(x,y) = x^{-1}y^{-1}xy$ the commutator operation, and let $G = G_0 \supseteq G_1 \supseteq G_2 \supseteq \cdots \supseteq G_n \supseteq \cdots$ be a central series in $G-$ that is the commutator subgroup (G_i,G_j) is contained in G_{i+j} for any i,j. Then $L = \oplus\frac{G_i}{G_{i+1}}$ is a Lie ring with the following operations:

- Addition
Define addition component-wise as follows: For $x,y \in G_i$ define $xG_{i+1} + yG_{i+1} = xyG_{i+1}$, where xy is the operation in the group G. It is easy to check that this addition is abelian for, $xG_{i+1}+yG_{i+1} = yG_{i+1}+xG_{i+1} \Leftrightarrow xyG_{i+1} = yxG_{i+1} \Leftrightarrow (yx)^{-1}(xy) \in G_{i+1} \Leftrightarrow x^{-1}y^{-1}xy \in G_{i+1}$. But since $x,y \in G_i$, $x^{-1}y^{-1}xy \in (G_i,G_i)$. Also $(G_i,G_i) \subseteq G_{2i}$ and $G_{2i} \subseteq G_{i+1}$ as $i \geq 1$. Thus $x^{-1}y^{-1}xy \in G_{i+1}$ and therefore addition is abelian componentwise. Hence addition defined is abelian.
- The bracket operation
The bracket operation be defined by $[xG_i,yG_j] = (x,y)G_{i+j}$, which is extended linearly so that bi-linearity is trivially satisfied.

  For anti-commutativity:
  $[xG_i,yG_j]+[yG_j,xG_i]$
  $= (x,y)G_{i+j}+(y,x)G_{j+i}$

$$= x^{-1}y^{-1}xyG_{i+1} + y^{-1}x^{-1}yxG_{i+1}$$
$$= x^{-1}y^{-1}xyy^{-1}x^{-1}yxG_{i+j}$$
$$= G_{i+j}, \text{ which is the zero element.}$$

Hence $L = \oplus \frac{G_i}{G_{i+1}}$ is a Lie ring.

5. Definition of modules and Lie modules

Definition 5.1. Let M be an additive abelian group, then M is called a *left* $K-$ *module* for the ring K if there is a binary product from

$$K \times M \to M \qquad (\lambda, m) \to \lambda m$$

such that
(i) $\lambda(m_1 + m_2) = \lambda m_1 + \lambda m_2$
(ii) $(\lambda_1 + \lambda_2)m = \lambda_1 m + \lambda_2 m$
(iii) $\lambda_1(\lambda_2 m) = (\lambda_1\lambda_2)m$
(iv) $1 \cdot m = m$
where m, m_1, m_2 are in M and $\lambda, \lambda_1, \lambda_2$ are in K, while 1 denotes the multiplicative identity of K.

Similarly **right** $K-$ **module** is defined.

Remark 5.1. If K is a commutative ring and that M is a left $K-$ module then we can turn M into a right $K-$ module simply by defining $x\lambda = \lambda x$. Conversely every right module can be regarded as a left $k-$ module, thus all modules over commutative rings are virtually two sided and distinction between left and right disappears

Definition 5.2. Let L be a Lie algebra and a vector space V is called $L-$ *module* if there is a operation $L \times V \to V \qquad (x, v) \to xv$ if it satisfies following conditions:
(i) $(ax + by) \cdot v = a(x \cdot v) + b(y\dot{v})$,
(ii) $x \cdot (av + bw) = a(x \cdot v) + b(x \cdot w)$,
(iii) $[x, y] \cdot v = x \cdot y \cdot v - y \cdot x \cdot v \qquad$ for all $x, y \in L, b, w \in V, ab \in F$.

Definition 5.3. An $R-$module M is said to be *finitely generated over* R if there is a finite subset X of M such that M is the submodule generated by X, i.e., if $X = \{x_1, x_2, \cdots, x_r\}$, then (assume $1 \in R$ and M is unitary) we have $M = \{\sum_{i=1}^{r} a_i x_i | a_i \in R\}$.

Definition 5.4. Let M be an $R-$module. A subset B of M is said to be *linearly independent* over R if for any finite subset $\{b_1, b_2, \cdots, b_r\} \subseteq B, \sum_{i=1}^{r} a_i b_i = 0$ with $a_i \in R$, then $a_i = 0$, for all i.

Definition 5.5. An $R-$modules M is called a free if M has a basis B, i.e., a linearly independent subset B of M such that M is spanned by B over R, i.e., every element $x \in M$ can be written uniquely as $x = \sum_{b \in \lambda_b} \lambda_b \cdot b, \lambda_b \in R, \lambda_b = 0$ except for finitely many $\lambda_b{}'s$, i.e., x is finite linear combination of elements in B, the scalars being unique for x.

Example 5.1.

(1) For any ring R with 1, the left $R-$module R is free with basis $\{1\}$ *or* $\{u\}$, u any unit in R. In fact, an element $b \in R$ is linearly independent if and only if b is not a right zero-divisor in R, i.e., $ab = 0 \Rightarrow a = 0$. Furthermore, $\{b\}$ is an $R-$basis of R if and only if b has a left inverse and is not a right zero-divisor.

(2) $R^n = R \times R \cdots \times R$, is a free $R-$module if R has 1. The set $B = \{(1,0,...,0),$ $(0,1,0,...,0)$, ..., $(0,0,...,1)\}$ is an $R-$ basis for R^n, called the standard basis of R^n.

(3) Direct sum of free modules is a free module. For, suppose M and N are free $R-$modules with bases A and B respectively. Now $M \oplus N = M \times N$ is free $R-$module because $(A \times \{0\}) \cup (\{0\} \times B)$ is an $R-$basis for $M \times N$. More generally, for any family of free $R-$modules, $\{M_i | i \in I\}$, with basis A_i, $M = \oplus_{i \in I} M_i$ is a free module with a basis $A = \cup_{i \in I} A_i$.

Example 5.2. Example of a non free module.
Any finite abelian group is not free as a module over Z, the ring of integers. In fact, any abelian group M which has a non-trivial element of finite order cannot be free as a module over Z. For, suppose M is free. Say B is a basis for M over Z. Let $0 \neq x \in M$ be such that $nx = 0$ for some $n \in N$, $nx \neq 0$ for $m < n$ and $n \geq 2$. Now we have $x = n_1 b_1 + n_2 b_2 + \cdots + n_r b_r$ for some $b_1, b_2, \cdots, b_r \in B$ and $n_1, n_2, \cdots, n_r \in Z$.

Hence $nx = n[n_1 b_1 + n_2 b_2 + \cdots + n_r b_r]$ imply that $nn_1 = 0, nn_2 = 0, \cdots, nn_r = 0$ (by linear independence of B) imply that $n_1 = 0, n_2 = 0, \cdots, n_r = 0$ (since $n \neq 0$, since Z is an integral domain), i.e., $x = 0$, a contradiction.

6. Projective Modules and Vector Bundles

Definition 6.1. Let K be a ring. A K-module P is said to be K-projective if given any diagram

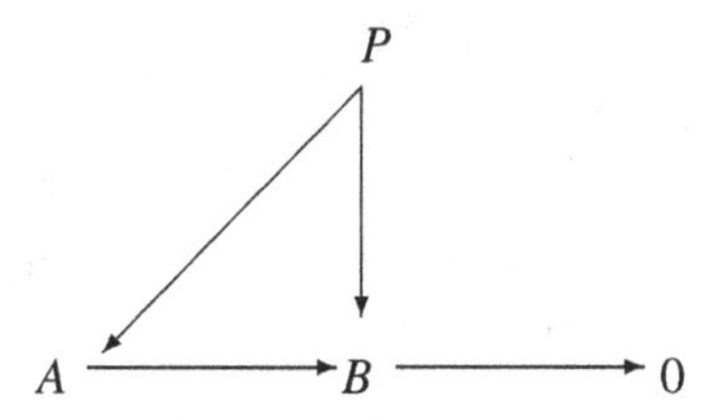

over K, in which the row is exact, it is always possible to find a K-homomorphism $p \to A$ such that $PAB = PB$.

Now we recall some results from Ref. 40.

Let $\xi = (\xi, p, X)$ be a bundle. A **section** S of ξ over a subset $U \subset X$ is a continuous map $S : U \to \xi$ such that $pS(x) = x$. A bundle ξ is locally trivially $\Rightarrow \forall x \in X$, there exists an open set U containing x, a fibre $F = R^n$ over R and a homomorphism $\phi : p^{-1}(U) \to U \times R^n$, such that

$$\phi_x : p^{-1}(x) \to R^n \text{ is a vector space isomorphism.}$$

Let $e_1 = (1,0,...,0),...,e_i = (0,...0.1,...0),...e_n = (0,0,...,1)$ be standard basis of $R^n = F_x$. Define $S_i : U \to R^n$, by $S_i(x) = e_i$, for all x in U.

Then $S_1,...,S_n$ are continuous being constant on U and they are sections of ξ over U such that $S_1(x),...,S_n(x)$ form a K base for F_x, for all $x \in X$. We will say that $S_1,...,S_n$ form a local base at x. Any section of ξ over U can be written as $S(y) = \sum_{i=1}^{n} a_i(y)S_i(y)$, where $a_i(y) \in K$. Note that S is continuous $\Leftrightarrow$ each $a_i(y)$ is continuous. This is immediate for the local base $e_1,...e_n$ which we get from the definition of vector bundle. If $S_1,...,S_n$ is another local base, $S_i(y) = \sum_{j=1}^{n} a_{ij}(y)e_j(y)$ and $y \to a_{ij}(y)$ is a continuous map $U \to GL(n,K)$. The result then follows from the fact that $A \to A^{-1}$ is a continuous map in $GL(n,K)$.

Similarly, if $S_1,...,S_n$ is a local base for ξ at $x, t_1,...,t_n$ is a local base for η at x, and $f : \xi \to \eta$, then near x, $f(S_i(y)) = \sum_{j=1}^{n} a_{ij}(y)t_j(y)$ and f is continuous $\Leftrightarrow$ each $a_{ij}(y)$ is continuous. If $f : \xi \to \eta$ is one-one and onto, the fact that $A \to A^{-1}$ in $GL(n,K)$ is continuous shows that f^{-1} is continuous. Thus, such an f must be a homeomorphism.

Let $C(X) = C_K(X)$ be the ring of continuous K-valued functions on X. If ξ is a K vector bundle over X. Let $\Gamma(\xi)$ be the set of all sections of ξ over X.

If $S_1, S_2 \in \Gamma(\xi)$, define $(S_1 + S_2)(x) = S_1(x) + S_2(x)$.

If $S \in \Gamma(\xi)$ and $a \in C(X)$, define $(aS)(x) = a(x)S(x)$.

Then $\Gamma(\xi)$ becomes a $C(X)$ module[SRG].

If ξ is trivial, $\Gamma(\xi)$ is obviously a free $C(X)$ module on n generators, $n =$ dimension of standard fibre.

Lemma 6.1. *Let X be normal. Let U be a neighborhood of x in X, and let S be a section of a vector bundle ξ over U. Then there is a section S' of ξ over X such that S' and S agree in some neighborhood of x.*

Proof. A space X being normal, there exist neighborhoods V, W of x such that $\overline{V} \subset U$, $\overline{W} \subset V$.

For X is normal $\Rightarrow X$ is regular. We know that X is regular if and only if for given a x in X and a neighborhood U of x there exists an open set V of x such that $\overline{V} \subseteq U$.

Then $A = X - U$ a closed set and $x \notin A$. Then X being regular there exist disjoint open sets V and W such that $x \in V$ and $A \subset W$.

For, all $y \in A \subseteq W$, W is a neighborhood of y.

Since $V \cap W = \phi \Rightarrow y$ is not a limit point of V implies $y \notin \overline{V} \Rightarrow \overline{V} \cap A = \phi \Rightarrow \overline{V} \subseteq A^c = U$ thus $\overline{V} \subset U$.

Given U a neighborhood of x, there exist an open set V such that $\overline{V} \subset U$.

Again V is a neighborhood of x similarly there exist an open set W such that $\overline{W} \subset V$.

Let ω be a real-valued function on X such that $\omega|\overline{W} = 1$, $\omega|X - V = 0$.

(By Uryson's lemma: X be a normal space. Let A and B be disjoint closed subsets of X. Let $[a,b]$ be a closed interval in the real line. Then there exist a continuous map $f : X \to [a,b]$ such that $f(x) = a$ for all $x \in A$ and $f(x) = b$ for all $x \in B$.)

Now define

$$s'(y) := \begin{cases} w(y)s(y), & \text{if } y \in V \\ 0, & \text{if } y \notin V \end{cases}.$$

□

Corollary 6.1. *Let X be normal. For any $x \in X$ there are elements $S_1,...,S_n \in \Gamma(\xi)$ which form a local base at x.*

Corollary 6.2. *Let X be normal. If $f,g:\xi \to \eta$ and $\Gamma(f) = \Gamma(g): \Gamma(\xi) \to \Gamma(\eta)$, then $f = g$.*

Proof. Given $e \in E(\xi)$, with $P(e) = x$, there is a section S over a neighborhood U of x with $S(x) = e$. By Lemma 3, there is a section $S' \in \Gamma(\xi)$ with $S'(x) = e$. Now $f(e) = fS'(x) = (\Gamma(f)S')(x) = (\Gamma(g)S')(x) = g(e)$. □

Lemma 6.2. *Let X be normal. Let $S \in \Gamma(\xi)$. Suppose $S(x) = 0$. Then there are elements $S_1,...,S_k \in \Gamma(\xi), a_i,...,a_k \in C(X)$ such that $a_i(x) = 0$ for $i = 1,...,k$ and $S = \sum a_i S_i$.*

Proof. Let $S_1,...,S_n \in \Gamma(\xi)$ be a local base at x (Corollary 6.1.1). Let $S(y) = \sum b_i(y)S_i(y)$ near $x, b_i(y) \in K$. Let $a_i \in C(X)$ be such that a_i and b_i agree in a neighborhood of x. These exist by Lemma 6.1 applied to $X \times K$. Then $S' = S - \sum a_i S_i$ vanishes in a neighborhood U of x. Let V be a neighborhood of x such that $\overline{V} \subset U$. Let $a \in C(X)$ be zero at x and 1 on $X - V$. Then $S = aS' + \sum a_i S_i$. But $a(x) = 0$ and $a_i(x) = b_i(x) = 0$. □

We will now show that if X is compact Hausdorff, the $C(X)$-modules which can occur as $\Gamma(\xi)$ for some ξ are exactly the finitely generated projective modules[SRG].

Lemma 6.3. *Let X be a compact Hausdorff. Let ξ be any K-vector bundle over X. Then there is a trivial vector bundle ζ (i.e $\zeta = X \times K^n$) and an epimorphism $f:\zeta \to \xi$.*

Proof. For each $x \in X$, choose a set of sections $S_1,...,S_k \in \Gamma(\xi)$ which form a local base over some neighborhood U of x. A finite number of the U cover X. Therefore, there are a finite number of sections $S_1,...,S_n \in \Gamma(\xi)$ such that $S_1(x),...,S_n(x)$ span ξ_x for every x in X. Let ζ be the trivial bundle with with the total space of $\zeta = X \times K^n$. Then $\Gamma(\zeta)$ is a free $C(X)$- module on n generators $e_1,...,e_n$. Map $\Gamma(\eta) \to \Gamma(\xi)$ by $e_i \to S_i$. By Theorem 1[SRG], this is induced by a map $f:\zeta \to \xi$. Since $f(e_i) = S_i, S_i(x) \in \text{ im } f$. Therefore f is onto. □

Corollary 6.3. *If X is compact Hausdorff, any K-vector bundle over X is a direct summand of a trivial K-vector bundle ζ.*

Proof. Let $f:\zeta \to \xi$ as in Lemma 6.3. Let $\eta = \ker f$. By Proposition 1[SRG], η is a subbundle of ζ. By Proposition 2[SRG], $\zeta = \eta \oplus \xi'$. Clearly $\xi' \approx \xi$. □

Corollary 6.4. *If X is compact Hausdorff and ξ is any K-vector bundle over X, then $\Gamma(\xi)$ is a finitely generated projective $C(X)$-module.*

Proof. By Corollary 6.1.3, $\Gamma(\xi)$ is a direct summand of $\Gamma(\zeta)$ which is a finitely generated free $C(X)$-module. □

7. Main Theorem

Here we show that a Lie ring P over $C(X)$ is isomorphic to the Lie ring of the set of all sections of ξ denoted $\Gamma(\xi)$, where ξ is a Lie bundle, if and only if P is a finitely generated projective module over $C(X)$, where X is a compact Hausdroff space.

Lemma 7.1. *Let ξ be a Lie bundle over X, then the set of all sections of $\xi, \Gamma(\xi)$ forms a Lie ring over $C(X)$ of all continuous real valued functions on X.*

Proof. Let $S_1, S_2 \in \Gamma(\xi)$. Then the map $X \xrightarrow{(S_1,S_2)} \xi \oplus \xi \xrightarrow{\Theta} \xi$ defined by $x \rightarrow (S_1(x), S_2(x)) \rightarrow \theta_x(S_1(x), S_2(x))$ is continuous.

Therefore, $[\ ] : \Gamma(\xi) \times \Gamma(\xi) \rightarrow \Gamma(\xi)$

$$(S_1, S_2) \rightarrow [S_1, S_2]$$

where $[S_1, S_2](x) = \theta(S_1(x), S_2(x))$ is continuous.

Here $\theta_x(S_1(x), S_2(x)) = [S_1(x), S_2(x)]_x \in \xi_x$.

Then $[S_1, S_2] \in \Gamma(\xi)$. Thus $\Gamma(\xi)$ forms a Lie ring over $C(X)$ since each ξ_x is a Lie algebra over R. □

To prove the next lemma we need the following results.

Remark 7.1. Let $I_x = \{\alpha | \alpha(x) = 0, \alpha \in C(X)\}$ be the maximal ideal of $C(X)$ attached to x in X, then $C(X)/I_x$ is a field and isomorphic to the field of real numbers R. For $\phi : C(X) \rightarrow R, \phi(f) = f(x)$ is an onto ring homomorphism then $\text{Ker}\phi = \{f | \phi(f) = 0\} = \{f | f(x) = 0\} = I_x$. Then by the first fundamental theorem of homomorphism of ring, $C(X)/I_x \cong R$.

Lemma 7.2. *Let $I_x = \{\alpha | \alpha(x) = 0, \alpha \in C(X)\}$ be the maximal ideal of $C(X)$ attached to x in X, then the map $\theta : \Gamma(\xi)/I_x\Gamma(\xi) \rightarrow \xi_x$ be given by*

$$\theta(\overline{S}) = \theta(S + I_x\Gamma\ (\xi)) = S(x)$$

is linear, and bijective and θ preserves the Lie product.

Proof. Define a map,

$$\theta : \frac{\Gamma(\xi)}{I_x\Gamma(\xi)} \rightarrow \xi_x$$

by $\theta(S + I_x\Gamma(\xi) = S(x)$.
map θ is one-one

$$\text{Ker}\theta = \{S + I_x\Gamma(\xi) \mid \theta(S + I_x\Gamma(\xi)) = S(x) = 0\}$$

then by (Swan), if X is normal, $S \in \Gamma(\xi)$ such that $S(x) = 0$ then there are elements $S_1, S_2, \cdots, S_k$ in $\Gamma(\xi)$ and $a_1, a_2, \cdots, a_k$ in $C(X)$ such that $a_i(x) = 0$ for all i, and $S = \sum a_i S_i$.

Therefore $S \in \sum I_x \Gamma(\xi)$

Thus $\text{Ker}\theta = I_x\Gamma(\xi)$

$\therefore \theta$ is one-one.

map θ is onto
Let $e \in \xi_x$ be arbitrary.
Then

$$e = \sum \alpha_i e_i$$

define

$$S_i(x) = e_i$$

$$a_i(x) = \alpha_i \qquad \text{for all } x \text{ in } X.$$

$$e = \sum a_i(x) S_i(x).$$

Consider,

$$S = \sum a_i S_i \Rightarrow S + I_x\Gamma(\xi) \in \frac{\Gamma(\xi)}{I_x\Gamma(\xi)}$$

such that

$$\theta(S + I_x\Gamma(\xi))$$

$$= S(x)$$

$$= \sum a_i(x) S_i(x)$$

$$= e.$$

Thus θ is onto. **map θ preserves the Lie product:**

$$\begin{aligned}\theta[\overline{S}_1, \overline{S}_2] = \theta([S_1, S_2] + I_x\Gamma(\xi)) &= [S_1, S_2](x) \\ &= \theta_x(S_1(x), S_2(x)) \text{ by definition} \\ &= [S_1(x), S_2(x)] \in \xi_x \\ &= [\theta\overline{S}_1, \theta\overline{S}_2].\end{aligned}$$

Thus the Lemma. □

Theorem 7.1. *Let ξ, η be two Lie bundles over a normal base space X and $F : \Gamma(\xi) \to \Gamma(\eta)$ be a Lie ring isomorphism over $C(X)$. Then there is a unique Lie bundle isomorphism $f : \xi \to \eta$ such that $F = \Gamma(f)$, where $\Gamma(f)(S) = foS$.*

Proof. The Lie ring isomorphism
$F : \Gamma(\xi) \to \Gamma(\eta)$ induces an isomorphism

$$F_x : \frac{\Gamma(\xi)}{I_x\Gamma(\xi)} \to \frac{\Gamma(\eta)}{I_x\Gamma(\eta)}$$

$$F_x(S + I_x\Gamma(\xi)) = (FS)(x) = F_xS(x) = FS + I_x\Gamma(\eta).$$

Since by the above lemma there exist an isomorphism $\phi : \frac{\Gamma(\xi)}{I_x\Gamma(\xi)} \to \xi_x, \theta : \frac{\Gamma(\eta)}{I_x\Gamma(\eta)} \to \eta_x$. Thus, $\xi_x \xrightarrow{\phi^{-1}} \frac{\Gamma(\xi)}{I_x\Gamma(\xi)} \xrightarrow{F_x} \frac{\Gamma(\eta)}{I_x\Gamma(\eta)} \xrightarrow{\theta} \eta_x$ is a Lie algebra isomorphism.
Say $f_x = \theta F_x \phi^{-1} : \xi_x \to \eta_x$.
The totally of these yield a morphism $f : \xi \to \eta$ defined by $e \to f_x(e), e \in \xi_x$. Then f is well-defined.
f is continuous
Let $S_1, ..., S_n \in \Gamma(\xi)$ be a local base at x. If $e \in \xi$ and $p(e)$ is near x, we have $e = \sum a_i(e)S_i(p(e))$, where the $a_i(e)$ are continuous real valued functions.
Now, $f(e) = \sum a_i(e)fS_i(p(e))$.

Since $fS_i = F(S_i), fS_i$ is a continuous section of η. Now all terms in the sum are continuous in e, thus, f is continuous.

f is Unique
If $f, g : \xi \to \eta$ and $\Gamma(f) = \Gamma(g) : \Gamma(\xi) \to \Gamma(\eta)$, then $f = g$.
Given $e \in \xi$, with $p(e) = x$, there is a section S over U of x with $S(x) = e$. By Lemma, there is a $S' \in \Gamma(\xi)$ with $S'(x) = e$, since X is normal. Now

$$\begin{aligned} f(e) &= fS'(x), \text{ since } S'(x) = e \\ &= (\Gamma(f)S')(x) = (\Gamma(g)S')(x), \text{ given} \\ &= g(e) \text{ for all e} \in \xi, \text{ thus } f \text{ is unique.} \end{aligned}$$

Obviously f preserves the Lie product restricted to each fibre ξ_x.

Thus $\xi \cong \eta \Leftrightarrow \Gamma(\xi) \cong \Gamma(\eta)$ as Lie rings and finitely generated projective modules over $C(X)$ if ξ and η are Lie bundles over the compact Hausdorff base space X. Given $\theta : \xi \to \eta$ Lie bundle morphism, then the map $\Gamma(\theta) : \Gamma(\xi) \to \Gamma(\eta)$ defined by $\Gamma(\theta)S = \theta oS$ is bijective, linear and preserves the Lie product. □

Theorem 7.2. *Let X be a compact Hausdroff space then a Lie ring P over $C(X)$ is isomorphic to a Lie ring of the form $\Gamma(\xi)$, where ξ is a Lie bundle if and only if P is finitely generated projective $C(X)-$module over $C(X)$.*

Proof. Suppose P is of the form $\Gamma(\xi)$, where ξ is a Lie bundle over X, then P is finitely generated projective module from Swan's theorem[SRG].

Suppose now that P is a finitely generated projective module, then there exists a vector bundle ξ with $\Gamma(\xi)$ isomorphic to P as modules over $C(X)$, again from Swan's theorem[SRG].

Now we will give a Lie bundle structure so that the Lie ring $\Gamma(\xi)$ (Lie ring structure induced from ξ) is isomorphic to P as Lie rings.

Suppose $[\ ]: P \times P \to P$ is the given Lie ring structure. Therefore

$$[\ ]: \frac{P}{I_xP} \times \frac{P}{I_xP} \to \frac{P}{I_xP}$$

is a Lie algebra structure for each x in X and we know that $\theta_x : \frac{P}{I_xP} \to \xi_x$ is given by $\theta_x(S + I_xP) = S(x)$ is a vector space isomorphic.

$$\begin{array}{ccc} \frac{P}{I_xP} \times \frac{P}{I_xP} & \xrightarrow{[\ ]_x} & \frac{P}{I_xP} \\ \Big\downarrow \theta_x \times \theta_x & & \Big\downarrow \theta_x \\ \xi_x \times \xi_x & \xrightarrow{\ominus} & \xi_x \end{array}$$

$(\theta_x \times \theta_x)\ominus_x = \theta_x[\ \ ]_x$, then obviously on each fibre ξ_x, $\ominus_x : \xi_x \times \xi_x \to \xi_x$ induces a Lie algebra structure over ξ_x. The totality of these yield a map $\ominus : \xi \oplus \xi \to \xi$.

We will now prove that the map $\ominus$ is continuous.

If $S_1, S_2 \in \Gamma(\xi), \ominus(S_1, S_2) = [S_1, S_2]$, where $\ominus(S_1, S_2)(x) = \ominus_x(S_1(x), S_2(x)) = \ominus_x[\overline{S}_1, \overline{S}_2] = [S_1, S_2]_x = [S_1, S_2](x)$, for all x in X.

$\ominus$ is continuous.

Let $\{S_i\}$ be a local base at $x \in X$ and $e = (e_1, e_2) \in \xi \oplus \xi$ such that $q(e) = p_1(e) = p_2(e)$ is near x. Then $(e_1, e_2) = (\sum_i a_i(e_1) S_i(p(e_1)),\ \sum_j b_j(e_2) S_j(p(e_2)))$, where $a_i(e_1), b_j(e_2) \in C(X)$.

Therefore, $\ominus(e_1, e_2) = \ominus(\sum_i a_i(e_1) S_i(p(e_1))),\ \sum_j b_j(e_2) S_j(p(e_2)) = \sum_i a_i(e_1)$ $\sum_j b_j(e_2) \ominus_x (S_i(p(e_1)), S_j(p(e_2)))$.

Hence $\ominus$. Thus is continuous, since $\ominus(S_i, S_j) = [S_i, S_j]$ is a continuous section of ξ. ξ is a Lie bundle. Then by the very construction, the Lie ring $\Gamma(\xi)$ is isomorphic to the given Lie ring P. □

8. Lie algebra bundles of finite type-[RGBSK]

Definition 8.1. A vector bundle over an arbitrary space X is of finite type if there is a finite partition S of 1 on X (that is a finite set S of nonnegative continuous functions on X whose sum is 1) such that the restriction of bundle to the set $\{x \in X | f(x) \neq 0\}$ is trivial for each f in S.

Example 8.1. Any vector bundle over a compact Hausdorff space X is of finite type.
Now we proceed to give a bijection between Lie algebra bundles of finite type over a general topological space X and finitely generated projective Lie rings over the ring of continuous functions on X.

Theorem 8.1. *For every Lie ring P which is also finitely generated projective $C(X)$−module, there is a Lie algebra bundle of finite type over X with $\Gamma(\xi) \cong P$.*

Proof. Since P is a finitely generated projective module over $C(X)$, P is isomorphic to the column space of a square hermitian idempotent matrix $e = e^2 = e^*$ over $C(X)$ and every finitely generated projective module P over $C(X)$ gives an vector bundle ξ over X whose fibre at x is just $e(x)F^N$[41].

Now we give a Lie algebra bundle structure on ξ. Let $I_x = \{\alpha \in C(X) | \alpha(x) = 0\}$ be the maximal ideal of $C(X)$ attached to $x \in X$. Then P/I_xP is isomorphic to $e(x)(F^N)$ given by the mapping $G_x : P/I_xP \to e(x)F^N$ defined by $G_x[e(f_1, f_2, \cdots, f_N) + I_xP] = e(x)(f_1(x), f_2(x), \cdots, f_N(x))$ which is an isomorphism of vector spaces.

Given two elements $e(x)(s)$, $e(x)(t) \in e(x)F^N$ we can define the multiplication

$$\theta_x(e(x)(s), e(x)(t)) = G_x(G_x^{-1}(e(x)(s)) * G_x^{-1}(e(x)(t))),$$

where "$*$" is the Lie multiplication on P. Hence $e(x)(F^N)$ has the structure of a Lie algebra as it inherits the Lie multiplication which we denote by θ_x from P and is having a vector space structure over F. Now let us define $\theta : \xi \to \xi$ as $\theta(u, v) = \theta_x(u, v)$, if u, v belong to $e(x)(F^N)$.

The continuity of θ follows obviously. □

Now we proceed to prove the converse of the above Theorem.

Theorem 8.2. *If ξ is Lie algebra bundle of finite type over the base space X, then $\Gamma(\xi)$ is a Lie ring and finitely generated projective $C(X)-$module.*

Proof. Suppose that P is of the form $\Gamma(\xi)$, where ξ is a Lie algebra bundle of finite type over X, then P is a Lie ring by above remark and also finitely generated projective $C(X)-$module[41]. □

Theorem 8.3. *The functor Γ from the category of Lie algebra bundles over X and the category of finitely generated projective $C(X)-$modules which are also Lie rings is an equivalence.*

Proof. From above Theorems, Γ induces a bijective map of the isomorphism classes of the objects in these categories. Further, if $\varphi : \xi_1 \to \xi_2$ is a Lie algebra bundle isomorphism, then we define $\Gamma(\varphi) : \Gamma(\xi_1) \to \Gamma(\xi_2)$ by $\Gamma(\varphi)(S) = \varphi \circ S$, which is a Lie ring isomorphism. Finally, given $\omega : P_1 \to P_2$ Lie ring isomorphism, there exist square hermitian idempotent matrices $e_1 = e_1^2 = e_1^*$ and $e_2 = e_2^2 = e_2^*$ over $C(X)$ such that P_1 and P_2 are isomorphic to the column space of e_1 and e_2 respectively. But corresponding Lie algebra bundle of finite type are $\xi_1 = \bigcup_{x \in X} e_1(x)F^N$ and $\xi_2 = \bigcup_{x \in X} e_2(x)F^N$. Define $\varphi : \xi_1 \to \xi_2$ by $\varphi|_{e_1(x)F^N} = e_2(x)F^N$. □

Acknowledgements

Author is thankful to the SERB/DST, New Delhi, India for the financial assistance SR/S4/MS:856 /13. I am thankful to Ranjitha Kumar, Ajay Kumar K. and B. Madhu for their help.

Bibliography

1. Albert, A. A., Power associate rings, transactions of the American Mathematical Society 64, (1948) 552–593.
2. Armand Borel, Linear algebraic groups, New York, Benjamin W.A, 1969.
3. Atiyah, M. F., K-Theory, W.A. Benjamin, Inc., New York, Amsterdam, 1967.
4. Bourbaki, N., Lie groups and Lie algebras-I, Addison-Wesley Publ., Co., 1975.
5. Chevalley, C., Theory of Lie Groups,Princeton, Princeton University Press, 1946.
6. Chindambara, C. and B. S. Kiranagi, On Cohomology of Associative Algebra Bundles, *J.* Ramanujan, *Math. Soc.*, **9**:1–12, 1994.
7. Coppersmith, D., A family of Lie algebras not extendible to a family of Lie group, *Proceedings of American Mathematical Society*, Vol. 66, No. 2 (1977), pp. 365–366.
8. Douady, A. and Lazard, M., Espace *fibrés algèbres* de Lie et en groupes, *Invent. Math.*, **1**:133–151, 1966.
9. Goodearl, K. R., Cancellation of Low rank vector bundles, *Pacific J. Math.*, **114**, 1984.
10. Greub, W., Halperin, S. and Vanstone, R., Connections, Curvature and Cohomology, Vol. 2, Academic Press, New York, 1973.
11. Gerd Rudolph and Matthias Schmidt, Differential Geometry and Mathematical Physics Part 1. Manifolds Lie Groups and Hamiltonian systems, 2013 Springer.
12. Husemoller, D., Fibre Bundles, McGraw-Hill, New-York 1966.
13. Humpreys, J. E., Introduction to Lie Algebras and Representation Theory, Springer-Verlag, 1972.
14. Jacobson, N., Lie Algebras, Wiley Interscience, New York, 1962.
15. Kiranagi, B. S., Lie Algebra Bundles, *Bul. Sci. Math.*, 2^e Series, **102**:57-62, 1978.
16. Kiranagi, B. S., Splitting Theorem for Lie Algebra bundles, *J. Madras Univ. Sec. B.*, **45**(1):108–115, 1982.
17. Kiranagi, B. S., Semisimple Lie Algebra Bundles, *Bull. Math. Sci. Math.* R.S. Roumanie, **27**(75):253–257, 1983.
18. Kiranagi, B. S., Lie Algebra Bundles and Lie Rings, *Proc. Nat. Acad. Sci. India*, **54**(A):38–44, 1984.
19. Kiranagi, B. S., Rings, Modules and Algebra Bundles, Anamaya Publishers, New Delhi, India, 2005.
20. Kiranagi, B. S. and Prema, G., Cohomology of Lie Algebra Bundles and its applications, *Ind. J. Pure and Appl. Math.*, **16**(7):731–735, 1985.
21. Kiranagi, B. S. and Prema, G., On Complete Reducibility of Bi-module bundles, *Bull. Math. Sci. Math.* R. S. Roumanie. Nouv. Ser., **33**(81):249–255, 1989.
22. Kiranagi, B.S. and Prema, G., A Decomposition Theorem of Lie Algebra Bundles, *Comm. Algebra*, **18**(6):1869–1877, 1990.
23. Kiranagi, B. S. and Rajendra, R., Revisiting Hochschild Cohomology for Algebra bundles, *J. Algebra Appl.*, Vol. 7, No. 6, (2008), pp. 685–715.
24. Kiranagi, B. S., Prema, G. and Chidambara, C., Rigidity Theorem for Lie Algebra Bundles, *Comm. Algebra*, **20**(6):1549–1556, 1992.
25. Kiranagi, B. S. and Prema, G., A Wedderburn principal Decomposition theorem for Jordon Bundles, *Algebra Appl.*, S. Narosa Publishing House, New Delhi, 179–183, 2001.
26. Kiranagi, B. S., Prema, G. and Kumar, R., On the radical bundle of a Lie algebra bundle, *Proc. Jangjeon Math. Soc.*, **15** (2012), No. 4, pp. 447–453.
27. Kiranagi, B. S., Ranjitha Kumar and Prema, G., On completely semisimple Lie algebra bundles, *J. Algebra Appl.*, Vol. 14, No. 2, (2015).
28. Kurosh, A. G., The current state of the theory of rings and algebras, Uspekhi Matematichheskikh Nauk VI, **2** (1951), 3–15.
29. Kyriazis, A., On topological algebra bundles, *J. Austr. Math. Soc.*, **43**, 398–419, (1987).

30. Leland Graeme McInnes, Pro-finite Lie rings and p-adic Lie algebras is accepted in partial fulfillment of the requirements for the degree of Doctor of Philosophy, 2007.
31. Musili, C., Introduction to Rings and Modules Narosa publishing House Pvt. Ltd., 2006.
32. Northcott, D. G., An Introduction to Homological algebra, Cambrige University press, 1996.
33. Prema, G. and Kiranagi, B. S., On Complete Reducibility of Module bundles, *Bull. Austral. Math. Soc.*, **28**:401–409, 1983.
34. Prema, G. and Kiranagi, B. S., Lie Algebra Bundles Defined by Jordon Algebra Bundles, *Bull. Math. Sci. Math.* R. S. Roumanie, **31**(79):255–264, 1987.
35. Ranjitha K., Prema, G. and Kiranagi, B. S., Lie Algebra Bundles of Finite Type, Acta Universitatis Apulensis No. 39, 2014, pp. 151–160
36. Rajendra, R., Kiranagi, B. S. and Ranjitha, K., On pullback Lie Algebra Bundles, *Adv. Studies Contemp. Math.*, **22**(4), (2012), pp. 521–524.
37. Serre, J. P., Faisceaux Algebriques Coherents, *Ann. Math.*, **6**(2):197–278, 1995.
38. Shirshov, A. I., Some problems in the theory of ring that are nearly associative, Uspekhi Matematichheskikh Nauk XIII, **6** (1958), 3–20.
39. Steenrod, N. E., Topology of Fibre Bundles, Princeton Mathematical Series, Princeton, 1951.
40. Swan, R. G., Vector bundles and Projective modules, *Trans. Am. Math. Soc.*, **105**:264–277, 1962.
41. Vaserstein, L. N., Vector Bundles and Projective Modules, *Trans. Amer. Math. Soc.*, **294**(2):749–755, 1986.

Study of theory of nearrings and Goldie structures, topological aspects, wreath sums etc.

Helen K. Saikia, K. C. Chowdhury

Department of Mathematics, Gauhati University, Guwahati, India-781014
E-mail: hsaikia@yahoo.com, khaninchowdhury1@gmail.com

1. A Brief Introduction

The research work along the track mentioned above started or began with the very strong hold of reverend B. K. Tamuli, the senior talent of the department of Mathematics of Gauhati University. Under his supervision, M. N. Barua did his research work and opened the door of research in Near-ring theory. At the very outset M. N. Barua brought to forefront the two main aspects along the line, viz. the wreath sum of near-rings and the near-ring of quotients. Taking into account of these developments, the followers of the later period could develop various aspects of near-ring theory along the line. M. N. Barua's approach paved a new path and along the same path in later period Graigner and J. R. Clay could establish what these famous near-ringers justified the utility of unusual module structure available in a near-ring group and in real sense, this useful tool helped M. N. Barua that he wanted and lead towards the expected development in the construction of near-ring of right quotients.

After M. Baruah, we took the responsibility of carrying out our research work along the same line with a varied taste in the field of B. K. Tamuli. Our main interest in Goldie distinctiveness in various algebraic structures took a new turn in the research field. We took our interest basically along the line of algebraic structures with finiteness conditions on annihilating substructures and their fuzzyfications. Keeping this in note we could justifiably give a new turn to the module theory (as in the case of Artinian Noetherian module) to a completely new algebraic structure with the notion of so called Goldie module (or in some context Goldie M-Groups). The decomposition of the zero of a Goldie module into closed submodules and the Artin Rees Analogue in a Goldie Module are two main feature at the first hit to such a structure. Another new connotation in the field of Near-ring theory is that of another notion, renowned as the notion of a Goldie near-ring. The very asymmetric notion of a near-ring naturally classified the study of Goldie structures into two classes. Our first interest in the right Goldie structure in a right near-ring to some restricted analogous of well-known Goldie theorems lead to right near-ring of right quotients. Another interestingly elegant study carried out at this stage was a radical structure which leads us to show the existence of a near-ring of quotients in substructures of it.

In 1994, we along with A. Masum discussed left Goldie near-rings and their parts having ACC or DCC on their subalgebraic structures. We established some useful results on left Goldie near-rings with their parts satisfying chain conditions on their subalgebraic structures. One of the most important results on singular left K-subsets modulo maximal annihilators of a left Goldie near ring led to the concept of cyclic structures of an ideal I of a left Goldie near ring with I satisfying DCC on its right K-subsets. We also obtained sufficient conditions for coincidence of complete near-ring of left quotient with classical near-ring of left quotients of right near-ring. This result is then led to obtain a sufficient condition on a minimal ideal with DCC on its right K-subsets to be a left K-group over a near-ring with DCC on its K-subgroup which is an extension of an epimorphic image of K. Left quotient near-ring structure was found in case of a left Goldie near-ring with left essential DCC which is also inherited by the near-ring K/P where P is a maximal left annihilator in some special cases. Near-ring K modulo a left annihilator of a countable left ideal I with DCC on its K-subgroups inherits the same character as with I.

We, along with A. Masum, carried out as our research work is on what may be called Finite spanning dimensional N-group. This is actually some sort bifurcation of what P. Fleury and B. Satyanarayana did in their respective attempts. The primary s-p decomposition of zero of such an N-group with Goldie characteristic was justifiably established with very ground level examples like the integers etc. In some special cases, we established a couple of interesting results for a class of commutative rings with unity with FGD (Finite Goldie Dimension).

In 1997, we have established some results on substructures with certain minimal conditions of a near-ring with chain conditions. We as well as to deal with some interesting properties of such a near-ring with some radical character. The inheritance of DCC of a near-ring modulo the left annihilator of a minimal countable left ideal with DCC on its left N-subgroups is established. An invariant subnear-ring, minimal as a left N-subgroup and is with DCC on its right N-subsets is, in some sense, a left near-ring group over a near-ring which is an extension of an epimorphic image of N. Also some important results on s-A(N) the sum of all left ideals of N with DCC on left N-subgroups. In some cases, s-A(N) is not contained in any minimal strongly prime ideal of N. In case, s-A(N) is a minimal countable invariant subnear-ring with DCC on left N-subgroups of the minimal strongly prime ideals, s-A(N) is direct summand of N.

Along with R. Kataki, we again considered the branch court of Goldie characters. In the perspective of the set of elements annihilated by the essential near-ring subgroups of the near-ring seems to behave in a natural way so far the idea of rank is concerned. The role of so called Ore condition to get some well behaved natural structure of such a rank is noticeable. Another developed algebraic observation is that of near-rings or near-ring groups with partial ordering lead the researcher to multifaceted possible developments for various types of structure theory. Near-rings in which its inherited subgroups form a chain, reveal some elegant dealings. With certain radical character, the near-ring establishes various fascinating decompositions leading to sequential types. The serial structure of a near-ring may also have something to look at from the annihilating point of view. It reveals some facts relating to its non-singularity character. All such observation intuitively motivates one to list the ideals of such subgroups generated by the idempotents. In turn this insists one to study the so-called modulo essential descending chain condition on a near-ring. And this comes out as a necessity from the restricted principal sequence condition of the intersection of left integral powers of its radical.

Later, G. Das, another scholar, joined in our research group who did at least some work on so-called semiring structure and develop the associated prime classification of such a structure, of course in the same line that is not giving up the Goldie characteristics. This research work enables to give exposure to multifaceted openings of what near-ring and near-ring groups structures were showing together with some topological vigour. More than one near-ring structure is possible from the wreath sum of two near-rings, as is verified through a number of justifiable examples, lead us to the notion of wreath sum of near-ring groups. We could justify that in such a wreath sum structure too more than one near-ring group structures are available from the same definite near-ring groups. And possibly this is what we may claim an up-level credit. This work deals with a generalized semi-prime character, the strongly semi-prime near-rings and so-called pseudo-strongly, semi-prime near-ring group play an important role. In this period (during 2004) new notion was developed to module theory, or so to say near-ring group theory. And this is the notion of nilpotency in a module of a near-ring group, though so called product in such an algebraic structure seems to be unavailable. And this very notion helped the whole group of the later period gave an impetus for their relevant work. The usual notions of nilpotency and strongly semi-prime characters have been replaced by so-called its pseudo character in case of the N-groups. The above mentioned so-called pseudo character helps us in establishing important results on non-singular N-group which plays an significant role in establishing the descending chain conditions on annihilators of subsets of it. Pseudo-strongly semi-prime duo ACC N-group leads one with some restriction to get the trivial annihilator of an element of the N-group. All these explained and obtained are in reference to the N-group where its topological structure were dealt with investigated some properties of topological structures on some generalization of what has been achieved on near-rings with ACC on annihilators having no infinite independent family of right (or left) ideals as observed by A. Oswald and our group. The so-called pseudo character on nilpotency and strongly semi-primeness, discussed, lead the group satisfactorily towards their goal giving well-designed results on N-groups rather than the near-ring structure. The generalization of the idea of boundedness of Beidleman and Cox on topological near-rings, E-boundedness etc. together with the notion of topologically nilpotent sets seems to play important role on an N-group E with finite number of elements (e's) belonging to it with zero annihilators $[Ann(e) = 0]$, which occur as a necessity of N-group E in above context. It is interesting to note the relevancy and elegancy of the result obtained, as the same were determined with obliging explanation on such topological N-groups that their discrete character is in association with the E-boundedness with zero radical or open character of the radical with E-boundedness. Some interesting results regarding the discrete character of the N-group were observed in case of locally compact groups if the near-ring is without unity. At this juncture one may recall that in the definition of topological ring, Kaplansky insisted that addition and multiplication be continuous on the product space; however, as defined by Beidleman *et al.* the authors found that co-ordinate wise continuity is all that is necessary in many cases. Some careful observations have elegantly revealed what our group have attempted and carried out with some sort of rare and alarming beauty, hitherto the so-called continuity of such structures are concerned. Keeping aside the concrete so-called topological aspects of what has been explained above, the group dare to review this aspect of above type of algebraic structure from more or less algebraic point of view in a broaden court-yard with a view to play the same game in a more sophisticated country of algebra. For the moment they leave available topological nomenclature, however instead, embrace some abstract familiar algebraic way of approach. Undoubtedly everything were justified with sufficient examples if and when necessary. The supposed pseudo continuity in such a so-called algebraic space is carried out by internal and external compositions of a group or a near-ring group respectively to give some general view of some topological properties including boundedness, connectedness and Hausdorff character etc. It is observed that near-ring groups with ACC on annihilators of subsets of the group in the attached near-ring found to be well behaved so far the so-called space biased algebraic structure is concerned with some so-called pseudo quality on nilpotency as well as strongly semi-prime character. This has involved for the proper development of what we have attempted for.

The notion of boundedness of Beidleman and Cox are playing some shaman character with the algebraic space as the authors are claiming for! Together with these, a near-ring group with so-called Goldie character has been playing an interesting and elegant worthwhile game where the group having finite number of elements (e's) belonging to the group with zero annihilator, which occurs as a necessity of such a near-ring group. The justification has properly been accommodated with a sufficient number of examples so as to congregate our endeavor.

The discrete character of such an algebraic biased space is playing a generously subjective role with a very deep insight, which seems to include so many aspects even in some cases, the orientable and non orientable space relating to Klein's example. (And this one may call as one milestone so far the research activity of this group is concerned...!!!) Extending the idea of boundedness already available with the help of the notion of (so-called) S_E-nilpotent set, we get some elegant results in case of such a space biased near-ring groups with ACC on annihilators. It is interesting to note the relevancy and elegancy of the results obtained, as the same may be determined with accommodating justification on such a space biased near-ring groups that their so-called ps(pseudo)-discrete character is in association with the so-called algebraic boundedness with zero-radical or so-called ps-openness of the radical with same character.

Later on B. De joined in our research group who did some important results on near-ring groups with irreducible substructures.

In the later period, J. Choudhury and K. Mishra have joined our group to continue our research work on near-rings and associated substructures. In this period, we introduced the notion of quasi I-bounded near-ring and investigated certain characteristics of such near-ring structures. Certain interesting properties of a quasi I-bounded near-ring with chain conditions on its substructures are established. We also introduced the notion of weakly quasi injective N-groups and investigated many properties.

On the other hand, we along with N. Hazarika extended the notion of honest submodules to near-ring groups. We have introduced honest N-subgroups, e-honest N-subgroups of an N-group E and investigate various characteristics of these subgroups. We have also extended the concepts of quasi-injective modules and their endomorphism rings to near-ring groups. The near-ring character of the set of endomorphism of quasi-injective N-groups under certain conditions were derived. This led us to a near-ring group structure which motivates us to study various characteristics of the structure.

Another researcher P. Das has joined in our group and her idea gives a new turn specifically establishing the important and interesting question whether the wreath sum of two Goldie near-ring groups again a Goldie? Through praiseworthy examples we could justifiably established the negation to this outstanding question. Here we discuss the structure of wreath sum of near-rings and as an extension that of near-ring groups. Some examples of wreath sum of near-rings lead us to the notion of what we have proposed. The same underlying set of wreath sum (of two near-rings) may have different explanation so far the near-ring group structure is concerned. This happens due to the variance in the definition of external product appeared in the picture as observed from the examples taken in to consideration. We could also here show the inheritance of so-called Noetherian character of wreath sum of two near-rings, however, on the contrary non-inheritance of the Goldie character even though the component near-rings are Goldie. It has been noted that such near-ring group structures arise from the same underlying set of wreath sum. One may choose a specific subset from the underlying set showing non-isomorphic algebraic character in the chosen near-ring group structure. Amazingly it is exhibited that the wreath sum of two Goldie near-rings bears Goldie like character though not Goldie exactly. Keeping aside the notion of continuity in such a topological structure, one dares to investigate the finitely ascending and descending parts Goldie structures using this notion as a topological tool. In this case the so-called minimal near-rings endowed with some sort of ideal like topology gives rise to some results on sum of its left Artinian ideals. In other words, we discuss here some topological relevance of what can be thought as an Artinian (Noetherian) part of a Goldie Near-ring. Moreover we here gives a semi-local description of what is known as the well known Uryshon's Lemma in classical topology. Giving up the usual notion of open and closeness in a topological space where a topology is, we deal with a pair with an arbitrary subset of power sets of and introduce the notion of a pseudo-continuous map, considering only the continuous character of a topological map. Dealing with such a pseudo(ps)-continuous function of the space we obtained some results up to so-called Uryshon's Lemma in classical topology which reveals that the theorem demands in contrast to global character of the space in the context to carry out analogous notion of the same that we mention here as a semi-global one. The so-called global nature is required subject to some types of continuous function in particular. Hence the results thus obtained may thus be regarded that some sort of global character is revealed through continuous embedding of such a space. And possibly this is the real vigour of what is examined in our so-called semi-global Uryshon's Lemma.

One among of our research team to be mentioned is the work of D. Nath. Her work deals with mainly on the near-ring groups of quotients. And in her discussion along this line she specifically dealt with the Goldie near-ring group as coined by us. We successfully developed the theory on Goldie near-ring groups specifically on the basis of what her fore-runner M. N. Barua did and this of course tallies with at least some aspects that Grainger carried on. To be more specific, the notion of unusual module structure. For proper development of near-ring group quotients the module structure in case of near-ring group appear s essential and this is what actually very closely noticeable. Toward the end, we gave some insight to the topological vigour that seems to be alive in any algebraic sentient. And as a result of such an insight he praise worthily justified his claim that is termed as the topological relevance in some algebra oriented paper.

On the other hand, we along with L. K. Barthakur introduced the fuzzy aspects of near-ring subgroups and ideals of near-ring groups. Here, we have investigated some characteristics of fuzzy subgroups and ideals of strongly regular near-rings. In 2003 we have introduced some fuzzy substructures of near-rings. Some interesting results are established using the concepts of fuzzy prime N-subgroup, fuzzy irreducible N-subgroup and fuzzy primary N-subgroup of N. In 2005, we have introduced and characterized fuzzy completely prime ideals of a near-ring in terms of fuzzy points, and also investigated many related properties. We have also introduced the fuzzy concept of annihilators of fuzzy subsets of near-ring groups. This concept leads to a structure that is termed as fuzzy prime N-subgroup. It was proved that the factor N-group induced by a fuzzy ideal is a fuzzy prime N-group if and only if the fuzzy ideal is a prime ideal of N. They finally concluded that using the fuzzy concepts of annihilator and prime N-subgroup, one may expect to obtain the fuzzy aspects of Goldie theorems in near-rings and primary decomposition of fuzzy ideals of near-rings with ACC on fuzzy annihilators.

2. Some Important Results

All the results are available in our publications.

Definition 2.1. An N-group E is said to be an N-group E with finite spanning dimension-1 (or in short fsd-1) if for every strictly descending sequence of ideals $M_0 \supsetneq M_1 \supsetneq M_2 \supsetneq \ldots$, there is an i such that $M_j \subset_s E$ for each $j \geq i$.

Definition 2.2. The collection $\mathcal{P} = \{P | P = Ann(M),\ for\ some\ prime\ N - subgroup\ M\ of\ E\}$ is defined as the family of associated strictly primes of E. An N-group E is primary if $\mathcal{P}$ is singleton. Clearly, a prime N-group is primary. But the converse is not true. For example, $E = \mathbb{Z}_9$ is an N-group (when $N = \mathbb{Z}$) which is not prime, because $Ann(E) = (9)$ and for its N-subgroup $G = \{0,3,6\}$, $Ann(E) = (3)$ and $Ann(E) \neq Ann(G)$. But E has only one prime N-subgroup G.

Lemma 2.1. *(Theorem 3.1 in Ref. 1) Let E be an fsd-1 N-group with ACC. If* $\mathcal{P} = X \cup Y$, *where* $X \cap Y = \phi$, *then there exists a supplement* $E' \underset{\sigma\pi}{\subset} E$ *such that* $\mathcal{P}(E') = X$ *and* $\mathcal{P}(\frac{E}{E'}) = Y$.

Lemma 2.2. *(Lemma 2.2.(i) in Ref. 1). If E_1 and E_2 are N-groups, then $\mathcal{P}(E_1 \oplus E_2) = \mathcal{P}(E_1) \cup \mathcal{P}(E_2)$.*

Theorem 2.1. *Let E be an fsd-1 N-group with ACC as above, then there exists a s-p decomposition of 0 in E.*

Proof. Let $\mathcal{P}(E) = \{P_1, P_2, ..., P_t\} = \{P_1, P_2, ..., \hat{P}_i, ..., P_t\} \cup \{P_i\}$. Then by Lemma 2.3, we have supplements $E_1, E_2, ..., E_t$ such that $\mathcal{P}(E_i) = \{P_1, P_2, ..., \hat{P}_i, ...P_t\}$ and $\mathcal{P}(\frac{E}{E_i}) = \{P_i\}, i = 1, 2, ..., t$.

Since each $\frac{E}{E_i}$ is primary and $\mathcal{P}(\frac{E}{E_i}) \neq \mathcal{P}(\frac{E}{E_j})$ for $i \neq j$, also since $\mathcal{P}(E_1 \cap E_2 \cap ... \cap E_t) \subseteq \mathcal{P}(E_1) \cap \mathcal{P}(E_2) \cap ... \cap \mathcal{P}(E_t) = \phi$, we have $\mathcal{P}(E_1 \cap E_2 \cap ... \cap E_t) = \phi$ and so $E_1 \cap E_2 \cap ... \cap E_t) = 0$. If for some i, $E_1 \cap E_2 \cap ... \cap \hat{E}_i \cap ... \cap E_t = 0$, then the homomorphism $\alpha : E \longrightarrow \bigoplus_{j \neq i} \frac{E}{E_j}$, $e \longrightarrow (e + E_1, e + E_2, ..., e + \hat{E}_i, ..., e + E_t)$ is an embedding, for $\bigcap_{j \neq i} E_j = 0$. Hence $\mathcal{P}(E) \subseteq \mathcal{P}(\bigoplus_{j \neq i} \frac{E}{E_j}) = \bigcup_{j \neq i} \mathcal{P}(\frac{E}{E_j})$ from Lemma 2.4 for each i, that is $\mathcal{P}(E) \subseteq \{P_1, P_2, ..., \hat{P}_i, ..., P_t\}$ which is not true. Thus $\bigcap_{j \neq i} E_j \neq 0$. □

We now recall a couple of results from Ref. 2.

Lemma 2.3. *Let K have no finite independent family of left K-subgroups and $x \in K$ is such that $l(x) = 0$ then Kx is an essential left K-subgroup of K.*

Lemma 2.4. *Let K be a strongly semiprime strictly left Goldie near-ring such that weakly essential left K-subgroups are essential K subgroups. If any finite intersection of left annihilators are distributively generated then a weakly essential left K-subgroup has a non zero divisor.*

We now show that the concept of complete near-ring of left quotients coincides with that of the classical near-ring of left quotients.

Theorem 2.2. *Let K be a strongly semiprime strictly left Goldie d.g.n.r. such that weakly essential left K-subgroups are essential and the non-zero divisors are distributive. Then the complete near-ring of left quotients w.r.t. S (a semigroup of some distributive non-zero divisors of K) coincides with the classical near-ring of left quotients of K.*

Proof. If $a \in S, b \in K$ then Ka is an essential left K-subgroup, by Lemma 2.6 and $A = \{k \in K : kb \in Ka\}$ is an essential left K-subgroup of K and hence weakly essential. So it contains a non-zero divisor, say a_1 (by Lemma 2.7). Thus $a_1 b = b a_1$ for some $b \in K$. Hence left Ore condition is satisfied w.r.t. S. So by the result of Tiwari and Seth[3], the complete near-ring of left quotients coincides with classical near-ring of left quotients w.r.t. S. □

Bibliography

1. K. C. Chowdhury, A. Masum and H. K. Saikia, FSD N-groups with ACC on Annihilators, *Indian J. Pure Appl. Math.* **24**(12) (1993), 747–755.
2. A. Masum, H. K. Saikia and K. C. Chowdhury, On left Goldie Near-Rings and its parts having minimum conditions, *Indian J. Pure Appl. Math.* **25**(11) (1994), 1155–1162.

3. K. Tiwari and V. Seth, Classical near-ring of left and right quotients, *Acad. Progress Math.* **12** (1978), 115–123.

Near-algebras and gamma near-algebras

T. Srinivas

Department of Mathematics, Kakatiya University, Warangal-506 009, Telangana State, India
E-mail: thotasrinivas.srinivas@gmail.com

P. Narasimha Swamy

Department of Mathematics, GITAM University, Hyderabad Campus-502 329, Telangana State, India
E-mail: swamy.pasham@gmail.com

In 1933, P. Jordan proposed a quantum mechanical formalism in which the operators form only a near-algebra. In this regard, the investigation of near-algebras is interesting not only as an axiomatic question but also for physical reasons. A near-algebra is a near-ring which admits a field as a right operator domain. Near-algebras occur more naturally as mappings of linear spaces. The concept Gamma ring (or Γ-ring) (a generalization of a ring) was introduced by Nobusawa [5] and generalized by Barnes [1]. A generalization of both the concepts near-ring and the Γ-ring namely Γ-near-ring was introduced by Satyanarayana [6]. Brown [2], Yamamuro [11], Irish [3] and Srinivas [4,7,8,9] have studied the structure of near-algebras. Γ-near-algebra is a generalization of both the concepts of near-algebra and Γ-near-ring is introduced by Srinivas *et al.* [10]. Throughout this lecture, "near-algebra" we mean a "right near-algebra", and X denotes a field.

Keywords: Near-ring, Near-algebra, Γ-near-algebra.

1. Near-algebras

In this section, we present necessary fundamental definitions and results.

Definition 1.1. [3] A *right near-algebra* Y over a field X is a linear space Y over X on which a multiplication is defined such that
(i) Y forms a semigroup under multiplication,
(ii) multiplication is right distributive over addition (i.e. $(a+b)c = ac+bc$ for every $a,b,c \in Y$) and
(iii) $\lambda(ab) = (\lambda a)b$ for every $a,b \in Y$ and $\lambda \in X$.

Definition 1.2. A subset M of a near-algebra Y over a field X is said to be a *sub near-algebra* of Y if it satisfies the following four conditions:
(i) M is a Linear Subspace of Y,
(ii) $(M,\cdot)$ is a Sub-semigroup of Y.

Example 1.1. $T(V) = \{f \mid f : V \to V, V \text{ is a linear space}\}$ forms a near-algebra with the point wise operations of sum and scalar multiplication by $(f+g)(x) = f(x)+g(x)$, $(\lambda f)(x) = \lambda(f(x))$ and function composition as the multiplication by $(fg)(x) = f(g(x))$. Then $T_0(V) = \{f : f \in T(V), f(0) = 0\}$ is a sub near-algebra of $T(V)$.

Theorem 1.1. *A subset M of a near-algebra Y over a field X is a sub near-algebra of Y if and only if the following two conditions hold: (i) $x-y \in M$, (ii) $xy \in M$, $\lambda x \in M$ for every $x, y \in M$ and $\lambda \in X$.*

Definition 1.3. Let Y and Y' be two near-algebras over a field X. A mapping $\phi : Y \to Y'$ is called a near-algebra homomorphism if
(i) $\phi(x+y) = \phi(x) + \phi(y)$,
(ii) $\phi(\lambda x) = \lambda\phi(x)$ and
(iii) $\phi(xy) = \phi(x)\phi(y)$ for every $x, y \in Y$ and $\lambda \in X$.

A homomorphism which is one-one and onto is called an isomorphism.

Definition 1.4. Let Y and Y' be two near-algebras over a field X. Let $f : Y \to Y'$ be a near-algebra homomorphism. Then the kernel of f is denoted by $Kerf$ and is defined by $Kerf = \{x \in Y \mid f(x) = 0', 0' \text{ is the zero element in } Y'\}$.

Theorem 1.2. *If f is a homomorphism of a near-algebra Y into a near-algebra Y', then f is one-one if and only if $Kerf = \{0\}$.*

Definition 1.5. [3] A non-empty subset I of a near-algebra Y is called a near-algebra ideal (NA-ideal) of Y if
(i) I is a linear subspace of the linear space Y,
(ii) $ix \in I$ for every $x \in Y, i \in I$ and
(iii) $y(x+i) - yx \in I$ for every $x, y \in Y, i \in I$.
If I satisfies (i) and (ii), then I is called a right ideal of Y.
If I satisfies (i) and (iii), then I is called a left ideal of Y.

Theorem 1.3. *If I_1 and I_2 are two ideals of a near-algebra Y over a field X, then $I_1 \cap I_2$ is also an ideal of Y over a field X.*

Theorem 1.4. *If $f : Y \to Y'$ be a near-algebra homomorphism, then $Kerf$ is an ideal of Y.*

Theorem 1.5. *If I_1 and I_2 are two right ideals of a near-algebra Y over a field X. Then $I_1 + I_2 = \{i_1 + i_2 \mid i_1 \in I_1, i_2 \in I_2\}$ is a right ideal of Y over a field X.*

Proof. Let I_1 and I_2 be two right ideals of a near-algebra Y over a field X. Then it is clear that $I_1 + I_2 \subseteq Y$. Let $i, j \in I_1 + I_2$. Then $i = i_1 + i_2$ and $j = j_1 + j_2$, where $i_1, j_1 \in I_1$ and $i_2, j_2 \in I_2$. Let $\lambda, \mu \in X$. Then $\lambda i_1 + \mu j_1 \in I_1$ and $\lambda i_2 + \mu j_2 \in I_2$.

Consider $\lambda i + \mu j = \lambda(i_1 + i_2) + \mu(j_1 + j_2) = \lambda i_1 + \lambda i_2 + \mu j_1 + \mu j_2 = (\lambda i_1 + \mu j_1) + (\lambda i_2 + \mu j_2) \in I_1 + I_2$. Therefore $I_1 + I_2$ is a linear subspace of Y over a field X.

Let $x, y \in Y$, $i = i_1 + i_2 \in I_1 + I_2$, where $i_1 \in I_1$ and $i_2 \in I_2$. Then $ix = (i_1 + i_2)x = i_1x + i_2x \in I_1 + I_2$. Hence $I_1 + I_2$ is a right ideal of Y. □

Theorem 1.6. *Let $A(I)$ be the set $\{x \in Y \mid xI = 0\}$ for any non-empty subset I of a near-algebra Y. Then $A(I)$ is a left ideal of Y.*

Proof. Given that $A(I) = \{x \in Y \mid xI = 0\} = \{x \in Y \mid xi = 0 \text{ for every } i \in I\}$. Let $\lambda, \mu \in X$ and $x, y \in A(I)$. Then $xi = 0$ and $yi = 0$ for every $i \in I$. Since Y itself is a linear space, we get $\lambda x + \mu y \in Y$. For each $i \in I$ consider $(\lambda x + \mu y)(i) = (\lambda x)i + (\mu y)i = \lambda(xi) + \mu(yi) = \lambda 0 + \mu 0 = 0 + 0 = 0$. Therefore $\lambda x + \mu y \in A(I)$. Thus $A(I)$ is a linear subspace of Y.

Let $u, v \in Y$ and $x \in A(I)$. Then $xi = 0$ for every $i \in I$. It is clear that $v(u+x) - vu \in Y$. For each $i \in I$, consider $(v(u+x) - vu)i = (v(u+x))i - (vu)i = v((u+x)i) - (vu)i = v(ui + xi) - (vu)i = v(ui+0) - (vu)i = v(ui) - (vu)i = (vu)i - (vu)i = 0$. Therefore $v(u+x) - vu \in A(I)$. Thus $A(I)$ is a left ideal of Y. □

Note that, $A(I)$ is not a right ideal of Y.

Definition 1.6. Let I be an ideal of a near-algebra Y over a field X. Then for any element $x \in Y$ the set $x + I = \{x + i \mid i \in I\}$ is called a *left coset* of I in Y, generated by x. Similarly, the set $I + x = \{i + x \mid i \in I\}$ is called a *right coset* of I in Y, generated by x.

Since $(I, +)$ is a subgroup of the abelian group $(Y, +)$, we have by commutative property $x + i = i + x$ for every $x \in Y, i \in I$. Hence $x + I$ is called a *coset* of I in Y, generated by x.

Remark 1.1. Let Y be a near-algebra over a field X and I be an ideal of Y. Then we have the following statements:
(i) $0 + I = I$, 0 is the additive identity in Y.
(ii) $y + I = I$ for every $y \in I$.
(iii) Any two cosets of I in Y are either identical or disjoint. That is either $x + I = y + I$ or $(x+I) \cap (y+I) = \emptyset$.
(iv) If $x + I$ and $y + I$ are two cosets of I in Y, then $x + I = y + I \Leftrightarrow x - y \in I$.

Notation 1.1. Let I be an ideal of a near-algebra Y over a field X. Then the set of all cosets of I in Y is denoted by Y/I and defined by $Y/I = \{x + I \mid \text{for every } x \in Y\}$.

Theorem 1.7. *Let I be an ideal of a near-algebra Y over a field X. Then the set Y/I is a near-algebra over X with respect to the induced operations of addition, scalar multiplication and product of cosets defined by*

$$\begin{aligned} (x+I) + (y+I) &= (x+y) + I, \\ \lambda(x+I) &= \lambda x + I, \\ (x+I)(y+I) &= xy + I \end{aligned}$$

for every $x + I, y + I \in Y/I, \lambda \in X$. (This near-algebra is called a Quotient near-algebra.)

Theorem 1.8. *If I is an ideal of a near-algebra Y over a field X, then the quotient near-algebra Y/I is a homomorphic image of Y.*

Theorem 1.9. *Let Y and Y' be two near-algebras over a field X. Let $f : Y \to Y'$ be a homomorphism with Kernel I. Then $f(Y)$ is isomorphic to Y/I.*

Theorem 1.10. *Let Y and Y' be two near-algebras over a field X. Then the direct product $Y \times Y' = \{(x, x') \mid x \in Y, x' \in Y'\}$ is a near-algebra over a field X.*

Proof. A direct verification shows that $Y \times Y'$ is a linear space over a field X. Now we verify that $Y \times Y'$ is a semigroup under multiplication. Let $(x,x'),(y,y') \in Y \times Y'$, where $x,y \in Y$ and $x',y' \in Y'$. Then $(x,x')(y,y') = (xy,x'y') \in Y \times Y'$. That is $Y \times Y'$ is closed under multiplication. Let $(x,x'),(y,y'),(z,z') \in Y \times Y'$, where $x,y,z \in Y$ and $x',y',z' \in Y'$. Then

$$\begin{aligned}((x,x')(y,y'))(z,z') &= (xy,x'y')(z,z')\\ &= ((xy)z,(x'y')z')\\ &= (x(yz),x'(y'z'))\\ &= (x,x')(yz,y'z'),\\ &= (x,x')((y,y')(z,z')).\end{aligned}$$

Now let $(x,x'),(y,y'),(z,z') \in Y \times Y'$, where $x,y,z \in Y$ and $x',y',z' \in Y'$. Then

$$\begin{aligned}((x,x')+(y,y'))(z,z') &= (x+y,x'+y')(z,z')\\ &= ((x+y)z,(x'+y')z')\\ &= (xz+yz,x'z'+y'z')\\ &= (xz,x'z')+(yz,y'z')\\ &= (x,x')(z,z')+(y,y')(z,z').\end{aligned}$$

For any $(x,x'),(y,y') \in Y \times Y'$ and $\lambda \in X$, where $x,y \in Y$, $x',y' \in Y'$, we get

$$\begin{aligned}(\lambda(x,x'))(y,y') &= (\lambda x,\lambda x')(y,y')\\ &= ((\lambda x)y,(\lambda x')y')\\ &= (\lambda(xy),\lambda(x'y'))\\ &= \lambda(xy,x'y')\\ &= \lambda((x,x')(y,y')).\end{aligned}$$

This shows that all the postulates of a near-algebra are satisfied. Hence $Y \times Y'$ is a near-algebra over the field X. □

2. Γ-near-algebra

In this section, we introduce the notion of a Γ-near-algebra. Properties such as distributor element, ideal and homomorphism of the Γ-near-algebra will be discussed [10]. Before going for a formal definition of a Gamma near-algebra, let us consider a natural example:

Let V be a vector space over a field X and Z be a non-empty set. Let $M = \{f \mid f : Z \to V\}$ and $\Gamma = \{g \mid g : V \to Z\}$. Since $(V,+)$ is an abelian group then $(M,+)$ is also an abelian group. Let $\lambda \in X, f \in M$. Then $f : Z \to V$ implies that $\lambda f : Z \to V$. Thus $\lambda f \in M$. Let $\lambda,\mu \in X, f_1, f_2 \in M$ and 1 is the unity in X. For any $x \in Z$, we have that $f_1(x), f_2(x) \in V$. This implies that $f_1(x)+f_2(x) \in V$(that is $(f_1+f_2)(x) \in V$) and $\lambda f_1(x), \lambda f_2(x) \in V$. It is clear that $(\lambda(f_1+f_2))(x) = (\lambda f_1+\lambda f_2)(x)$, $((\lambda+\mu)f)(x) = (\lambda f+\mu f)(x)$, $((\lambda\mu)f)(x) = \lambda(\mu f)(x)$, $1(f(x)) = (1f)(x) = f(x)$.

Hence M is a linear space over the field X. Now for every $f_1, f_2, f_3 \in M$ and $g_1, g_2 \in \Gamma$, it is clear that $(f_1g_1f_2)g_2f_3 = f_1g_1(f_2g_2f_3)$ and $(f_1+f_2)g_1f_3 = (f_1g_1f_3)(x)+(f_2g_1f_3)(x)$. Let $\lambda \in X, f_1, f_2 \in M, g_1 \in \Gamma$. Then for every $x \in Z$, we have that $((\lambda f_1)g_1f_2)(x) = \lambda(f_1g_1f_2)(x)$. That is $(\lambda f_1)g_1f_2 = \lambda(f_1g_1f_2)$. Thus, if V is a linear space and Z is a non empty set, then $M = \{f \mid f : Z \to V\}$ is a linear space and there exists a mapping $M \times \Gamma \times M \to M$, where $\Gamma = \{g \mid g : V \to Z\}$ satisfying the following conditions: $(i)(f_1g_1f_2)g_2f_3 = f_1g_1(f_2g_2f_3)$, $(ii)(f_1+f_2)g_1f_3 = (f_1g_1f_3)(x)+(f_2g_1f_3)(x)$, and $(iii)(\lambda f_1)g_1f_2 = \lambda(f_1g_1f_2)$ for every $f_1, f_2 \in M, g \in \Gamma, \lambda \in X$. Hence M is a Γ-near-algebra over the field X.

Keeping these axioms in mind, we define the concept of a "gamma near-algebra".

Definition 2.1. Let M be a linear space over a field X and Γ be a non-empty set. Then M is said to be a Γ-*near-algebra* over a field X if there exist a mapping $M \times \Gamma \times M \to M$ (the image of (a, α, b) is denoted by $a\alpha b$) satisfying the following three conditions:
(i) $(a\alpha b)\beta c = a\alpha(b\beta c)$,
(ii) $(a+b)\alpha c = a\alpha c + b\beta c$,
(iii) $(\lambda a)\alpha b = \lambda(a\alpha b)$
for every $a, b, c \in M, \alpha, \beta \in \Gamma$ and $\lambda \in X$.

Definition 2.2. Let M be a Γ-near-algebra over a field X. Then M is said to be a *zero symmetric* Γ-*near-algebra* or Γ-*near-c-algebra* if $a\alpha 0 = 0$ for every $a \in M$ and $\alpha \in \Gamma$, where 0 is the additive identity in M.

Definition 2.3. Let M be a Γ-near-algebra over a field X. For $x, y, z \in M, \alpha \in \Gamma$, the element $x\alpha(y+z) - x\alpha y - x\alpha z$ is called the *distributor* of y and z with respect to x and is denoted by $[x, y, z]$.

Proposition 2.1. *Let M be a Γ-near-algebra over a field X. For any $x, y, z, w \in M, \alpha \in \Gamma$ and $\lambda \in X$, we have the following identities:*
(i) $[x, y, z] = [x, z, y]$
(ii) $[x+w, y, z] = [x, y, z] + [w, y, z]$
(iii) $\lambda[x, y, z] = [\lambda x, y, z]$
(iv) $[x, y, z]\alpha w = [x, y\alpha w, z\alpha w]$.

Proof. Let M be a Γ-near-algebra over a field X. Let $x, y, z, w \in M, \alpha \in \Gamma$ and $\lambda \in X$. Then
$(i)[x, y, z] = x\alpha(y+z) - x\alpha y - x\alpha z = x\alpha(z+y) - x\alpha z - x\alpha y = [x, z, y]$,
$(ii)[x+w, y, z] = (x+w)\alpha(y+z) - (x+w)\alpha y - (x+w)\alpha z = x\alpha(y+z) + w\alpha(y+z) - x\alpha y - w\alpha y - x\alpha z - w\alpha z = x\alpha(y+z) - x\alpha y - x\alpha z + w\alpha(y+z) - w\alpha y - w\alpha z = [x, y, z] + [w, y, z]$,
$(iii)[\lambda x, y, z] = (\lambda x)\alpha(y+z) - (\lambda x)\alpha y - (\lambda x)\alpha z = \lambda(x\alpha(y+z)) - \lambda(x\alpha y) - \lambda(x\alpha z) = \lambda(x\alpha(y+z) - (x\alpha y) - (x\alpha z)) = \lambda[x, y, z]$,
$(iv)[x, y\alpha w, z\alpha w] = x\alpha(y\alpha w + z\alpha w) - x\alpha(y\alpha w) - x\alpha(z\alpha w) = x\alpha((y+z)\alpha w) - (x\alpha y)\alpha w - (x\alpha z)\alpha w = (x\alpha(y+z))\alpha w - (x\alpha y)\alpha w - (x\alpha z)\alpha w = (x\alpha(y+z) - x\alpha y - x\alpha z)\alpha w = [x, y, z]\alpha w$. □

Definition 2.4. Let M be a Γ-near-algebra over a field X. A linear subspace L of M is said to be a *sub* Γ-*near-algebra* over a field X if there exist a mapping $L \times \Gamma \times L \to L$ satisfying the following three conditions:
(i) $(a\alpha b)\beta c = a\alpha(b\beta c)$,
(ii) $(a+b)\alpha c = a\alpha c + b\beta c$ and

(iii) $(\lambda a)\alpha b = \lambda(a\alpha b)$
for every $a, b, c \in L, \alpha, \beta \in \Gamma$ and $\lambda \in X$.

In other words, A linear subspace L of a Γ-near-algebra M over a field X is said to be a *sub Γ-near-algebra* over a field X if there exist a mapping $L \times \Gamma \times L \to L$ satisfying the following two conditions:
(i) $(a\alpha b) \in L$,
(ii) $(\lambda a) \in L$
for every $a, b, \in L, \alpha \in \Gamma$ and $\lambda \in X$.

Theorem 2.1. *A non-empty subset L of a Γ-near-algebra M over a field X is a sub Γ-near-algebra of M if and only if the following three conditions hold: (i) $x - y, \lambda x \in L$, (ii) $x\alpha y \in L$ and (iii) $(\lambda x)\alpha y = \lambda(x\alpha y)$ for every $x, y \in L, \lambda \in X$ and $\alpha \in \Gamma$.*

Definition 2.5. Let M be a Γ-near-algebra over a field X. Then the linear subspace I of a linear space M is called
(i) a *left ideal* if $y\alpha(x+i) - y\alpha x \in I$ for all $x, y \in M, \alpha \in \Gamma$ and $i \in I$,
(ii) a *right ideal* if $i\alpha x \in I$ for all $x \in M, \alpha \in \Gamma$ and $i \in I$ and
(iii) an *ideal* if it is both a left and a right ideal.

Definition 2.6. Let M and M' be two Γ-near-algebras over a field X. A mapping $\phi : M \to M'$ is called a Γ-*near-algebra homomorphism* if the following three conditions hold:
(i) $\phi(x+y) = \phi(x) + \phi(y)$,
(ii) $\phi(\lambda x) = \lambda\phi(x)$,
(iii) $\phi(x\alpha y) = \phi(x)\alpha\phi(y)$ for every $x, y \in M, \lambda \in X, \alpha \in \Gamma$.
We say that ϕ is a Γ-near-algebra isomorphism if ϕ is one-one, onto and homomorphism.

Proposition 2.2. *Let $\phi : M \to M'$ be a Γ-near-algebra homomorphism. Then (i) $\phi(0) = 0'$ (ii) $\phi(-x) = -\phi(x)$ (iii) $\phi(x-y) = \phi(x) - \phi(y)$ for every $x, y \in M$, where 0 and $0'$ are the zero elements in M and M' respectively.*

Theorem 2.2. *If $\phi : M \to M'$ is a Γ-near-algebra homomorphism, then the homomorphic image $\phi(M)$ is a sub Γ-near-algebra of M'.*

Proof. Given that M and M' are two Γ-near-algebras over a field X, and $\phi : M \to M'$ is a Γ-near-algebra homomorphism. The homomorphic image of M is $\phi(M) = \{x' \in M' : \phi(x) = x', x \in M\}$. To show $\phi(M)$ is a sub Γ-near-algebra of M', it is sufficient to prove (i) $x' - y', \lambda x' \in \phi(M)$ for every $x', y' \in \phi(M), \lambda \in X$, (ii) $x'\alpha y' \in \phi(M)$ for every $x', y' \in \phi(M), \alpha \in \Gamma$, (iii) $\lambda(x'\alpha y') = (\lambda x')\alpha y'$ for every $x', y' \in \phi(M), \alpha \in \Gamma, \lambda \in X$.

Let $x', y' \in \phi(M), \alpha \in \Gamma, \lambda \in X$. Then there exists $x, y \in M$ such that $x' = \phi(x), y' = \phi(y)$. Since M is a Γ-near-algebra, we get $x - y, \lambda x, x\alpha y \in M$ and $(\lambda x)\alpha y = \lambda(x\alpha y)$. Now (i) $x' - y' = \phi(x) - \phi(y) = \phi(x-y) \in \phi(M)$ and $\lambda x' = \lambda\phi(x) = \phi(\lambda x) \in \phi(M)$. (ii) $x'\alpha y' = \phi(x)\alpha\phi(y) = \phi(x\alpha y) \in \phi(M)$. (iii) $\lambda(x'\alpha y') = \lambda(\phi(x)\alpha\phi(y)) = \lambda(\phi(x\alpha y)) = \phi(\lambda(x\alpha y)) = \phi((\lambda x)\alpha y) = \phi(\lambda x)\alpha\phi(y) = (\lambda\phi(x))\alpha\phi(y) = (\lambda x')\alpha y'$. Hence $\phi(M)$ is a sub Γ-near-algebra of M'. $\square$

Definition 2.7. Let M and M' be two Γ-near-algebras over a field X. Let $\phi : M \to M'$ be a Γ-near-algebra homomorphism. Then the Kernel of ϕ is denoted by $Ker\phi$ and is defined by $Ker\phi = \{x \in M : \phi(x) = 0'\}$, where $0'$ is the zero element in M'.

Theorem 2.3. *If $\phi : M \to M'$ is a Γ-near-algebra homomorphism, then* $\mathrm{Ker}\phi$ *is an ideal of M.*

Theorem 2.4. *Let M and M' be two Γ-near-algebras over a field X. Let $\phi : M \to M'$ be a Γ-near-algebra homomorphism. If L is an ideal of M, then $\phi(L)$ is an ideal of $\phi(M)$.*

Proof. We know that $\phi(M) = \{x' \in M' : \phi(x) = x', x \in M\}$ and $\phi(L) = \{\phi(y) : y \in L \subseteq M, \phi(y) = y', y' \in \phi(M)\}$. Then $\phi(L) \subseteq \phi(M)$.

Let $a, b \in X; i', j' \in \phi(L)$. Then $i', j' \in \phi(M)$ and there exists $i, j \in L$ such that $\phi(i) = i', \phi(j) = j'$. Since L is an ideal in M, we get $ai + bj \in L$ for every $a, b \in X; i, j \in L$. Now $ai' + bj' = a\phi(i) + b\phi(j) = \phi(ai) + \phi(bj) = \phi(ai + bj) \in \phi(L)$(since $ai + bj \in L$). Therefore for every $a, b \in X; i', j' \in \phi(L), ai' + bj' \in \phi(L)$. Thus $\phi(L)$ is a linear subspace of $\phi(M)$.

Let $i' \in \phi(L), \alpha \in \Gamma, x' \in \phi(M)$. Then $i' \in \phi(M)$ and there exists $i \in L, x \in M$ such that $\phi(i) = i', \phi(x) = x'$. Since L is an ideal in M, we get $i\alpha x \in L$ for every $i \in L, \alpha \in \phi, x \in M$. Now $i'\alpha x' = \phi(i)\alpha\phi(x) = \phi(i\alpha x) \in \phi(L)$(since $i\alpha x \in L$). Thus $\phi(L)$ is a right ideal of $\phi(M)$.

Let $x', y' \in \phi(M), \alpha \in \Gamma, i' \in \phi(L)$. Then $i' \in \phi(M)$ and there exists $x, y \in M, i \in L$ such that $\phi(x) = x', \phi(y) = y', \phi(i) = i'$. Since L is and ideal in M, we get $y\alpha(x+i) - y\alpha x \in L$ for every $x, y \in L, \alpha \in \Gamma, i \in L$. Consider $y'\alpha(x'+i') - y'\alpha x' = \phi(y)\alpha(\phi(x) + \phi(i)) - \phi(y)\alpha\phi(x) = \phi(y)\alpha\phi(x+i) - \phi(y\alpha x) = \phi(y\alpha(x+i)) - \phi(y\alpha x) = \phi(y\alpha(x+i) - y\alpha x) \in \phi(L)$. Thus $\phi(L)$ is a left ideal of $\phi(M)$. Hence $\phi(L)$ is an ideal of $\phi(M)$. □

Definition 2.8. Let I be an ideal of a Γ-near-algebra M over a field X. Then for any element $x \in M$, the set $x + I = \{x + i : i \in I\}$ is called the *left coset* of I in M generated by x. Similarly, the set $I + x = \{i + x : i \in I\}$ is called the *right coset* of I in M generated by x.

Basic properties corresponding to these cosets are as usual.

Theorem 2.5. *Let I be an ideal of a Γ-near-algebra M over a field X. Then the quotient set $M/I = \{x + I : x \in M\}$ is a Γ-near-algebra over X with respect to the operations defined by*

$$\begin{aligned}
(x+I) + (y+I) &= (x+y) + I, \\
\lambda(x+I) &= \lambda x + I, \\
(x+I)\alpha(y+I) &= x\alpha y + I
\end{aligned}$$

for every $x, y \in M, \lambda \in X, \alpha \in \Gamma$. (This Γ-near-algebra is called a Quotient Γ-near-algebra.)

Theorem 2.6. *Let I be an ideal of a Γ-near-algebra M over a field X, then the quotient Γ-near-algebra M/I is the homomorphic image of M.*

Theorem 2.7. *Let M and M' be two Γ-near-algebras over a field X. Let $\phi : M \to M'$ be a Γ-near-algebra homomorphism with kernel I. Then $\phi(M)$ is isomorphic to M/I.*

Bibliography

1. W. E. Barnes, On the Γ-rings of Nobusawa, *Pacific Journal of Math.*, **18** (1966), 411–422.
2. H. Brown, Near-algebras, *Illinois J. Math.*, **12** (1968), 215–227.
3. J. W. Irish, Normed near-algebras and finite dimensional near-algebras of continuous functions, Doctoral thesis, University of New Hampshire, 1975.
4. P. Narasimha Swamy, Some aspects on Near-rings, Doctoral Dissertation, Kakatiya University, 2012.
5. Nobusawa, On a generalization of the ring theory, *Osaka Journal of Math.* **1** (1964), 81–89.
6. Pilz G., Near-rings, North-Holland pub., 1983.
7. Bh. Satyanarayana, Contribution to near-ring theory, Doctoral Dissertation, Acharya Nagarjuna University, 1984.
8. Bh. Satyanarayana and K. Syam Prasad, Near-rings, Fuzzy Ideals and Graph Theory, CRC Press (Taylor and Francis Group), England/New York, 2013 (ISBN: 978-1-4398-7310-6).
9. Bh. Satyanarayana and K. Syam Prasad, Discrete Mathematics and Graph Theory, Printice Hall of India, New Delhi, 2014 (ISBN: 978-81-203-4948-3).
10. T. Srinivas, Near-rings and application to function spaces, Doctoral Dissertation, Kakatiya University, 1996.
11. T. Srinivas and K. Yugandhar, A note on normed Near-algebras, *Indian J. Pure and Appl. Math.*, **20**(5), 433–438, May 1989.
12. T. Srinivas, A note on the radicals in a normed Near-algebra, *Indian J. Pure and Appl. Math.*, **21**(11), 989–994, Nov. 1990.
13. T. Srinivas, P. Narasimha Swamy, K. Vijaykumar, Gamma Near-Algebras, *International Journal of Algebra and Statistics*, Vol. 1:2 (2012), 107–117.
14. S. Yamamuro, On Near-algebras of mappings of Banach spaces, *Proc. Japan Acad.*, **41** (1965), 889–892.

Introducing the n–gen problem

S. D. Scott

Department of Mathematics,
University of Auckland, New Zealand
E-mail: ssco034@math.auckland.ac.nz

1. What is the n–gen problem?

The n–gen problem asks a question about any given finite transformation nearring. So straight away it is needful to define such a nearring. In order to do this it is necessary to say a little about nearrings and groups. Here nearrings will be taken as left distributive and zero–symmetric and most groups will be written additively. This additive convention also applies to N–groups (N a nearring). It is not taken as meaning groups being considered are abelian. Certain groups specified by the context remain multiplicative. For example if N is a nearring with identity then this is true for the group $u(N)$ of units of N. It is also true for an automorphism group of a group (the operation here is composition). However, the operation of an arbitrarily given group is $+$. Also, unless stated to the contrary, our nearrings have an identity.

A transformation nearring is obtained when we take a group V and all zero–fixing self–maps on V. Here under the operations of pointwise addition and composition this set of functions becomes a nearring. It is denoted by $M_0(V)$. As it has an identity consideration of the group $u(M_0(V))$ of units of $M_0(V)$ becomes possible. Elements of $M_0(V)$ in $u(M_0(V))$ are in fact precisely the bijections on V that fix 0. It will be seen below elements of $u(M_0(V))$ are important when we are considering how $M_0(V)$ may be generated (as a nearring) by a finite subset of $M_0(V)$. It is not difficult to see that for such a finite set to exist, V (thus $M_0(V)$) must be finite. It is this situation considerations are restricted to. From now on all groups are taken as finite.

The problem of how simply $M_0(V)$ can be generated is inspired by looking at symmetric groups. $M_0(V)$ is all zero–fixing functions of a finite group V into itself, while a symmetric group is all bijections on a finite set. However, the symmetric group on a set of m elements has a very elementary set of generators. It is generated by an element of order m and one of order 2. Thus, the question arises as to whether or not $M_0(V)$ is generated (as a nearring) in some elementary manner. This is indeed nearly always the case. It is a significant fact that, apart from the cyclic group of order three and the 4–group, $M_0(V)$ is generated by a single element. Some notation needs specifying. If α is in $M_0(V)$, then from now on $N(\alpha)$ will denote the subnearring of $M_0(V)$ (with or without identity) generated by α. According to the above there is nearly always an α in $M_0(V)$ with $N(\alpha) = M_0(V)$. However, a feature of an α with this property is that it is in $u(M_0(V))$. This in itself raises a question. Taking arbitrary elements of $u(M_0(V))$ what is the probability that $N(\alpha) = M_0(V)$ (ie. that α generates $M_0(V)$)? It has been shown by Aichinger, Mašulović, Pöschel and Wilson in [1] that this probability, with $|V|$ tending to infinity, is one.

The problem of how simply $M_0(V)$ is generated first made an appearance in [5] where it was shown the $M_0(V)$, apart from V cyclic of order three and V elementary two, are generated by a unit α with $\alpha^2 = 1$. This is a remarkable fact. Nearly all $M_0(V)$ (V finite) are generated by a unit of order two. There followed in 1997 in [14] the classification of $M_0(V)$ generated by a unit α with $\alpha^3 = 1$. Once again, apart from groups of an elementary nature, all $M_0(V)$ are so generated. The n–gen problem seeks to take this much further and asks for a complete classification of those V where $M_0(V)$ is generated by an element of $u(M_0(V))$ of order n (this is what it means for $M_0(V)$ to be n–gen). It is very similar to asking as to when $M_0(V)$ is a homomorphic image of the nearring generated by a β where the only defining relationship is $\beta^n = 1$. The unpublished (as yet) book [3] provides the entire solution of this very deep problem. In the next section some indication is given as to what was previously known.

2. Previous results

The nature of the n–gen problem was explained in the previous section. This one indicates progress up until the solution [3] provided. These results have been collected over about thirty years. Quite a number of them are found in unpublished manuscripts. Although this material is a foundation for much of [3] overall development in [3] is more general. It is true this [3] is often into such areas. It is also true that on the whole its perspective is much wider.

As indicated in the previous section the $n = 2$ situation is completely solved. This result initiated the study of $M_0(V)$ generation and in particular examination of when $M_0(V)$ is n–gen. The paper involved (see [5]) dates back to 1979. Its main theorem states:

The nearring $M_0(V)$ is 2–gen if, and only if, V is not a cyclic group of order three and not elementary two ($\{0\}$ included).

The answer as to how things are when $n = 3$ was given in 1997. Providing this involved much more work than the $n = 2$ case. Indeed, although small values of n can create problems, general theory seems to take over only when n becomes ≥ 11. The 3–gen solution supplied in [14] is as follows:

The nearring $M_0(V)$ is 3–gen if, and only if, V is not of order ≤ 3, not the 4–group, not the symmetric group of order six and not an elementary 2–group of order > 4 incongruent to 1 mod 3.

In progressing from $n = 2$ or 3 to $n = p$ a prime ≥ 5 an observation needs to be made. This is that it looks as if elementary two groups behave very differently from other groups. This is indeed the case. The theorem dealing with V elementary two is difficult to prove but is easily stated. The result (see [12]) we have here is:

If V is a non–zero elementary two group, then $M_0(V)$ is p–gen ($p \geq 5$ a prime) if, and only if, $|V| \equiv 1$ mod p.

The theorem dealing with p a prime ≥ 5 and V not elementary two is unlike that just given. In this situation it is needful to define (in terms of p) three distinct finite families of groups. The result in question states that $M_0(V)$ is p–gen if, and only if, V is not in their union. Its statement and proof can be found in [11].

All results mentioned above are for n a prime. Indeed, the prime case was completely covered before [3] was present. Although [5], [14], [11] and [12] supply this, the full picture first appeared in the 318 page manuscript [4]. The advantage of [11] and [12] (dealing with $p \geq 5$ a prime) is that they present a much more concise approach.

The case of n composite involves difficulties that are formidable. The distinction between V elementary two, and otherwise, again seems to be forced on us. As far as the first goes [5] seeks to give a complete solution. The main theorem of [6] is stated as an 'if, and only if' result. One implication, where the proof is nearly all of the manuscript, holds up, but the very brief attention given to the reverse is flawed.

In spite of the flawed nature of the proof of 1.1 of [6] much goes through. The theorem allows the statement that, when n is divisible by two distinct primes, $M_0(V)$ (V elementary two) is nearly always n–gen. In other words, apart from a finite collection of such groups (depending on n), $M_0(V)$ is n–gen. [6] also allows a complete classification in the case of n being a prime power. With n not being a prime power only specification of the finite set had, prior to [3], failed to be determined.

We come now to the previous, inroads made into solving the n–gen problem when n is composite and V in not elementary two. Here [13], although not directly concerned with this matter, does in fact lay foundations. [13] proves three theorems dealing with n divisible by 2, 3 or 5. In fact [8], [9] and [7] make use of [13] to handle all such n. What [8], [9] and [7] prove is that, for such a composite n, $M_0(V)$ (V not elementary two) is n–gen if, and only if, a straightforward numerical condition on n and $|V|$ is satisfied.

3. The size of the problem

The first excursion into n–gen considerations consisted of a short paper. It was the five pages of [5] that gave us the classification of 2–gen $M_0(V)$. At that time it looked as though this might be the end of the matter. It was nice to know that most finite transformation nearrings were generated so simply. However, the analogy between the simple nearring $M_0(V)$ and simple groups (non–abelian) could not be suppressed. The fact that many of these groups are generated by an element of order two and of order three, raised questions. One of these was: were most of the $M_0(V)$ generated by an element of $u(M_0(V))$ of order three and if so what were the exceptions? The problem involved seemed quite difficult. After a period of about 18 years from the publication of [5] a paper supplying the full answer to this appeared (see [14]). It was of considerable length (21 pages) compared with [5]. The arguments involved tended to display a higher order of complexity.

The matter of n–gen considerations rested with [14] for some time. However, in writing up a list of interesting nearring questions (see [10]) it occurred to me that [5] and [14] were indicative of something far more general. It looked as though, apart from a finite number of exceptions, $M_0(V)$ (V not elementary two) were p–gen ($p \geq 5$ a prime). It also looked as though the p–gen nature of elementary two groups could be dealt with. A conjecture covering both aspects of this problem was formulated and put into the list (see question six of [10]). Once again this matter was not immediately taken up. However, finding it a personal challenge, it was not that long before I gave it more serious consideration.

Question six of [10] asked if, apart from a finite collection of groups the $M_0(V)$ were p–gen ($p \geq 5$ a prime) if, and only if, V was not elementary two of order incongruent to $1 \ mod \ p$. At the time of formulating [10] this seemed plausible. It also seemed difficulties involved in answering it would be substantial. By 2000 something like a roughed out manuscript of 200 pages had evolved which looked as though it answered this. However, there were problems with this attempt and shortly after near completion it was abandoned. It did, however, provide a foundation for the completed manuscript [4]. This was a 318 page effort giving a complete answer to when $M_0(V)$ is p–gen (p a prime). Not only had question six of [11] been answered but the finite collection had been specified.

The solution as to which $M_0(V)$ were p–gen still presented a problem. The finite collection of groups they had was specified in terms of the number of prime order subgroups. Deciding when this integer was $> \frac{1}{2}|V| - 1$, would allow these groups to be determined. The answer to this was given in 2009. In [2] Tim Burness and myself provided a 31 page paper answering this. This more than resolved the p–gen problem but to some extent the n–gen problem had not really been touched on, only the situation where n is a prime.

The solution of the p–gen problem given in [4] was not entirely satisfactory. It looked as though more concise treatment could be obtained. Somewhat different ways of approaching this matter were possible. Although the p–gen solution appeared to be the end of things, there still seemed to be the matter of accessibility. In [11] and [12] I provided two much more concise approaches. The 27 page manuscript [11] dealt with V not elementary two, while the elementary two case was covered in the 15 page manuscript [12].

Over about four or five years leading up to 2012 I gave serious attention to the n–gen problem (n an integer ≥ 2). It was not until [11] and [12] were in place that I began to see a way to resolve this for V not elementary two. V elementary two and otherwise were distinct problems. In the case of the first, something like a complete solution was given in the 20 page manuscript [6]. However, the second was still very thorny territory but p–gen considerations provided a ray of hope. If n divisible by a prime in $\{2,3,5,7\}$ could be eliminated general theory might be able to proceed. The 31 page manuscript [13] paved the way for this. In [8] (8 pages) the case of n even was solved. The 14 page manuscript [9] dealt with n divisible by 3, while [7] (14 pages) dealt with n divisible by 5. The case of n divisible by 7 had, at that time, not been considered. [3] handles n divisible by 7 and much more. There, general theory takes over and the full solution of the n–gen problem is supplied.

What is packed into the two hundred and more pages of [3] is considerable. Virtually all of it is directed towards the one goal. There is much material directly concerned with this. Such material and that of a supportive nature, along with examples, eventually yield the complete solution.

4. The complexity of the problem

The problem that is presented to us is as follows: given a group V and integer $n \geq 2$, is there some method of deciding if $M_0(V)$ is n–gen? Having before us V and n does not generally ensure the n–gen nature of $M_0(V)$ is easily determined. Moving from what is given to a solution is, in general, very circuitous, evidenced by a number of deep results of seemingly unrelated areas that must be obtained along the way. They are aspects of number and group theory. As far as number theory goes both standard and rather unconventional questions arise. This can be seen in chapter two of [3]. The calculations expressing integers as the sum of two or more prime powers has a conventional flavour about it while calculations relating to the l–function tend to the opposite extreme. Also, group theory development in reality is, rather non–standard. Although requiring some such knowledge of features of V (a group) seems necessary. One of the problems faced there relates to the number of elements of order two of V. Another relating to the number of prime order subgroups of V, when this, $> \frac{1}{2}|V| - 1$, has only recently been solved. This knowledge is used in a key manner at times throughout [3].

Something like three reasons can be given why solving the n–gen problem can be expected to be difficult. These are outlined in this and the next two paragraphs. The nature of the n–gen problem explained at the beginning of this section does not readily allow movement from a given n and V to an understanding of whether $M_0(V)$ is n–gen. Here n is associated with certain elements of $u(M_0(V))$ and is, therefore, related to multiplicative structure. However, V is closely related to the additive structure of $M_0(V)$. In this regard there is not that much relationship between them. Addition and multiplication fit together as the operations of $M_0(V)$ determine. This seems to indicate that to solve the n–gen problem means grappling with particular elements of $M_0(V)$ and carrying out necessary computations within the nearring. In other words, at the outset, some overview of how a solution might proceed is hard to imagine. This indeed seems to be the case.

The second reason (related to the first) is that if we see addition and multiplication fitting snugly together, quite a bit of theory is obtained. This is in many respects rather superficial. Such facts as $M_0(V)$ being simple or a direct sum of minimal right ideals are features that all $M_0(V)$ share. It would seem they cannot be expected to shed light on n–gen considerations. Such considerations are really more difficult and look deep into the internal structure of the $M_0(V)$. It is the difference between individual $M_0(V)$ that hopefully will shine light on the n–gen problem. Attempting to solve it by looking at nearring structure is futile. Thus it would seem necessary to regard all the $M_0(V)$ as different and having their own brand of uniqueness.

The final reason that the n–gen problem looks to be complex is now explained. If we take an element α of $u(M_0(V))$ of order n, then the multiplicative subgroup it generates is a C_n (cyclic group of order n). The nature of C_n changes considerably, if n is replaced by $n-1$ or $n+1$. In fact, n, $n-1$ or $n+1$ can be a prime so the cyclic subgroup involved has no proper subgroups while the other two cyclic groups may have many. This perhaps provides us with reason to believe closely related n may have very different $M_0(V)$, which are n–gen. Indeed, it is borne out in practice that small changes in n give rise to diverse n–gen $M_0(V)$. The last chapter of [3] gives examples of this sort of thing. One example provided is where thirteen categories list the $M_0(V)$, when $n = 37$, which are not n–gen while n–gen $M_0(V)$ with $n = 36$ are described in terms of two. What this means is that the nature of n is strongly related to the problem's solution. Indeed, regarding the final solution, there is quite a big difference between prime and non–prime n.

In the above, reasons are given why solving our problem can be expected to be difficult, which turns out to be the case. Even developments depending on this are substantial, a lot of which are supplied in chapters two to four with more being introduced as [3] proceeds. Five onwards grapples more directly with the problem. There are at least three aspects to this, which are covered in the remainder of this section.

In dealing with groups which are not elementary two, small primes create difficulties. Chapters five to seven of [3] handle this. Five and six supply the needed amount of detailed material where quite specific matters are attended to. Out of this (chapter seven) comes the $\{2,3,5,7\}$ theorem which gives us the n–gen nature of $M_0(V)$, when n is divisible by one of these primes. In this way low primes are sorted.

Essentially, chapters two to seven of [3] allow the complex general theory of eight and nine to handle n with primes ≥ 11 dividing it. In this manner problems concerning groups which are not elementary two are resolved. The three chapters that follow involve quite intense investigations into elementary two groups. This is substantial enough to entirely deal with such groups. With twelve in place the resolution of major difficulties is complete. What remains is to put together notation and previous work that tie up the solution. Chapter thirteen of [3] not only handles this but supplies a presentation (along with examples) that gives the entire picture. Although lengthy in its statement, the solution of chapter thirteen of [3] is not really hard to comprehend.

Bibliography

1. E. Aichinger, D. Mašulović, R. Pöschel and J. S. Wilson, *Completeness of concrete near-rings*, J. Alg **279** (2004), 61–78.
2. T. C. Burness and S. D. Scott, *On the Number of Prime Order Subgroups of Finite Groups*, J. Austr. Math. Soc **87** (2009), 329–357.
3. S. D. Scott, *Entire solution of the n–gen problem*, book awaiting publication, 1–210.
4. S. D. Scott, *Generators of Finite Transformation Nearrings*, unpublished paper, 1–318.
5. S. D. Scott, *Involution Near–rings*, Proc. Edin. Math. Soc **22** (1979), 241–245.
6. S. D. Scott, *Particular n–gen $M_0(V)$*, preprint.
7. S. D. Scott, *Solving the n–gen problem when five divides n*, preprint.
8. S. D. Scott, *Solving the n–gen problem when n is even*, preprint.
9. S. D. Scott, *Solving the n–gen problem when three divides n*, preprint.
10. S. D. Scott, *Some Interesting Open Problems*, Near-ring Newsletter **16**, Jan. (1995), 160–161.
11. S. D. Scott, *The p–gen nature of $M_0(V)$ (I)*, Alg. Dis. Math **15** (2013), #2 237–268.
12. S. D. Scott, *The p–gen nature of $M_0(V)$ (II)*, submitted.
13. S. D. Scott, *Towards solving the n–gen problem*, preprint.
14. S. D. Scott, *Transformation Near–rings Generated by a Unit of Order Three*, Alg. Col. **4:4** (1997), 371–392.

A study of permutation polynomials as Latin squares

Vadiraja Bhatta G. R.

Department of Mathematics, Manipal Institute of Technology, Manipal University,
Manipal, Karnataka, India
E-mail: vadiraja.bhatta@manipal.edu

B. R. Shankar

Department of Mathematical and Computational Sciences, National Institute of Technology Karnataka,
Surathkal, Karnataka, India
E-mail: brs@nitk.ac.in

Permutation Polynomials over finite rings have many applications, including cryptography. The behaviour of bivariate permutation polynomials that form Latin squares over some finite rings is studied. The linear bivariate permutation polynomials for all finite rings and quadratic bivariate permutation polynomials for all rings of even order have been characterized. The Latin squares formed by some permutation polynomials are considered over a few small rings and studied with regard to few Latin square properties.

Keywords: Permutation polynomials, Latin squares.

1. Introduction

A Latin square of order (or size) n is an $n \times n$ array based on some set S of n symbols (treatments), with the property that every row and every column contains every symbol exactly once. In other words, every row and every column is a permutation of S. Also it can be thought of as a two dimensional analogue of a permutation. A Latin square can be viewed as a quadruple $(R, C, S; L)$, where R, C, and S are sets of cardinality n, L is a mapping $L : R \times C \rightarrow S$ such that for any $i \in R$ and $x \in S$, the equation, $L(i, j) = x$ has a unique solution $j \in C$, and for any $j \in C$, $x \in S$, the same equation has a unique solution $i \in R$. That is any two of $i \in R, j \in C, x \in S$ uniquely determine the third so that $L(i, j) = x$, i.e., the cell in row i and column j contains the symbol $L(i, j)$.

Using the concept of permutation functions we can define Latin squares as follows:

A function $f : S^2 \rightarrow S$ on a finite set S of size $n > 1$ is said to be a **Latin square** (of order n) if for any $a \in S$ both the functions $f(a, .)$ and $f(., a)$ are permutations of S.

Here, $f(a, .)$ determines the rows and $f(., a)$ determines the columns of the Latin square. Latin squares exist for all n, as an obvious example we can consider addition modulo n.

Example: A Latin square of order 5 over the set $\{a,b,c,d,e\}$ is below:

a	b	c	d	e
b	a	e	c	d
c	d	b	e	a
d	e	a	b	c
e	c	d	a	b

The terminology 'Latin square' originated with Euler who used a set of Latin letters for the set S.

Here, we study polynomials in two variables x and y (i.e., bivariate polynomials) over some finite rings Z_n. These type of permutation polynomials permute in two dimensions. Hence, these polynomials can form *Latin squares* over finite rings. The formation of different Latin squares using permutation polynomials can be done as follows:

(1) For all the coefficients in order in the ring Z_n, form $P(x,y) = a_{10}x + a_{01}y + a_{11}xy + a_{20}x^2 + a_{02}y^2$.
(2) For each polynomial $P(x,y)$, check check whether permutation of Z_n holds with respect to the variable x only.
(3) If 2 is true, then do the same with respect to the variable y.
If 2 is false, then go to the next polynomial in the list.
(4) When both 2 and 3 are true, the polynomial represents a Latin square.

Rivest[1] considered such polynomials *modulo* $n = 2^w$, where $w \geq 2$ and showed that orthogonal pairs of Latin squares do not exist[1]. Here we have considered them modulo n, $n \neq 2^w$ and to our surprise, found that there are many examples of orthogonal pairs of Latin squares. For $n = 2^w$ Rivest proved that a bivariate polynomial $P(x,\ y)$ can form a Latin square if and only if the 4 univariate polynomials $P(x,\ 0)$, $P(x,\ 1)$, $P(0,\ y)$ and $P(1,\ y)$ are permutation polynomials modulo n. Based on preliminary computations, if $n \neq 2^w$, we have found that a bivariate polynomial can fail to form a Latin square even when these 4 univariate polynomials are permutation polynomials. In a Latin square determined by $P(x,\ y)$, values of $P(x,\ 0)$, $P(x,\ 1)$, $P(0,\ y)$ and $P(1,\ y)$ are given by the entries of first two columns and first two rows.

2. Linear case

Theorem 2.1. *A bivariate linear polynomial $a+bx+cy$ represents a Latin square over Z_n if and only if one of the following equivalent conditions is satisfied:*
(i) both b and c are coprime with n,
(ii) $a+bx$, $a+cy$, $(a+c)+bx$ and $(a+b)+cy$ are all permutation polynomials modulo n.

Proof: For linear polynomials over any Z_n, we can observe that $a+bx+cy$ forms a Latin square if and only if $a+bx$, $a+cy$, $(a+c)+bx$, $(a+b)+cy$ are permutation polynomials. This is because, whenever b and c are both co-prime with n, all those 4 polynomials will be permutation polynomials and in those cases we can fill all the entries of the Latin squares by just looking at first row and first column. As these are all distinct elements in the first row and column, and polynomial $bx+cy$ having only two terms, the entries are got by just adding $a(mod\ n)$ to all entries of $bx+cy$. So Rivest's result holds in the linear case.

Example 2.1. The polynomial $3x+4y+2$ over Z_5 forms the following Latin square.

2	0	3	1	4
1	4	2	0	3
0	3	1	4	2
4	2	0	3	1
3	1	4	2	0

Corollary 2.2. *The Latin squares formed by bivariate linear permutation polynomials will always have all the rows and the columns as cyclic shifts.*

Proof: First, if we neglect the constant term a of the polynomial $P(x,y) = a+bx+cy$, the resulting square will have the first row and the first column due to permutations by "ax" and "by" respectively. That is, when $x = 0$, the values "by", for $y = 0,1,...,n-1$ are the entries of the first column and when $y = 0$, the values "ax", for $x = 0,1,...,n-1$ are the entries of the first row. The remaining entries are the sum of corresponding entries in the first row and first column. So, this square will have rows and columns in a cyclic way. The required Latin square by $P(x,y)$ is got by adding each entry in this square by the constant $a(mod\ n)$, so that it will again have the rows and columns in a cyclic way.

3. Quadratic case

We also try to characterize quadratic bivariate polynomials modulo $n \neq 2^w$, in this way. We represent a bivariate quadratic polynomial as $P(x,y) = a_{10}x + a_{01}y + a_{11}xy + a_{20}x^2 + a_{02}y^2$. Without loss of generality, we have ignored the constant term.

Remark 3.1. If a polynomial $P(x,y)$ represents a Latin square, then our 4 polynomials $P(x,0)$, $P(x,1)$, $P(0,y)$ and $P(1,y)$ will be obviously permutation polynomials, as they form the first two rows and first two columns of the Latin squares.

However, to our surprise, many quadratic polynomials failed to form Latin squares, even though the 4 polynomials $P(x,0)$, $P(x,1)$, $P(0,y)$ and $P(1,y)$ are permutation polynomials. The number of such polynomials over different rings Z_n are shown below.

Ring	No. of polynomials	Examples
Z_6	48	$1+5x+2y+2xy+3y^2$
Z_7	1,050	$x+y+xy$
Z_9	4,374	$x+y+xy+3y^2$
Z_{10}	1,440	$9x+9y+8xy$
Z_{11}	8,910	$10x+10y+10xy$
Z_{12}	768	$7x+7y+10xy+6x^2+6y^2$
Z_{13}	1,8876	$12x+12y+12xy$
Z_{14}	8,400	$13x+11y+6xy$
Z_{15}	3,720	$8x+14y+14xy$

From the data collected, we observed that in all cases where $P(x,y)$ formed a Latin square, the cross term xy was always absent. Hence we could formulate and prove two interesting results.

However, we need an interesting fact regarding orthomorphisms in proving the theorem. The definition as well as proof of the theorem quoted are given in the well-known text of J. H. Van Lint and R. M. Wilson, "A Course in Combinatorics", chapter 22, page 297. We include it for easy reference.

Definition 3.1.[5] An **orthomorphism** of an abelian group G is a permutation σ of the elements of G such that $x \mapsto \sigma(x) - x$ is also a permutation of G.

Theorem 3.1.[5] *If an abelian group G admits an orthomorphism, then its order is odd or its Sylow 2-subgroup is not cyclic.*

Proof: If the Sylow 2-subgroup is cyclic and nontrivial, then there is exactly one element z of G with order 2. When we add all elements of G, each element pairs with its additive inverse except z and 0; so the sum of all elements of G is z. But if σ were an orthomorphism, then

$$z = \sum(\sigma(x) - x) = \sum\sigma(x) - \sum x = z - z = 0,$$

a contradiction to the choice of z.

Theorem 3.2. *If $P(x,y)$ is a bivariate polynomial having* no cross term, *then $P(x,y)$ represents a Latin square if and only if $P(x,0)$ and $P(0,y)$ are permutation polynomials.*

Proof: $P(x,0)$ is the first column of the square and $P(0,y)$ is the first row. If $P(x,y) = f(x) + g(y)$, looking at first row and column, we can complete the square just as addition modulo n (which is a group). So, $P(x,y)$ will be a Latin square.

Theorem 3.3. *Let $n > 2$ be even and $P(x,y) = f(x) + g(y) + kxy$, where k is co-prime with n, be a bivariate quadratic polynomial, where $f(x)$ and $g(x)$ are permutation polynomials modulo n. Then $P(x,y)$ does not represent a Latin square modulo n.*

Proof: If $f(x)$ is a permutation polynomial then $f(x) + a$ is also a permutation polynomial for any constant $'a'$. So, we may assume that the constant term in $P(x,y)$ is 0. Now $f(x) +$

$g(y)$ always represents a Latin square whenever $f(x)$ and $g(y)$ are permutation polynomials, by the last theorem. When $x = c$, the cth row entries will be $P(c,0), P(c,1), ..., P(c,n-1)$, i.e., $f(c)+g(0)+0, f(c)+g(1)+kc, f(c)+g(2)+2kc, ..., f(c)+g(n-1)+(n-1)kc$. Let $f(c) = \theta$, a constant. Then, $\theta+0, \theta+kc, ..., \theta+(n-1)kc$ will be a permutation of $\{0,1,...,n-1\}$ if $g.c.d.(n,c) = 1$. So, let c be such that $g.c.d.(n,c) = 1$. Without loss of generality, we may ignore the constant θ in the sequence. Also $g(0), g(1), ..., g(n-1)$ is some permutation of $\{0,1,...,n-1\}$. The sum of these two permutations fails to be a permutation of Z_n, since there are no orthomorphisms of Z_n as n is even. Hence the cth row contains repetitions and $P(x,y)$ does not represent a Latin square.

However, there are plenty of quadratic bivariate polynomials which do form Latin squares. In the next section we try to characterize them.

4. Formation of Latin Squares and Coefficients of Permutation Polynomials

The permutation of elements of a finite ring varies with the order of the ring. So Latin square structure depends on n, the order of the ring. Also, they are dependent on the coefficients of the polynomials. If we write a quadratic bivariate $P(x,y) = a_{10}x + a_{01}y + a_{11}xy + a_{20}x^2 + a_{02}y^2$, then the numbers in the table of the last section can be explicitly given as the possible choices for the coefficients in $P(x,y)$.

Theorem 4.1. *If $P(x,y)$ represents a Latin square then $P(y,x)$ will represent the Latin square which is just a transpose of the former.*

The proof is obvious.

Corollary 4.1. *For a quadratic bivariate $P(x,y) = a_{10}x + a_{01}y + a_{11}xy + a_{20}x^2 + a_{02}y^2$, over any finite ring, the set of possible coefficients a_{10} and a_{01} are same. This is also true in case of the coefficients a_{20} and a_{02}.*

Proof: The coefficients a_{10} and a_{01} of x and y respectively in some $P(x,y)$ are nothing but the coefficients of y and x in $P(y,x)$.

Theorem 4.2. *If a bivariate polynomial $P(x,y)$, represents a Latin square over the finite ring Z_n, then the constant term a_{00} can be taken as $0,1,..., n-1$ to yield n equivalent Latin squares.*

Proof: If a constant k is added to the polynomial $P(x,y)$, then the values of the polynomial over the ring will be shifted by k *mod* n, i.e., the Latin square entries will be shifted cyclically by k. Thus, we get n equivalent squares.

5. Coefficients of the polynomials and order n of the rings

Also we observe that in the case of Z_n where n is a prime or a product of distinct odd primes, the coefficients of x^2, y^2 and xy are all zero or equivalently there are no quadratic bivariate polynomials which represent Latin squares. So, in these type of rings we find the number of polynomials that yield Latin squares is k^2, where k is the number of possible coefficients of x and y. When n is a prime number, all $n-1$ nonzero elements of Z_n occur as coefficients of both x and y. When n is a product of distinct odd primes, then all the $\varphi(n)$ nonzero elements of Z_n which are coprime with n occur as coefficients of both x and y.

We tabulate a few cases below:

n	**number of P(x,y)**	**set of possible values of a_{10} and a_{01}**
3	$4 = 2^2$	$\{1,2\}$
5	$16 = 4^2$	$\{1,2,3,4\}$
7	$36 = 6^2$	$\{1,2,3,4,5,6\}$
11	$100 = 10^2$	$\{1,2,...,10\}$
13	$144 = 12^2$	$\{1,2,...,12\}$
15	$64 = 8^2$	$\{1,2,4,7,8,11,13,14\}$
17	$256 = 16^2$	$\{1,2,...,16\}$
19	$324 = 18^2$	$\{1,2,...,18\}$
21	$144 = 12^2$	$\{1,2,4,5,8,10,11,13,16,17,19,20\}$
23	$484 = 22^2$	$\{1,2,...,22\}$
29	$784 = 28^2$	$\{1,2,...,28\}$

From the above table we can see that the number N of bivariate quadratic polynomials $P(x,y)$ with constant term zero which yield Latin squares is given by $N = (\varphi(n))^2$, if n is a prime or product of distinct odd primes.

From the table on page 25, we note the following: if n is of the form $2p$, p prime, the coefficient a_{11} of xy is *zero* for all polynomials that represent a Latin square. Also, we can note that the coefficients a_{20} and a_{02} of x^2 andy^2 are either *zero* or p. a_{20} and a_{02} depends on a_{10} and a_{01} respectively, which can be seen below.

$$a_{20}\ (or\ a_{02}) = \begin{cases} 0 \text{ if } a_{10} \text{ (or } a_{01}) \text{ is odd} \\ p \text{ if } a_{10} \text{ (or } a_{01}) \text{ is even.} \end{cases}$$

Example 5.1. $n = 6,10,14,22,26.$

When n is composite but not a product of distinct primes, we observe that the same set of elements of Z_n are possible coefficients a_{11}, a_{20} and a_{02}. Also, in case of the rings Z_n, where n is a power of a prime, Z_n can be partitioned into 2 classes; one is set of possible coefficients of x or y, and the other is that of coefficients of xy, or x^2, or y^2. And in this case, the numbers coprime to n can be coefficients of x and y, the numbers that are not coprime to n can be coefficients of xy, x^2 and y^2.

6. Latin Squares of Lower order

The special thing in a Latin square is the arrangements of rows and columns. Slight variation of the position of a symbol or repetition of a symbol may violate the Latin square structure. Already we observed that all bivariate polynomials modulo n can not represent a Latin square modulo n. However, such squares seem to exhibit some interesting patterns. We can, in some cases, get a Latin square of lower order by deleting some rows and columns in which entries have repetitions. Obviously, number of rows and columns deleted must be equal. In this section we reduce the order of the square to get a Latin square of lower order.

Example 6.1. The polynomial $5x+2y+2xy+3y^2$ over Z_6 will not form a Latin square as shown below.

0	5	4	3	2	1
5	0	1	2	3	4
4	1	4	1	4	1
3	2	1	0	5	4
2	3	4	5	0	1
1	4	1	4	1	4

The third and sixth rows as well as columns contain repetitions. In these rows and columns we see only the entries 1 and 4. Deleting these two rows and columns, we get a square of order 4×4, which is a Latin square over the set $\{0, 2, 3, 5\}$.

0	5	3	2
5	0	2	3
3	2	0	5
2	3	5	0

In the above example, it is to be noted that the polynomial $5x+2y+2xy+3y^2$ is such that $5x$ and $2y+3y^2$ both are permutation polynomials in x and y respectively. But the term $2xy$ prevents the formation of a Latin square. So, it is possible to delete the corresponding rows and columns to get a Latin square of order 4. Similarly, the bivariate $P(x,y) = 9x+9y+8xy$ over Z_{10} will give a 10×10 square which can be reduced to a Latin square of order 8×8 after deleting 2 rows and 2 columns, having only the entries 3 and 8.

$$P(2,y) = \begin{cases} 3 \text{ for all odd y} \\ 8 \text{ for all even y} \end{cases}$$

$$P(7,y) = \begin{cases} 8 \text{ for all odd y} \\ 3 \text{ for all even y.} \end{cases}$$

Similar expressions hold for $P(x,2)$ and $P(x,7)$, because $P(x,y)$ is a symmetric polynomial. So we delete the rows and columns corresponding to both x and y equal to 2 and 7.

But things will not work well in some other cases. If $P(x,y) = f(x)+g(y)+kxy$ is the form of a bivariate quadratic permutation polynomial, not representing a Latin square, then getting a lower order Latin square is possible only if $f(x)$ and $g(y)$ themselves are univariate permutation polynomials.

Example 6.2. The following is an example of a square over Z_9 got by $2x+7y+4xy+3x^2+5y^2$.

0	5	7	6	2	4	3	8	1
3	3	0	3	3	0	3	3	0
7	2	3	1	5	6	4	8	0
3	2	7	0	8	4	6	5	1
0	3	3	0	3	3	0	3	3
7	5	0	1	8	3	4	2	6
6	8	7	3	5	4	0	2	1
6	3	6	6	3	6	6	3	6
7	8	6	1	2	0	4	5	3

In this square we see that no column is a permutation of Z_9. But, there are a few rows which are permutations of Z_9. The rows corresponding to $y = 0, 2, 3, 5, 6$ and 8 are permutations. So, in this case we are not able to decrease the order to get a Latin square.

Remark 6.1. In the previous example, if we write the polynomial as $P(x,y) = f(x) + g(y) + kxy$ where $f(x) = 2x + 3x^2$ and $g(y) = 7y + 5y^2$ and $k = 4$, which is coprime with 9. The univariate polynomial $g(y) = 7y + 5y^2$ itself is not a permutation polynomial. Its values over the ring Z_9 are $\{0, 3, 7, 3, 0, 7, 6, 6, 7\}$. So, the first column itself has repetitions. Same is true with the other columns. That is why we are not able to delete equal number of rows and columns to get a Latin square of lower order.

Theorem 6.1. *The square formed by a quadratic bivariate polynomial $P(x,y) = f(x) + g(y) + kxy$, if not a Latin square, can be reduced to a Latin square of lower order by deleting equal number of rows and columns, if and only if both $f(x)$ and $g(y)$ are themselves univariate permutation polynomials over Z_n.*

Proof: From the Theorem 3.3, the presence of kxy prevents the square from being a Latin square. We can write the square to be sum of two squares L_1 and L_2, where L_1 is obtained by $f(x) + g(y)$ and L_2 by kxy. Also from the Theorem 3.2, L_1 is a Latin square, if and only if $f(x)$ and $g(y)$ are univariate permutation polynomials over Z_n. Obviously, the square L_2 is symmetric about the diagonal, as it is obtained by the symmetric polynomial kxy. So, in the resulting square formed by $P(x,y)$, the number of rows and columns which fail to be permutations of Z_n are same and the entries on these rows and columns are also the same. So, it can be be reduced to a square of lower order by deleting such rows and columns, if and only if both $f(x)$ and $g(y)$ are themselves univariate permutation polynomials over Z_n.

If we have two orthogonal Latin squares of order 4, both over the set, $\{1, 2, 3, 4\}$, the configuration of their superposition is as follows:

1	2	3	4
2	1	4	3
3	4	1	2
4	3	2	1

1	2	3	4
3	4	1	2
4	3	2	1
2	1	4	3

The orthogonal configuration is:

(1,1)	(2,2)	(3,3)	(4,4)
(2,3)	(1,4)	(4,1)	(3,2)
(3,4)	(4,3)	(1,2)	(2,1)
(4,2)	(3,1)	(2,4)	(1,3)

Rivest[1] proved that no two bivariate polynomials modulo 2^w, for $w \geq 1$ can form a pair of orthogonal Latin squares. This is because all the bivariate polynomials over Z_n, where $n = 2^w$, will form Latin squares which can be equally divided into 4 parts as shown below, where the $n/2 \times n/2$ squares A and D are identical and $n/2 \times n/2$ squares B and C are identical.

A	B
C	D

So, no two such Latin squares can be orthogonal.

Example 6.3. The following is a pair of Latin squares over Z_9 which are orthogonal to each other.

0	8	4	6	5	1	3	2	7
7	0	8	4	6	5	1	3	2
8	4	6	5	1	3	2	7	0
3	2	7	0	8	4	6	5	1
1	3	2	7	0	8	4	6	5
2	7	0	8	4	6	5	1	3
6	5	1	3	2	7	0	8	4
4	6	5	1	3	2	7	0	8
5	1	3	2	7	0	8	4	6

Latin square formed by
$5x + y + 3xy + 3x^2 + 6y^2$

0	5	7	6	2	4	3	8	1
2	4	3	8	1	0	5	7	6
7	6	2	4	3	8	1	0	5
6	2	4	3	8	1	0	5	7
8	1	0	5	7	6	2	4	3
4	3	8	1	0	5	7	6	2
3	8	1	0	5	7	6	2	4
5	7	6	2	4	3	8	1	0
1	0	5	7	6	2	4	3	8

Latin square formed by
$2x + 5y + 6xy + 3x^2 + 6y^2$

The two bivariate quadratic polynomials $x + 5y + 3xy + 6x^2 + 3y^2$ and $4x + 7y + 6xy + 3x^2 + 6y^2$ give two orthogonal Latin squares over Z_9. Also, $x + 4y + 3xy$ is a quadratic bivariate which gives a Latin square orthogonal to Latin square formed by $x + 5y + 3xy + 6x^2 + 3y^2$ over Z_9, but not to that of $4x + 7y + 6xy + 3x^2 + 6y^2$.

Remark 6.1. We have found many examples in which the rows or columns of the Latin square formed by quadratic bivariates over Z_n are cyclic shifts of a single permutation of $\{0, 1, 2, ..., \text{n-1}\}$. If two bivariates give such Latin squares, then corresponding to any one entry in one Latin square, if there are n different entries in n rows of the other Latin square, then those two Latin squares will be orthogonal. For instance, in the example below, the entries in the second square corresponding to the entry 0 in the first square are 0, 8, 7, 6, 5, 4, 3, 2, 1. The rows of the first square are all cyclic shifts of the permutation (0, 8, 4, 6, 5, 1, 3, 2, 7), not in order. Also the columns of the second square are the cyclic shifts of the permutation (0, 7, 2, 3, 1, 5, 6, 4, 8), not in order.

Example 6.4.

0	8	4	6	5	1	3	2	7
7	0	8	4	6	5	1	3	2
8	4	6	5	1	3	2	7	0
3	2	7	0	8	4	6	5	1
1	3	2	7	0	8	4	6	5
2	7	0	8	4	6	5	1	3
6	5	1	3	2	7	0	8	4
4	6	5	1	3	2	7	0	8
5	1	3	2	7	0	8	4	6

Latin square formed by
$5x + y + 3xy + 3x^2 + 6y^2$

0	4	2	3	7	5	6	1	8
7	8	3	1	2	6	4	5	0
2	0	1	5	3	4	8	6	7
3	7	5	6	1	8	0	4	2
1	2	6	4	5	0	7	8	3
5	3	4	8	6	7	2	0	1
6	1	8	0	4	2	3	7	5
4	5	0	7	8	3	1	2	6
8	6	7	2	0	1	5	3	4

Latin square formed by
$7x + 4y + 6xy + 6x^2 + 3y^2$

Instead of looking for an orthogonal mate of Latin square formed by some other polynomial we looked at the mirror image of the square itself.

Example 6.5. The Latin square formed by the polynomial $4x + 7y + 6xy + 6x^2 + 3y^2$ over Z_9 and its mirror image are given below:

0	1	5	3	4	8	6	7	2
1	8	0	4	2	3	7	5	6
8	3	1	2	6	4	5	0	7
3	4	8	6	7	2	0	1	5
4	2	3	7	5	6	1	8	0
2	6	4	5	0	7	8	3	1
6	7	2	0	1	5	3	4	8
7	5	6	1	8	0	4	2	3
5	0	7	8	3	1	2	6	4

2	7	6	8	4	3	5	1	0
6	5	7	3	2	4	0	8	1
7	0	5	4	6	2	1	3	8
5	1	0	2	7	6	8	4	3
0	8	1	6	5	7	3	2	4
1	3	8	7	0	5	4	6	2
8	4	3	5	1	0	2	7	6
3	2	4	0	8	1	6	5	7
4	6	2	1	3	8	7	0	5

These two are orthogonal to each other. Over the ring Z_7, here is a Latin square formed by the polynomial $3x + 4y$, with its mirror image, which are orthogonal to each other:

0	3	6	2	5	1	4
4	0	3	6	2	5	1
1	4	0	3	6	2	5
5	1	4	0	3	6	2
2	5	1	4	0	3	6
6	2	5	1	4	0	3
3	6	2	5	1	4	0

4	1	5	2	6	3	0
1	5	2	6	3	0	4
5	2	6	3	0	4	1
2	6	3	0	4	1	5
6	3	0	4	1	5	2
3	0	4	1	5	2	6
0	4	1	5	2	6	3

But this is not always true. We have the example below (6.6): in the ring Z_8, the Latin square formed by the polynomial $x+5y+4xy+2x^2+6y^2$ is not orthogonal with its mirror image.

Example 6.6.

0	3	2	5	4	7	6	1
3	2	5	4	7	6	1	0
2	5	4	7	6	1	0	3
5	4	7	6	1	0	3	2
4	7	6	1	0	3	2	5
7	6	1	0	3	2	5	4
6	1	0	3	2	5	4	7
1	0	3	2	5	4	7	6

1	6	7	4	5	2	3	0
0	1	6	7	4	5	2	3
3	0	1	6	7	4	5	2
2	3	0	1	6	7	4	5
5	2	3	0	1	6	7	4
4	5	2	3	0	1	6	7
7	4	5	2	3	0	1	6
6	7	4	5	2	3	0	1

Here there are 32 distinct pairs, each appearing twice. Also all the pairs of the form (a,b), where one of a and b is odd and the other is even are appearing and all pairs that appear are of this form.

Theorem 6.2. *For odd n, Latin square over Z_n formed by a bivariate permutation polynomial $P(x,y)$ is orthogonal with its mirror image.*

Proof: Each entry in the Latin square of odd order formed by $P(x,y)$ is uniquely determined by the pair (x,y). Let (x',y') be the corresponding cell index in the mirror image. So, in the mirror image of the square, for each fixed row (i.e., for each fixed y'), the entry in the cell (x',y') is given by $P(x,y)$, where x is such that, $x+x'=n-1(mod\ n)$ and $y=y'$. So, in the two squares (i.e., the Latin square and its mirror image, we can have a bijection between the triples $(x,y,P(x,y))$ and $(x',y',P(x,y))$. So, there are n^2 distinct pairs of corresponding entries. Hence, these two squares are orthogonal.

Remark 6.2. In case of rings Z_n, where n is even, we know from the previous theorem, the Latin squares formed by bivariate permutation polynomials have four parts with diagonally opposite pair of parts being same. Mirror images of such squares are also of the same kind. So, in the first $n/2$ rows we can get $n/2$ distinct pairs of corresponding entries. In the last $n/2$ rows, these pairs will repeat in the same order. So, these two pair of squares are not orthogonal.

Bibliography

1. Ronald L. Rivest, Permutation Polynomials modulo 2^w, *Finite Field and Applications*, **7**(2), 287–292, 2001.
2. Appa G., Magos D., Mourtos I., An LP-based proof for the non existance of a pair of orthogonal Latin squares or order 6, *Operations Research Letters*, **32**, 2004, 336–344.
3. Appa G., Magos D., Mourtos I., Searching for mutually orthogonal Latin squares via integer and constrain programming, *European Journal of Operational Research*, **173**, 2006, 519–530.
4. Vadiraja Bhatta G. R. and Shankar B. R., Permutation Polynomials modulo $n, n \neq 2^w$ and Latin Squares, *International Journal of MathematicalCombinatorics*, Vol. 2, 58–65, 2009.

5. Van Lint J. H. and Wilson R. M., A Course in Combinatorics, second edition, *Cambridge University Press, Cambridge, New York*, 2001.
6. Wallis W. D., Introduction to Combinatorial Designs, second edition, *Chapman and Hall/CRC*, 1998.

List of participants

(1) **G. L. Booth,** Nelson Mandela Metropolitan University, South Africa, geoff.booth.nmmu.ac.za.
(2) **J.H. Meyer,** University of the Free State, South Africa, meyerjh@ufs.ac.za
(3) **Timothy H. Boykett,** Johannes Kepler University, Austria, tim.boykett@jku.at
(4) **N. J. Groenewald,** Nelson Mandela Metropolitan University, South Africa, Nico.Groenewald@nmmu.ac.za
(5) **Wen-Fong Ke,** National Cheng Kung University, Taiwan, wfke@mail.ncku.edu.tw
(6) **Stefan Veldsman,** Nelson Mandela Metropolitan University, South Africa, Stefan.veldsman@nmmu.ac.za
(7) **Gary F. Birkenmeier,** University of Louisiana at Lafayette, USA, gfb1127@louisiana.edu
(8) **S. D. Scott,** University of Auckland, New Zealand, ssco034@math.auckland.ac.nz
(9) **I.B.S. Passi,** Indian Institute of Science Education and Research, Mohali India, ibspassi@yahoo.co.in
(10) **Sudhir R. Ghorpade,** Indian Institute of Technology Bombay, India, srg@matth.iitb.ac.in.
(11) **Ambedkar Dukkipati,** Indian Institute of Science, India, ad@csa.iisc.ernet.in
(12) **Shankar B. R.,** National Institute of Technology Karnataka, India, shankarbr@gmail.com
(13) **Bijan Davvaz,** Yazd University, Iran, davvvaz@yazd.ac.ir
(14) **Ayman Badawi,** American University of Sharjah, UAE, abadawi@aus.edu
(15) **S. Juglal,** Nelson Mandela Metropolitan University, South Africa, suresh.juglal@nmmu.ac.za
(16) **Rajan A. R.,** University of Kerala, India, arrunivker@yahoo.com
(17) **Kent Neuerburg,** Southeastern Louisiana University, USA, kneuerburg@southeastern.edu
(18) **Mark Farag,** Fairleigh Dickinson University, USA, mfarag@fdu.edu
(19) **Sushma Singh,** Indian Institute of Technology Patna, India, sushmasingh@gmail.com.
(20) **P. K. Sharma,** D. A. V. College, Jalandhar, India, pksharma@davjalanhaar.com
(21) **Pasham Narasimha Swamy,** GITAM University, India, swamy.pasham@gmail.com

(22) **Nagaiah Thandu,** Kakatiya University, India, nagaiahphd4@gmail.com
(23) **Sharad Vitthal Gaikwad,** Shivaji University, India, sharadvg@gmail.com
(24) **Kuncham Ravi Kiran,** Acharya Nagarjuna University, India, ravikirankuncham@gmail.com
(25) **Om Prakash,** Indian Institute of Technology, Patna India, om@iitp.ac.in
(26) **Bhavanari Satyanarayana,** Acharya Nagarjuna University, India, bhavanari2002@yahoo.co.in
(27) **Deepak Shetty,** Moodlakatte Institute of Technology, India, deepak14_shetty@yahoo.com
(28) **Gampa Veera Nagaraju,** Acharya Nagarjuna University, India, gampanagaraju.2929@gmail.com
(29) **Shaik Mohiddin Shaw,** Narasaraopeta Engineering College, India, mohiddin_shaw26@yahoo.co.in
(30) **Thota Srinivas,** Kakatiya University, India, thotasrinivas.srinivas@gmail.com
(31) **Hamsa Nayak,** Manipal University, India, nayakhamsa@gmail.com
(32) **Deiborlang Nongsiang,** North-Eastern Hill University, India, ndeiborlang@yahoo.in
(33) **Ginjupalli Ayyappa,** Acharya Nagarjuna University, India, GinjupalliAyyappa@gmail.com
(34) **Jagadeesha B.,** St. Joseph Engineering College, Mangalore, India, jagadeesh1_bhat@yahoo.com
(35) **Kavitha Koppula,** Manipal University, India, kavithag.koppula@gmail.com
(36) **Dhananjaya Reddy,** Yogi Vemana University, India, djreddy65@gmail.com
(37) **C. Meera,** Bharathi Women's College, Chennai, India, eya65@rediffmail.com
(38) **Mallikarjuna Bavanari,** CEERI Chennai, India
(39) **Rekha Rani,** N. R. E. C. College, India
(40) **Ayush Kumar Tewari,** National Institute of Science Education and Research, India, ayush.t@niser.ac.in
(41) **Tumurukota Venkata Pradeep Kumar,** Acharya Nagarjuna University, India, Pradeeptv5@gmail.com
(42) **Mohana K. S.,** Mangalore University, India, krish305@gmail.com
(43) **Y. Venkateshwara Reddy,** Acharya Nagarjuna University, India, yvreddy47@gmail.com
(44) **K. Pushpalatha,** Acharya Nagarjuna University, India, kpushpamphil@gmail.com
(45) **Chodi Setty Seshaiah,** Andhra Loyola College, India, ch.seshaiah@gmail.com
(46) **Aditi Sharma,** Bangalore University, India, aditisharma1729@gmail.com
(47) **Kontham Vijay Kumar,** Kakatiya University, India, vijay.kntm@gmail.com
(48) **Mahender Kumar,** Jawaharlal Nehru University, India, kumar17@gmail.com
(49) **Vinay Madhusudanan,** Manipal Institute of Technology, India, vinay.m2000@gmail.com
(50) **Soumitra Das,** North-Eastern Hill University, India, soumitrad330@gmail.com

(51) **Sudhakara G.,** Manipal Institute of Technology, India, sudhakara.g@manipal.edu
(52) **Oviya V.,** Bharathiyar University, India, oviyavisu@gmail.com
(53) **Poornima S.,** Bharathiyar University, India, spoorni26@gmail.com
(54) **Puligadda Jyothi,** LITS-JNTU Hyderabad, India, puligaddajyothi@gmail.com
(55) **Devanaboina Srinivasulu,** Acharya Nagarjuna University, India, gktsrinu@gmail.com
(56) **Ambashree K.,** Manipal University, India, ambashreek51@gmail.com
(57) **Tamil Selvan,** Bharathiyar University, India, tamilmath94@gmail.com
(58) **Rajeshwari M.,** Bharathiyar University, India, rajijas@gmail.com
(59) **R. Nandhini,** Bharathiyar University, India, nandhupriya3124@gmail.com
(60) **Srikanth Prabhu,** Manipal Institute of Technology, India, srikanth.prabhu@manipal.edu
(61) **Vadiraja G. R. Bhatta,** Manipal Institute of Technology, India, vadiraja.bhatta@manipal.edu
(62) **Raplang Nongsiej,** North-Eastern Hill University, India, rap890002@gmail.com
(63) **Vismith,** Brahmavar Udupi, India, mvismitha2@gmail.com
(64) **Deepika Bhat,** Saibrakatte, Udupi India, deepikabhat9686@gmail.com
(65) **Deepak D.,** National Institute of Technology Calicut, India, deepakd1987@gmail.com
(66) **Shijina V.,** National Institute of Technology Calicut, India, shijichan@gmail.com
(67) **Gayathri Varma,** National Institute of Technology Calicut, India, gvarma29@gmail.com
(68) **Jupudi Lakshmi Rama Prasad,** P. B. Siddhartha College, India, jlrprasad@gmail.com
(69) **Prathima J.,** Manipal Institute of Technology, India, prathima.amrutharaj@manipal.edu
(70) **Indira K. P.,** Manipal Institute of Technology, India, indira.kp@manipal.edu
(71) **Sujatha H. S.,** Manipal Institute of Technology, India, sujatha.jayaprakash@manipal.edu
(72) **Baiju T.,** Manipal Institute of Technology, India, baijutmaths@gmail.com
(73) **Sowmya,** Manipal University, India, sankudru@gmail.com
(74) **Monica Rao,** Manipal University, India, raomonica94@gmail.com
(75) **Aishwarya S.,** Manipal University, India, aishwarya93s@gmail.com
(76) **Chaitra Shetty,** Manipal University, India, chaitra.shetty7@gmail.com
(77) **Abhishek N.,** Manipal University, India, abhigowdan@gmail.com
(78) **B. Sandesh,** Manipal University, India, sandesh.aithal@gmail.com
(79) **Shwetha Nayak,** Manipal University, India
(80) **Anusha Kumari,** Manipal University, India, anushavpalan@gmail.com
(81) **Sowmya K.,** Manipal University, India, soumya.bhat706@gmail.com
(82) **Ramya Chandran,** Manipal University, India, ramyachandran30@yahoo.com

(83) **Varalakshmi Alapati,** Manipal University, India, Lakshmi.alapati@manipal.edu
(84) **Goli Haveesh,** Hyderabad, India, haveeshgoli4@gmail.com
(85) **Sangeeta Hemant Joshi,** University of Mumbai, India, sangeetajoshi@yahoo.com
(86) **Anand Biradar,** Manipal University, India, anandbiradar1991@gmail.com
(87) **Rajashekhar Choudhari,** Manipal University, India, choudhariraj3@gmail.com
(88) **M. Raja Pavan Kumar,** Acharya Nagarjuna University, India, rajapavan127@gmail.com
(89) **S. Parameshwara Bhatta,** Mangalore University, India
(90) **B. S. Kiranagi,** University of Mysore, India, bskiranagi@gmail.com
(91) **Mary Lynn Scott,** Accompanying Scott, New Zealand, ml.scott@xtra.co.nz
(92) **Marie Therese Bieber,** Austria
(93) **Satya Gyanasri,** Vignan's Nirula Engineering College, Guntur
(94) **Lakshmi,** Manipal Institute of Technology, Manipal University
(95) **Rashmitha,** Manipal Institute of Technology, Manipal University
(96) **Chaithra,** Manipal Institute of Technology, Manipal University
(97) **Ashwith,** Manipal Institute of Technology, Manipal University
(98) **Shruthi D. Nayak,** Manipal Institute of Technology, Manipal University
(99) **B. Narayana Rao,** Rtd. Head Master, Warangal District, Telangana
(100) **K. Manjunath Prasad,** Department of Statistics, Manipal University
(101) **S. Balasubramanyam,** Karur, Tamilnadu
(102) **P. Arunaswathi Vyjayanthi,** Tamilnadu
(103) **Kuncham Syam Prasad,** Department of Mathematics, MIT, Manipal University, syamprasad.k@manipal.edu
(104) **Kedukodi Babushri Srinivas,** Department of Mathematics, MIT, Manipal University, babushrisrinivas.k@manipal.edu
(105) **Harikrishnan Panackal,** Department of Mathematics, MIT, Manipal University, pk.harikrishnan@manipal.edu

Dr Vinod Bhat
Vice-Chancellor, Manipal University

MANIPAL INSTITUTE OF TECHNOLOGY
(A constituent institute of Manipal University)
DST - DRDO - ISRO - KSTA Sponsored
24th International Conference on
Nearrings, Nearfields and Related Topics
05-12, JULY 2015
Organized by
Department of Mathematics, MIT, Manipal University

Inagural session

$C(N) \subseteq hr \langle C(N) \rangle \subseteq$

R comm ring with 1 ≠ 0
M R-module
N submodule of M

Conference tour

MANIPAL INSTITUTE OF TECHNOLOGY
(A constituent institute of Manipal University)
DST - DRDO - ISRO - KSTA Sponsored
24th International Conference on
Nearrings, Nearfields and Related Topics
05-12, JULY 2015
Organized by
Department of Mathematics, MIT, Manipal University

MANIPAL INSTITUTE OF TECHNOLOGY
(A constituent institute of Manipal University)
DST - DRDO - ISRO - KSTA Sponsored
24th International Conference on
Nearrings, Nearfields and Related Topics
05-12, JULY 2015
Organized by
Department of Mathematics, MIT, Manipal University

MANIPAL INSTITUTE OF TECHNOLOGY
(A constituent institute of Manipal University)
DST - DRDO - ISRO - KSTA Sponsored
24th International Conference on
Nearrings, Nearfields and Related Topics
05-12, JULY 2015
Organized by
Department of Mathematics, MIT, Manipal University

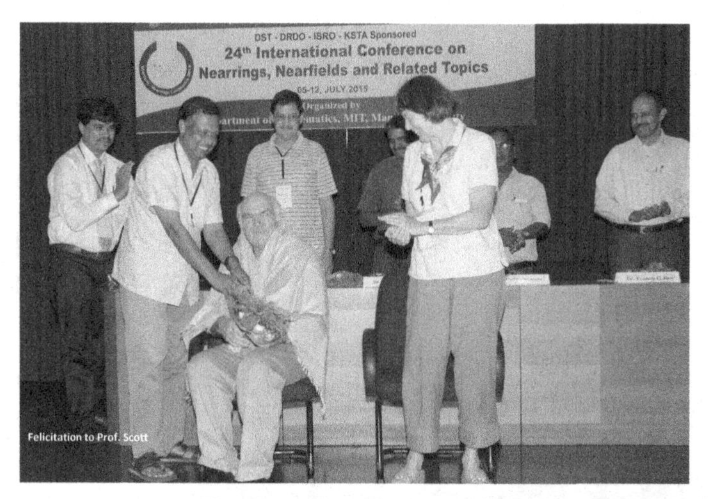

Felicitation to Prof. Scott

MANIPAL INSTITUTE OF TECHNOLOGY
DST - DRDO - ISRO - KSTA Sponsored
24th International Conference on
Nearrings, Nearfields and Related Topics
05-12, JULY 2015
Organized by
Department of Mathematics, MIT, Manipal University

MANIPAL INSTITUTE OF TECHNOLOGY
DST - DRDO - ISRO - KSTA Sponsored
24th International Conference on
Nearrings, Nearfields and Related Topics
05-12, JULY 2015
Organized by
Department of Mathematics, MIT, Manipal University

MANIPAL INSTITUTE OF TECHNOLOGY
DST - DRDO - ISRO - KSTA Sponsored
24th International Conference on
Nearrings, Nearfields and Related Topics
05-12, JULY 2015
Organized by
Department of Mathematics, MIT, Manipal University
Validictory session

MANIPAL
DST - DRDO - ISRO - KSTA
24th International
Nearrings
05-12, JULY 2015

WILEY
CRC Press
Roland

Some preceding photographs

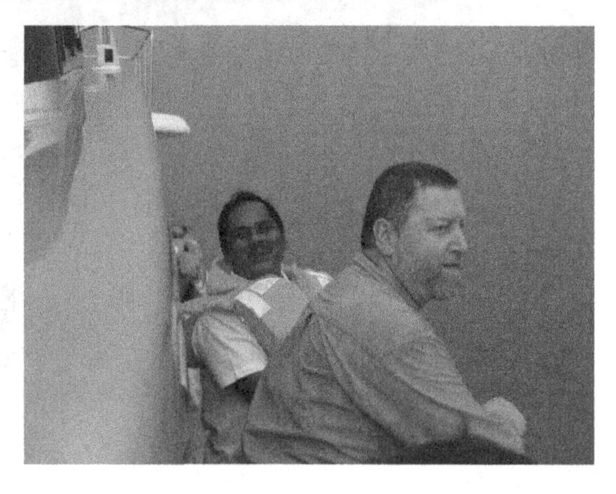

NBHM - DST - DRDO - CSIR Sponsored
National Workshop
puter Applications based on Modern Algebra

http://www.algebra.uni-linz.ac.at/Nearrings/conferences.html

Nearring conferences

(1) **1968** 5–8 December, Oberwolfach, Germany
(2) **1972** 30 January–3 February, Oberwolfach, Germany
(3) **1976** 27 June–3 July, Oberwolfach, Germany
(4) **1978** 6–12 August, Edinburgh UK
(5) **1980** 12–19 April, Oberwolfach, Germany
(6) **1981** 13–19 September, San Benetto del Tronto, Italy
(7) **1983** 1–6 August, Harrisonburg, USA
(8) **1985** 7–11 January, Nagarjuna University, India
(9) **1985** 4–10 August, Tübingen, Germany
(10) **1987** 2–8 August, Middlesbrough, UK
(11) **1989** 5–11 November, Oberwolfach, Germany
(12) **1991** 14–20 July, Linz, Austria
(13) **1993** 18–24 July, Fredericton, Canada
(14) **1995** 30 July–6 August, Hamburg, Germany
(15) **1997** 14 July–18 July, Stellenbosch, South Africa
(16) **1999** 12 July–16 July, Edinburgh, U. K.
(17) **2001** 22 July–28 July, Harrisonburg, VA, USA
(18) **2003** 27 July–3 August, Hamburg, Germany
(19) **2005** 31 July–6 August, Chi-Tou, Taiwan
(20) **2007** 23 July–27 July, Linz, Austria
(21) **2009** 26 July–1 August, Abbey Vorau (Styria, Austria)
(22) **2011** 25–29 July, Southeastern Louisiana University, Louisiana USA
(23) **2013** 07–13 July, University of the Free State, Bloemfontein, South Africa
(24) **2015** 05–12 July, Manipal Institute of Technology, Manipal University, India

Printed in the United States
By Bookmasters